# Environmental Health Impacts of Transport and Mobility

# Environmental Science and Technology Library

VOLUME 21

# Environmental Health Impacts of Transport and Mobility

Edited by

## P. Nicolopoulou-Stamati
*National and Kapodistrian University of Athens,*
*Medical School, Department of Pathology,*
*Athens, Greece*

## L. Hens
*Vrije Universiteit Brussel,*
*Human Ecology Department,*
*Brussels, Belgium*

and

## C.V. Howard
*University of Liverpool,*
*Department of Human Anatomy and Cell Biology,*
*Liverpool, United Kingdom*

 Springer

A C.I.P. Catalogue record for this book is available from the Library of Congress.

ISBN-10  1-4020-4304-X (HB)
ISBN-13  978-1-4020-4304-8 (HB)
ISBN-10  1-4020-4307-4 (e-book)
ISBN-13  978-1-4020-4307-9 (e-book)

Published by Springer,
P.O. Box 17, 3300 AA Dordrecht, The Netherlands.

*www.springer.com*

*Printed on acid-free paper*

Desktop publishing by Vu Van Hieu.

**Editorial Statement**

It is the policy of AREHNA, EU-SANCO project. to encourage the full spectrum of opinion to be represented at its meetings. Therefore it should not be assumed that the publication of a paper in this volume implies that the Editorial Board are fully in agreement with the contents, though we have tried to ensure that contributions are factually correct. Where, in our opinion, there is scope for ambiguity we have added footnotes to the text, where appropriate.

# TABLE OF CONTENTS

# FOREWORD

The day I became the Flemish Minister for Mobility in 1999, my region of Flanders was one of the poorest students of the "EU class" for traffic safety. This was in spite of a variety of measures taken by previous governments, to improve traffic and safety because of unsafe driving behaviour and to reduce the emissions of traffic-related pollutants. In this context, I focussed on one specific aspect of traffic safety: the number of traffic deaths and injuries on the roads. My intention was to raise a broad societal debate on mobility. It seemed to me that the most complex issue was to convince those people who are most intimately related to their cars, to leave their vehicles at home. After decades with slogans such as "My car, my freedom" it is politically impossible to withdraw this "freedom" on the basis of ecological arguments. I often noticed that although the majority of the population wants a cleaner environment, the same majority is convinced that 'pollution is caused by others'. Moreover, the most important environmental problems tend to appear only in the long term. In contrast, the numerous, mainly young, victims on the road, among them the dead and the seriously injured youngsters from 'weekend' accidents are immediately visible. In 2001, Flanders totalled 848 road-accident deaths, 7725 seriously injured persons and 39070 minor injuries, from a total population of 6 million people. These are only the recorded figures. Every year, Flanders records 490,000 people who are involved in one way or another in a traffic accident. Furthermore, the perception of the threat is even larger, because many people have a 'near miss' experience. Every year, this involves some 60 million people. More precisely, every Flemish citizen has the impression to be almost involved in an accident ten times each year.

All this shows that traffic offers a clear problem of public health. For young people, traffic is even the single most important cause of death in Flanders. Apart from the enormous grief this causes for all those who are involved, the cost for society is tremendous. EU assessments show that the societal cost for one person killed in an accident is €1.5 million. Using this calculation, the cost of roadside deaths to Flanders totals €1.3 billion.

Although we performed and still perform poorly, there is also good news. During the past sixteen years, the number of accidents has declined by 5.8%, the number of traffic deaths by 20.3% and the number of serious injuries by 44.1%, despite the increase in traffic density.

To make sure the number of victims will decline further, drastic measures were necessary to achieve a change in the mentality of both the drivers and the authorities. A carrot and stick approach was used. This means that the authorities needed to make it clear that there are no acceptable arguments to excuse reckless driving.

Therefore I decided to install wanted automatic speed cameras at all accident black spots in Flanders. By the end of 2004, over 350 crossroads in the region will have such a traffic camera. Installing these devices in Belgium is complex, because the political responsibilities are shared amongst the regional and federal governments. Therefore I threatened that if these targets were not realised, I would consider resigning. In this way, I managed as a Flemish minister to involve the federal government in this project. Installing traffic cameras is not an easy decision but it leads to rewarding results. In contrast to what people presumed, the number of fines did not increase but decreased. That was the result of drivers who decreased speed at the crossroads. Moreover, it had a most beneficial effect on the number of traffic injuries. Police statistics show that the number of accidents in locations with automatic cameras decreased by between 20 to 50% and that the number of victims decreased proportionally. In the City of Leuven, where the speed limit on the ring road was reduced to 50 km/h and the junctions have been furnished with automatic cameras, the number of accidents was reduced by 19% in one year. In Hasselt, my home town, automatic cameras were installed in January 2003. During that year, the number of victims decreased by 35%. A stricter policy does not necessarily mean that the citizens are dissatisfied. European statistics from 2002 showed that 60% of Belgians either agreed or strongly agreed with the statement that speeding offences should be punished more strictly. This rating came out second of seven EU countries, after Finland where the rating was 80%. Moreover, almost 80% of the Belgians are more or less in favour of installing automatic cameras to control speed and the 'jumping' of red lights at junctions. These figures are also higher than comparative data for other countries.

The policy should also contain an element of reward for citizens. Therefore, we started official discussions with the insurance companies. Less traffic accidents with reduced risks for the insurers. In theory, it should be possible that car premiums should become 20 to 30% cheaper. A mobility policy should not, therefore, be solely repressive but can also include positive incentives.

Positive steps were also taken at a second level, that of public transport. My opinion is that public transport should be free for everyone. In my home town Hasselt, where I am the Mayor, I introduced free public bus transport in 1997. The city pays a yearly contribution to the public transport company, which allows all inhabitants to use the bus free of charge. Between 1997 and 2002, the number of people using the bus in Hasselt increased twelve-fold. My political opponents call free public transport a 'populistic utopia'. Meanwhile, all over Flanders there are quite a number of groups who use buses and trams for free. The introduction of free public transport needs to be effected step by step, because the transport company is unable to adapt its infrastructure instantaneously. Therefore, I decided to start with offering free buses to pensioners and subsequently to young people below the age of twelve. For students, I introduced cheap bus passes and also adults can use buses cheaply with a simplified pass. The number of users of the Flemish Transportation Company increased dramatically during the period of the last government. In 1998 the number of travellers was about 250 million; by 2002 this number had increased to 318

million and in 2003 over 360 million travellers were registered, which is almost 1
million per day.

This cannot only be explained because people travel cheaply or for free, but also
because we improved the infrastructure. This was possible because the demand for
public transport increased; because only after an increase in demand, you will be
able to find a wider public consensus to invest in public transport. A parliament does
not refuse to invest if the buses are overcrowded.

Moreover, communities and employers have the possibility to make deals with the
public transport company. These deals allow their employees to have free bus
transport. We intend to apply the same system on the railway. Employers will have
the possibility to offer free train travel to their employees. The employer pays part of
the cost and the authorities pay for the remainder.

I linked free public transport with a decree on basic mobility rights. Basic mobility
refers to the right of each citizen to be mobile. Those who don't have a car must
have the possibility to travel easily by public transport. That means that accessibility
and frequency of public transport are legal rights in Flanders.

When it comes to public transport, creativity also helps. In 2003, I suggested people
could obtain a 3-year free bus pass for the whole family if they gave up their vehicle.
In particular for households with two cars, this is a stimulus towards a more
sustainable mobility. In the beginning, the idea was considered ridiculous. However,
in 2003 over 10,000 Flemish people sold their cars in exchange for a free bus pass.
These are modest interventions but they urge people to think about mobility.

A public transport company also allows authorities to set up experiments with
alternative energy and sustainable technology. Experiments with particle traps on
buses showed that we could limit emissions by 90%. In cities in particular, these
interventions can be effective in reducing impacts on health, although the reduction
of the total impact on the environment remains limited. The public traffic company
did also set up experiments using biodiesel, which is less toxic than regular diesel
oil. Unfortunately, here too the benefits for the environment are limited.

Free public transport is effective. It removes cars from the roads. Also the
introduction of very low shipping taxes has removed trucks from the roads. In the
past, to use the Flemish inland waters, one had to pay by kilometre and by
displacement. This is contradictory to the more environmentally-friendly character
of inland shipping as compared to road transport. Since the almost complete removal
of shipping taxes, we have succeeded in reducing truck movements by 50,000. The
tonnage/km carried on Flemish waterways increased by 55%. This is equal to
approximately 400,000 truck movements per year.

Traffic injuries can also be limited by persuading car drivers to drive more slowly.
The higher the speed at the moment of an accident, the more serious the
consequences are. Car passengers have a twofold higher chance of dying in a

collision at 80 km/h than at 30 km/h. Decreasing speed on busy roads is most effective when it comes to accident reductions. Therefore, Flemish communities can now propose to reduce speed on regional roads from 90 km/h to 70 km/h. By September 2003, almost half of the Flemish communities applied for such a permission. The effects of these measures will become apparent in terms of traffic safety and noise. Also, more speed controls on highways will contribute to safety. There, the effects will also involve a reduction of air pollution.

The measures we have introduced in Flanders during the past few years are insufficient to solve all traffic problems, but they are the starting point of an important change in mentality and have undoubtedly removed a number of cars and trucks from the roads. In 2003, the number of kilometres on Flemish regional roads decreased slightly. Between 2001 and 2002, the energy consumption of Flemish traffic similarly decreased slightly. For NOx and small particles, the decrease between 1995 and 2002 was limited respectively to 22 and 12%. But, for instance, the decrease during the same period for VOCs, lead and SOx is more pronounced, with a 44%, 96% and 73% reduction respectively. This was mainly the result of stricter requirements for fuel and new vehicles. However, in fact, this decrease is still insufficient. Health problems resulting from existing pollutants are still serious. It is estimated that some sixty Flemish people die each year as a consequence of air pollution produced by car and truck traffic, but thousands of people suffer from associated respiratory diseases.

I don't believe in short-term solutions, because this would mean that you would have to ban the use of motorised vehicles. As stated before, I'm convinced there is no societal basis for such measures. Authorities, however, can implement stimulating measures, promote cleaner energy sources and create an environment in which sustainable technology can develop faster and be supported by fiscal measures. But the most important step we had to take towards better transport safety had to do with reducing the number of traffic victims. In doing this, we established a broad societal support that was beneficial for the public debate on mobility.

**Steve Stevaert**
*Former Minister for Mobility of the Flemish Region.*
*Chairman of the Flemish Socialist Party.*

# LIST OF CONTRIBUTORS

W. BABISCH
*Federal Environmental Agency*
*P.O. BOX 330022*
*D-14191 Berlin*
*GERMANY*

F. BALLESTER
*Valencian School of Studies for Health*
*Unit of Epidemiology and Statistics*
*Calle Juan de Garay 21*
*E -46017 Valencia*
*SPAIN*

P. BEHRAKIS
*National and Kapodistrian University*
*of Athens*
*Department of Experimental*
*Physiology*
*Mikras Assias 75 Goudi*
*GR-11527 Athens*
*GREECE*

K. BOISON
*Arizona State University*
*Environmental Technology*
*Management*
*College of Technology and Applied*
*Sciences*
*Mesa, Arizona 85212*
*USA*

C. DORA
*WHO*
*20 Avenue Appia*
*1211 Geneva 27*
*SWITZERLAND*

K. ELEFTHERATOS
*National and Kapodistrian University*
*of Athens*
*Department of Geology*
*University of Athens*
*15784 Athens*
*GREECE*

L. HENS
*Vrije Universiteit Brussel*
*Human Ecology Department*
*Laarbeeklaan 103*
*B-1090 Brussel*
*BELGIUM*

C.V. HOWARD
*University of Liverpool*
*Developmental Toxico-Pathology*
*Research Group*
*Department of Human Anatomy and*
*Cell Biology*
*Sherrington Buildings*
*Ashton Street*
*L69 3GE Liverpool*
*UNITED KINGDOM*

H.-P. HUTTER
*Institute for Environmental Health*
*Medical University of Vienna*
*Kinderspitalgasse 15*
*1095 Vienna*
*AUSTRIA*

I.S.A. ISAKSEN
*Institute of Geophysics,*
*University of Oslo,*
*P B 1022, Blindern*
*Olso 3*
*NORWAY*

D. KALTSAS
Endocrine Division
Hippocrateion Hospital
Vassilisis Sofias Avenue 108
GR-115 27 Athens
GREECE

P. LAMMAR
Vrije Universiteit Brussel
Human Ecology Department
Laarbeeklaan 103
B-1090 Brussel
BELGIUM

S. LIVADAS
Endocrine Division
Hippocrateion Hospital
Vassilisis Sofias Avenue 108
GR-115 27 Athens
GREECE

B. MARTIN
Institute of Sports Sciences
Federal Office of Sports
Hauptstrasse 243
CH-2532 Magglingen
SWITZERLAND

E. MARTIN-DIENER
Institute of sports Sciences
Federal Office of Sports
Hauptstrasse 243
CH-2532 Magglingen
SWITZERLAND

U. MÄDER
Institute of sports Sciences
Federal Office of Sports
Hauptstrasse 243
CH-2532 Magglingen
SWITZERLAND

H. MOSHAMMER
Institute for Environmental Health
Medical University of Vienna
Kinderspitalgasse 15
1095 Vienna
AUSTRIA

V. MOUNTFORD
University of Liverpool
Developmental Toxico-Pathology
Group
Department of Human Anatomy and
Cell Biology
Sherrington Buildings
Ashton Street
L69 3GE Liverpool
UNITED KINGDOM

J.A. NEWBY
University of Liverpool
Developmental Toxico-Pathology
Group
Department of Human Anatomy and
Cell Biology
The Sherrington Buildings
L69 3GE Liverpool
UNITED KINGDOM

P. NICOLOPOULOU-STAMATI
National and Kapodistrian University
of Athens
Medical School - Dept. of Pathology
75 Mikras Asias Street,
11527, Athens
GREECE

L.W. OLSON
Arizona State University
Environmental Technology
Management
College of Technology and Applied
Sciences
Mesa, Arizona 85212
USA

A.D. PAPAYANNIS
*National Technical University of*
*Athens*
*School of Applied Mathematical and*
*Physical Sciences*
*Physics Department*
*Heroon Polytechniou 9*
*15780 Zografou*
*GREECE*

W.F. PASSCHIER
*Universiteit Maastricht*
*Department of Health Risk Analysis*
*and Toxicology*
*PO BOX 616*
*6200 MD Maastricht*
*THE NETHERLANDS*

W. PASSCHIER-VERMEER
*TNO - Environment and Geosciences*
*Department of Environment and*
*Health*
*P.O.BOX 6041*
*2600 JA Delft*
*THE NETHERLANDS*

F. RACIOPPI
*WHO Regional Office for Europe*
*European Centre for Environment and*
*Health*
*Via Francesco Crispi 10*
*00187 Rome*
*ITALY*

T. ROUSSOU
*Metropolitan Hospital*
*Ethnarhou Makariou 9 & EL.*
*Venizelou 1*
*18547, Neo Faliro*
*GREECE*

L. SCHMIDT
*SOMO*
*Social Sciences Mobility Research and*
*Consultancy*
*Sonnenweg 5*
*1140 Vienna*
*AUSTRIA*

S. STOYANOV
*University of Chemical Technology*
*and Metallurgy*
*Head of Ecology Center*
*8, Kl. Ohridski blvd,*
*1756 Sofia*
*BULGARIA*

E. TERLEMESIAN
*·University of Chemical Technology*
*and Metallurgy*
*Ecology Center*
*8, Kl. Ohridski blvd,*
*1756 Sofia*
*BULGARIA*

G. TOLIS
*Endocrine Division*
*Hippocrateion Hospital*
*Vassilisis Sofias Avenue 108*
*GR-115 27 Athens*
*GREECE*

C.S. ZEREFOS
*Director of the National Observatory*
*of Athens, P.O. BOX 20048, Thissio,*
*11810 Athens,*
*GREECE*
*Director of the Laboratory of*
*Atmospheric Environment, Foundation*
*for Biomedical Research of the*
*Academy of Athens, Soranou Efesiou 4,*
*11527 Athens,*
*GREECE*

# LIST OF FIGURES

# LIST OF TABLES

# LIST OF BOXES

# EFFECTS OF MOBILITY ON HEALTH

P. NICOLOPOULOU-STAMATI
*National and Kapodistrian University of Athens*
*Medical School, Department of Pathology*
*75 Mikras Asias Street, 11527, Athens, GREECE*

## Summary

The current organization of society requires mobility as an essential part of its structure. Complex activities are interwoven in such a way that problems deriving from the actual process of constant movement cannot be separated as they are so tightly interconnected.

This book examines the health effects of mobility by addressing the major issues related to the subject and analyzing their consequences. Thus, air pollution, noise, sedentarism and its related endocrine problems are addressed. Moreover, by integrating information from different geographical regions it aims at enhancing the understanding of the issue. Furthermore, currently available scientific knowledge is presented in a comprehensive way that is useful for non-expert advisors.

## 1. Introduction

Society increasingly demands the relocation of persons and products. Strong networks of land, air and water routes are built and intensely used. Mobility that has been for thousands of years more or less local in scope, can now be found on a global scale, due to the evolution and implementation of technology. The principal consequences of this extension are air pollution, increased noise, accidents and climatic changes (Van Leeuwen, 2002; Goldberg *et al.*, 2003).

Air pollution is largely attributed to engine exhausts. The contamination of clean air has become a major problem in urban areas. In most European cities the relative contribution of vehicle exhausts to urban air pollutants is increasing. Furthermore, emissions from aircraft around airports and ships along sea lanes and river routes constitute additional sources of pollution (Künzli, 2002; Zhang and Lioy, 2002).

*P. Nicolopoulou-Stamati et al. (eds), Environmental Health Impacts of Transport and Mobility, 1-7.*
© 2005 *Springer. Printed in the Netherlands.*

The air pollutants that are most often addressed are carbon dioxide ($CO_2$), carbon monoxide (CO), nitrogen oxides ($NO_X$), particulate matter (PM), polycyclic aromatic compounds, benzene and ozone. Needless to say the list is not exhaustive. Lead is probably the best studied example (Katsougianni, 2003; Englert, 2004; Smuts, 2001).

Noise is another major issue related to mobility. The increase in transport has directly increased the exposure of populations to health impairing levels of noise, that affect people physically, mentally and socially (Tarnopolsky et al., 1980; Stansfeld et al., 2000; Haines et al., 2002).

Rapid transport facilities carry with them the risk of accidents. Human error, considered to be the main cause of accidents, is clearly related to the high speed attainable by modern vehicles (Patel et al., 2002).

Climatic change and its consequent health effects is also related to increased mobility (Patz et al., 2000; Beniston, 2002; Sutherst, 2004).

Moreover, there are a number of less extensively addressed issues related to mobility. Endocrine-related problems, resulting from mobility and contributing to obesity is one of them.

All the above-mentioned issues raise health considerations and they need to be addressed and incorporated in the development of transport policies (Dora and Racioppi, 2003).

## 2.    Health effects related to mobility: a complex problem

Health effects related to mobility have obvious different aspects. Examples entail the effects on the respiratory system and/or the auditory system, and body injury.

The impact of air pollution on pulmonary function is mainly reflected in the increase of asthma. There is accumulating evidence that the number of children requiring asthma medication and hospital admittance has increased over recent decades. The prevalence ranges between 4% and 27% in different geographical regions (Teague and Bayer, 2001). Air pollution is also related to allergies and lung cancer. Information was drawn mainly from epidemiological studies (Brunekreef and Holgate, 2002; Vedal, 2002; Baldacci and Viegi, 2002; Annesi-Maesano et al., 2003). More recent scientific evidence suggests that health effects deriving from air pollution also impair the cardiovascular system and the bone marrow. These health effects strongly relate to the pathology of particles. Particulate matter consisting of organic and non–organic compounds can enter the bloodstream and reach the myocardium and the bone marrow (Fuji et al., 2002; Suwa et al., 2002). Mobility related pollutants at very low concentrations affect health. Extra mortality and hospital admissions are related to exposure of low concentrations of $SO_2$, $NO_2$ and ozone (Kim et al., 2004).

Exposure to air pollution is usually accompanied by exposure to noise. Noise can also affect the abovementioned cardiovascular system by increasing the catecholamine secretion and thus affecting blood pressure. Hypertension is related to noise (Rosenlund *et al.*, 2001). Increased catecholamine secretion is also related to psychological disorders, such as irritation, depression, sleeping disturbances and stress. Psychological disorders can predispose to accidents (OECD, 1998), with most serious consequences (Di Gallo *et al.*, 1997).

## 3. Integration of local information to planning

WHO air guidelines have provided a general basis for the development of an air quality framework in the EU (WHO, 2000a). Noise and accidents are also important for transport policies. However, wide geographical variations occur in levels of measurable pollutants, noise effects and accident rates related to mobility. Urban areas and rural areas tend to be separated into different studies. Long-term effects and short-term effects have also been studied separately. Despite the fact that this plethora of data has been incorporated into simulation models there remains a deficiency in the quality of current health indicators which would permit the drawing of concrete conclusions (Mason, 2000; Mindell *et al.*, 2004). However, research selected from different countries and European regions allows one to compare data and perhaps provides a common basis to harmonize policies. This book contains information from Bulgaria, Greece, Austria, Flanders and the UK in the form of country reports.

## 4. Why this book?

This book aims at overviewing the hazardous side-effects from increased mobility: air pollution, noise and hidden problems such as sedentarism and endocrine dysfunctions. Furthermore, it aims at identifying gaps in policies by focusing on problems that have received less attention and providing scientific documentation that may be included in applied policies.

Although currently transport is essential for economic and social development, the adverse health effects necessitate policy interventions (WHO, 2000b; Tiwari, 2003; Dora, 1999; OECD, 1998; Sharpe, 1999). How this policy interventions could be accomplished is addressed by different papers in this book. Wider knowledge of the mobility-related problems is provided, together with integration of information, linking available knowledge and experience between European countries aligned with the EU common strategy on Environment and Health (EU, 2003). Non expert advisors will find reviewed information that might contribute to their policy advice and will assist them to develop sectoral policies maximizing prevention from mobility-related adverse effects.

Because of the complexity of the problem, no lines can be drawn to separate the effects attributed to different causes. An example is how air pollution and noise may affect the cardiovascular system (Peters *et al.*, 2004; Pope *et al.*, 2004; Babisch *et al.*, 2005; De Paula Santos *et al.*, 2005). It is essential that decision-makers should have a most complete understanding of the whole issue.

It is understood that further research will be needed to clarify the precise relation between health effects and mobility. Nevertheless, the papers collected in this book show that there currently exists a considerable body of comprehensive information. This can form the basis for addressing some of the most pertinent questions that need to be answered concerning the relation between health effects and mobility.

## 5.   Conclusion

The aim of this book is to meet the urgent need to address the health effects deriving from mobility in their entirety and to raise awareness of the cost to society consequent upon current practices. It attempts to provide non-expert advisors with peer-reviewed information that may serve to aid decision-making and policy development that will lead to an amelioration of the current impact on health. Moreover, it tries to provide a better understanding of the health mechanisms related to a number of externalities arising from human mobility patterns, in order to provide a sounder scientific foundation on which further specific knowledge may be based in the future. Risk-reduction policies can be successful only if citizens and decision-makers are well informed which will in turn lead to the political pressure necessary to  implement change. This capacity-building of decision makers is an essential element in reinforcing sustainable development. Public health and environmental professionals have to develop a common language to accomplish their fundamental mission - the protection of public health.

## 6.   Acknowledgements

All papers in this book have been peer reviewed. The editors are most indebted to the colleagues who reviewed various chapters of this book: Balis D. (Aristotle University of Thessaloniki), Barnaba F. (Istituto di Scienze dell'Atmosfera e del Clima), Berry B. (Berry Environmental Ltd), Bluhm G. (Karolinska Hospital), Botteldooren D. (Universiteit Gent), Brauer M. (University of British Columbia), Carroll L. (University of Alberta), Colbeck I. (University of Essex), Daniëls S. (Limburgs Universitair Centrum), De Bondt R. (Vrije Universiteit Brussel), De Geus B. (Vrije Universiteit Brussel), Dewit J. (Vrije Universiteit Brussel), Geurts K. (Limburgs Universitair Centrum), Griefahn B. (Universität Dortmund), Hermans E. (Limburgs Universitair Centrum), Hoet P. (Katholieke Universiteit Leuven), Howard V. (University of Liverpool), Klijnen E. (Vlaamse Stichting Verkeerskunde), Koppe J. (Universiteit Amsterdam), Lammar P. (Vrije Universiteit Brussel), Maynard M. (Department of Health UK), Miermans W. (Limburgs Universitair Centrum), Moshammer H. (Medical University of Vienna), Nath B. (European Centre for Pollution Research ), Nemery B. (Katholieke Universiteit Leuven), Nyssen M. (Vrije Universiteit Brussel), Rylander R. (Göteborg

University), Schoeters G. (Vlaams Instituut voor Technologisch Onderzoek), Skanberg A. (Göteborg University), Torfs R. (Vlaams Instituut voor Technologisch Onderzoek), Vanlaar W. (Belgisch Instituut voor de Verkeersveiligheid), Vesentini L. (Provinciale Hogeschool Limburg), Wets G. (Limburgs Universitair Centrum), Willems B. (Vrije Universiteit Brussel), Witters H. (Vlaams Instituut voor Technologisch Onderzoek), Zanis P. (Academy of Athens), Ziomas I. (National Technical University of Athens).

The editors of this book wish also to thank the EU-SANCO and the Municipality Dikaiou Kos for their support to the A.R.E.H.N.A. project (www.arehna.di.uoa.gr), which provided the main scientific bases of this book. The help of Prof. Costas Evangelides who was in charge of the final language review is most sincerely appreciated.

## References

Annesi-Maesano, I., Agabiti, N., Pistelli, R., Couilliot, M.F., and Forastiere, F. (2003) Subpopulations at increased risk of adverse health outcomes from air pollution, *Eur. Respir. J. Suppl.* 40, 57s-63s.

Babisch, W., Beule, B., Schust, M., Kersten, N., and Ising, H. (2005) Traffic Noise and Risk of Myocardial Infarction, *Epidemiology* 16 (1), 33-40.

Baldacci, S., and Viegi, G. (2002) Respiratory effects of environmental pollution: epidemiological data, *Monaldi. Arch. Chest. Dis.* 57 (3-4), 156-160.

Banatvala, J. (2004) Unhealthy airports, *Lancet* 364 (9435), 646-648.

Beniston, M. (2002) Climatic Change: possible impacts on human health, *Swiss Med. Wkly.* 132 (25-26), 332-337.

Brunekreef, B., and Holgate, S.T. (2002) Air pollution and health, *Lancet* 360 (9341), 1233-1242.

De Paula Santo, U., Braga, A.L., Giorgi, D.M., Pereira, L.A., Grupi, C.J., Lin, C.A., Bussacos, M.A., Zenetta, D.M., do Nascimento Saldiva, P.H., and Filho, M.T. (2005) Effects of air pollution on blood pressure and heart rate variability: a panel study of vehicular traffic controllers in the city of Sao Paulo, Brazil, *Eur. Heart J.* 26 (2), 193-200.

Di Gallo, A., Barton, J., and Parry-Jones, W.L. (1997) Road traffic accidents: early psychological consequences on children and adolescents, *Br. J. Psychiatry.* 170, 358-362.

Dora, C. (1999) A different route to health: implications of transport policies, *BMJ* 318 (7199), 1686-1689.

Dora, C., and Racioppi, F. (2003) Including health in transport policy agendas: the role of health impact assessment analyses and procedures in the European experience, *Bull World Health Organ.* 81 (6), 399-403. Epub. 2003 Jul 25.

Englert, N. (2004) Fine particles and human health–a review of epidemiological studies, *Toxicol. Lett.* 149 (1-3), 235-242.

European Environment and Health Strategy (2003) *Communication from the Commission to the council*, the European Parliament and the European Economic and Social Committee, Brussels 11.6.2003 COM(2003), 338 final.

Fujii, T., Hayashi, S., Hogg, J.C., Mukae, H., Suwa, T., Goto, Y., Vincent, R., and Eeden, S.F. (2002) Interaction of alveolar macrophages and airway epithelial cells following exposure to particulate matter produces mediators stimulate the bone marrow, *Am. J. Respir. Cell Mol. Biol.* 27 (1), 34-41.

Goldberg, M.S., Burnett, R.T., and Stieb, D. (2003) A review of time-series studies used to evaluate the short-term effects of air pollution on human health, *Rev. Environ. Health* 18 (4), 269-303.

Haines, M.M., Stansfeld, S.A., Head, J., and Job, R.F. (2002) Multilevel modelling of aircraft noise on performance tests in schools around Heathrow Airport London, *J. Epidemiol. Community Health* 56 (2), 139-144.

Katsouyianni, K. (2003) Ambient air pollution and Health, *Br. Med. Bull.* 68, 143-156.

Kim, J.J., Smorodinsky, S., Lipsett, M., Singer, B.C., Hodgson, A.T., and Ostro, E. (2004) Traffic-related Air Pollution Near Busy Roads: The East Bay Children's Respiratory Health Study, *Am. J. Respir. Crit. Care Med.* (in print).

Kunzli, N., Kaiser, R., Medina, S., Studnicka, M., Chanel, O., Filliger, P., Herry, M., Horak, F. Jr., Puybonnieux-Texier, V., Quenel, P., Schneider, J., Seethaler, R., Vergnaud, J.C., and Sommer, H. (2002) Public-health impact of outdoor and traffic-related air pollution: a European assessment, *Lancet* 356 (9232), 795-801.

Mason, C. (2000) Transport and Health: en route to a healthier Australia? *Med. J. Aust.* 172 (5), 230-232.

Mindell, J., Sheridan, L., Joffe, M., Samson-Barry, H., and Atkinson, S. (2004) Health impact assessment as an agent of policy change: improve the health impacts of the mayor of London's draft transport strategy, *J. Epidemiol. Community Health* 58 (3), 169-174.

OECD (1998) *Safety of vulnerable road users*, http://www.oecd.org/topic/0,2686,en_2649_37433_1_1_1_1_37433,00.html.

Patel, D.R., Greydanus, D.F., and Rowlet, J.D. (2002) Romance with the automobile in the 20th Century: implication adolescents in a new millennium, *Adolesc. Med.* 11 (1), 127-139.

Patz, J.A., Engelberg, D., and Last, J. (2000) The effects of changing weather on public health, *Annu. Rev. Public Health* 21, 271-307.

Peters, A., von Klot, S., Heier, M., Trentinaglia, I., Hormann, A., Wichmann, H.E., Lowel, H., and Cooperative Health Research in the Region of Augsburg Study Group (2004) Exposure to traffic and the onset of Myocardial infarction, *N. Engl. J. Med.* 351 (17), 1721-1730.

Pope, C.A. 3rd, Hansen, M.L., Long, R.W., Nielsen, K.R., Eatough, N.L., Wilson, W.E., and Eatough, D.J. (2004) Ambient particulate air pollution, heart variability, and blood markers of inflammation in a panel of elderly subjects, *Environ. Health Perspect.* 112 (3), 339-345.

Rosenlund, M., Berglind, N., Pershagen, G., Jarup, L., and Bluhm, G. (2001) Increased prevalence of hypertension in a population exposed to aircraft noise, *Occup. Environ. Med.* 58 (12), 769-773.

Sharpe, M. (1999) An unhealthy road, *J. Environ. Monit.* Apr.1(2), 23N-25N.

Smuts, M. (2001) Hazardous air pollutants: inside and out, *Public Health Rep.* 116 (1), 58-60.

Stansfeld, S., Haines, M., and Brown, B. (2000) Noise and health in the urban environment, *Rev. Environ. Health.* 15 (1-2), 43-82.

Sutherst, R.W. (2004) Global change and human vulnerability to vector-borne diseases, *Clin. Microbiol. Rev.* 17 (1), 136-173.

Suwa, T., Hogg, J.C., Vincent, R., Mukae, H., Fujii, T., and Van Eeden, S.F. (2002) Ambient air particulates stimulate alveolar macrophages of smokers to promote differentiation of myeloid precursor cells, *Exp. Lung Res.* 28 (1), 1-18.

Tarnopolsky, A., Watkins, G., and Hand, D.J. (1980) Aircraft noise and mental health: I. Prevalence of individual symptoms, *Psychol. Med.* 10 (4), 683-698.

Teague, W.G., and Bayer, C.W. (2001) Outdoor air pollution, Asthma and other concerns, *Pediatr. Clin. North Am.* 48 (5), 1167-1183, ix.

Tiwari, G. (2003) Transport and land-use policies in Delhi, *Bull. world Health Organ.* 81 (6), 444-450

Van Leeuwen, F.X. (2002) A European perspective on hazardous air pollutants, *Toxicology* 181-182, 355-359.

Vedal, S. (2002) Update on the health effects of outdoor air pollution, *Clin. Chest. Med.* 23 (4), 763-775, vi.

WHO (2000a) *Air quality guidelines for Europe*, Copenhagen, Denmark, Regional Office for Europe. Second Ed., Chapt.2.

http://www.euro.who.int/air/activities/20020620_1.

WHO (2000) *Transport, Environment and Health* (89), I-II, 1-81.

Zhang, J.J., and Lioy, P.J. (2002) Human exposure assessment in air pollution systems, *Scientific World Journal* 2, 497-513.

# TRAFFIC, NOISE AND HEALTH

W. BABISCH
*Federal Environmental Agency*
*P.O. BOX 330022*
*D-14191 Berlin, GERMANY*

## Summary

Traffic noise causes a lot of concern in the population. It annoys, disturbs sleep and can cause cardiovascular problems in chronically noise-exposed subjects. Approximately 50 million people in the European community are exposed to sound levels from road traffic at home, that are suspected of increasing the risk of cardiovascular disorders. The noise effects hypothesis is based on the general stress model. The mechanism includes arousal of the sympathetic and endocrine system. Heart rate, blood pressure, stress hormones and classical biological risk factors of ischaemic heart disease (IHD) are affected by the noise. In epidemiological studies, subjects who live in noise exposed areas show a higher prevalence of high blood pressure and IHD (including myocardial infarction). It is estimated that approximately 3 per cent of IHD cases in the general population may be attributed to traffic noise.

## 1. Introduction

It is common experience that noise is unpleasant and affects the quality of life. It disturbs and interferes with activities of the individual including concentration, communication, relaxation and sleep (WHO, 2000a and 2000b; Schwela, 2000). Besides the psychosocial effects of community noise, there is concern about the impact of noise on public health, particularly regarding cardiovascular outcomes (Suter, 1992; Passchier-Vermeer and Passchier, 2000; Stansfeld *et al.*, 2000a). Non-auditory health effects of noise have been studied in humans for a couple of decades using laboratory and empirical methods. Biological reaction models have been derived which are based on the general stress concept (Henry and Stephens, 1977; Ising *et al.*, 1980; Lercher, 1996). The noise-hypothesis is nowadays well established, and large-scale epidemiological studies have been carried out for a long time (Babisch, 2000). The epidemiological evidence of the long-term effects of

*P. Nicolopoulou-Stamati et al. (eds), Environmental Health Impacts of Transport and Mobility, 9-24.*

environmental noise on health is constantly increasing (Babisch, 2002; Babisch 2004). Studies suggest, that transportation noise is associated with adverse cardiovascular effects, in particular ischaemic heart disease. Other important health endpoints that have intensively been investigated in relation to chronic noise exposure are disrupted sleep (Ouis, 1999; Passchier-Vermeer, 2003a and 2003b), mental health (Stansfeld *et al.*, 2000b), and effects on the endocrine system (Ising and Braun, 2000; Babisch 2003). This article focuses on the impact of ambient noise on the prevalence and incidence of hypertension and ischaemic heart diseases (including myocardial infarction) as severe health endpoints.

## 1.1.    *Traffic noise level*

The A-weighted long-term average sound pressure level is used to describe the noise exposure at the facades of the people's homes ($L_{Aeq}$). A distinction is often made between the exposure during the day (6-22 hr) and the night (22-6 hr). To assess an overall indicator of the noise exposure, a weighted average was usually calculated ($L_{dn}$), giving a 10 dB(A) penalty to the night period. The new directive of the European Union on the assessment and management of environmental noise considers a weighted long-term average ($L_{den}$) of the sound pressure levels during day (e. g. 7-19 hr), evening (e. g. 19-23 hr, penalty 5 dB(A)) and night (e. g. 23-7 hr, penalty 10 dB(A)) and an un-weighted night-time noise indicator ($L_{night}$) (Directive 2002/49/EC, 2002).

In Table 1 the distributions of residential noise due to road and rail traffic in Germany are given (Umweltbundesamt, 2001). Based on model calculations, approximately 16 per cent of the German population live in 'noisy zones' (WHO Regional Office Europe, 2000) where the road traffic noise level outdoors exceeds 65 dB(A) during the day and 55 dB(A) during the night. For such noise levels considerable annoyance occurs (WHO European Centre for Environment and Health, 1995), and the cardiovascular risk tends to increase (Babisch, 2002). Only a few European countries have assessed the noise exposure completely. In European Union countries approximately 13 per cent of the population are exposed to such levels at the facades of their houses due to road traffic (EEA, 1999). For the whole of Europe the estimate is approximately 20 per cent (Schwela, 2000).

**Table 1**. Noise exposure from road and rail traffic in Germany (Umweltbundesamt, 2001).

| Sound level, outdoors [dB(A)] | Percentage of the population [%] | | | |
|---|---|---|---|---|
| | Road traffic day | Road traffic night | Rail traffic day | Rail traffic night |
| >45 - 50 | 16.4 | 17.6 | 12.4 | 15.5 |
| >50 - 55 | 15.8 | 14.3 | 14.9 | 10.8 |
| >55 - 60 | 18.0 | 9.3 | 10.4 | 6.2 |
| >60 - 65 | 15.3 | 4.2 | 6.2 | 2.7 |
| >65 - 70 | 9.0 | 2.9 | 2.3 | 0.9 |
| >70 - 75 | 5.1 | 0.2 | 0.7 | 0.4 |
| >75 | 1.5 | 0.0 | 0.1 | 0.1 |

## 1.2.    *Noise annoyance*

Figure 1 shows the relationships between the noise level (outdoors) and the percentage of 'highly' annoyed people due to road, air and rail traffic. The dose-response curves are taken from Miedema and Vos (1998). They are derived from meta-analyses considering a large number of social surveys that were carried out over the recent decades in different countries. For the same noise level, aircraft noise is more annoying than road traffic noise and railway noise. However, there are limitations regarding the universal applicability of the curves in different environmental and cultural settings (Ouis, 2002; Diaz *et al.*, 2001; Finegold and Finegold, 2003). Therefore it was suggested by a WHO working group on noise and health indicators that countries should assess their individual dose-response curves (WHO, 2003).

From the results of nationwide representative surveys carried out regularly in Germany, it is known that road traffic noise is the predominant source of annoyance for the population. Table 2 gives the results of the year 2000 (Ortscheid and Wende, 2002). Eighteen percent of the population are 'highly' annoyed by road traffic noise (categories 'extremely' and 'very' on a five-point scale). Annoyance due to noise from neighbours (6.5 percent) and air traffic (5.7 percent) follow next in the ranking.

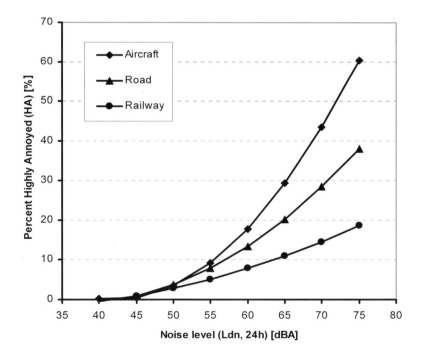

**Figure 1**. Relationship between sound level and noise annoyance (Miedema and Vos, 1998).

Table 2. Noise annoyance in Germany (Ortscheid and Wende, 2002).

| Noise source | Percentage of disturbed or annoyed subjects (%) | | | | |
|---|---|---|---|---|---|
|  | Extremely | Very | Moderately | Slightly | Not at all |
| Rail | 1.7 | 3.1 | 7.3 | 11.3 | 76.6 |
| Road | 6.4 | 11.6 | 19.4 | 26.2 | 36.3 |
| Industry | 1.4 | 3.0 | 7.4 | 16.4 | 71.7 |
| Air | 2.0 | 3.7 | 9.1 | 17.7 | 67.5 |
| Neighbour | 2.2 | 4.3 | 10.7 | 22.3 | 60.4 |

## 2.    Noise and stress

The auditory system is continuously analysing acoustic information, which is filtered and interpreted by different cortical and sub-cortical brain structures. The limbic system, including the hippocampus and the amygdala, plays an important role in the emotional processing pathways (Spreng, 2000). It has a close connection to the hypothalamus that controls the autonomic nervous system and the hormonal balance of the body. In laboratory studies, changes in blood flow, blood pressure and heart rate were found as well as increases in the release of stress hormones including the catecholamines adrenaline and noradrenaline, and the corticosteroid cortisol (Babisch, 2003; Berglund and Lindvall, 1995; Maschke et al., 2000). Such changes also occur during sleep without involvement of cortical structures due to the capacity of the amygdala to learn (plasticity), particularly with respect to adverse sound stimuli (Spreng, 2000 and 2004).

Noise is an unspecific stressor that arouses the autonomous nervous system and the endocrine system. The generalised psycho-physiological concept given by Henry and Stephens can directly be applied to noise-induced stress reaction (Henry and Stephens, 1977). The stress-mechanism as such is genetically determined. It may be modified by experience and environmental factors. Its biological function is to prepare the organism to cope with a demanding stressor. Any arousal of the sympathetic and endocrine system is associated with changes in physiological functions and the metabolism of the organism, including blood pressure, cardiac output, blood lipids (cholesterol, triglycerides, free fatty acids, phosphatides) and carbohydrates (glucose), electrolytes (magnesium, calcium), blood clotting factors (thrombocyte aggregation, blood viscosity, leukocyte count) and others (Friedman and Rosenman, 1975; Lundberg, 1999; Cohen et al., 1995). In the long term functional changes and dysregulation due to changes of physiological set points may occur, thus increasing the risk of manifest diseases. Since many of the above factors are known to be classical cardiovascular risk factors, the hypothesis has emerged that chronic noise exposure increases the risk of hypertension, arteriosclerosis and ischaemic heart disease (Suter, 1992).

Figure 2 shows the reaction schema used in epidemiological noise research for hypothesis testing (Babisch, 2002). It simplifies the cause-effect chain i.e.: sound - annoyance (noise) - physiological arousal (stress indicators) - (biological) risk factors - disease - and mortality (the latter is not explicitly considered in the graph).

The mechanism works 'directly' through synaptic nervous interactions and 'indirectly' through the emotional and the cognitive perception of the sound. It should be noted that the 'direct' pathway is relevant even at low sound levels particularly during sleep, when the organism is at its nadir of arousal. The objective noise exposure (sound level) and the subjective noise exposure (annoyance) may independently serve as exposure variables in the statistical analyses of the relationship between noise and health endpoints.

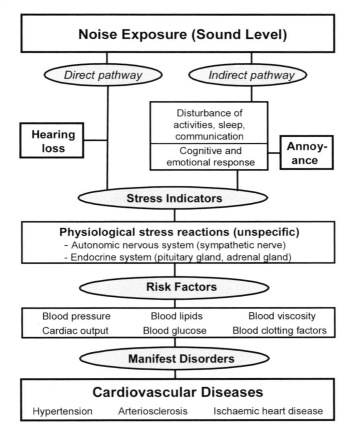

**Figure 2**. Noise effects reaction scheme (Babisch, 2002).

Principally, the effects of environmental noise cannot be extrapolated from results of occupational noise studies. The two noise environments cannot be merged into one sound energy-related dose-response model (e. g., a simple 24 hour average noise level measured with a dose-meter). Noise effects do not only depend on the sound intensity but also on the frequency spectrum, the time pattern of the sound and the individuals' activities, which are affected. For example, it may very well be that a truck driver reacts little to the sound of his engine, but is affected more if disturbed by traffic noise at home or during sleep although the exposure levels are much lower. Therefore, epidemiological studies carried out under real-life conditions are needed to assess the impact of a specific noise source on the health outcomes and provide the basis for a quantitative risk assessment. Other noise sources might act as confounders and/or effect modifiers on the association of interest. It was shown that the effects of road traffic noise (at home) were stronger in subjects that were also exposed to high noise levels at work (Babisch *et al.*, 1990).

## 3.    Epidemiological traffic noise studies

A number of epidemiological studies have been carried out in children and adults regarding changes of mean blood pressure, hypertension and ischaemic heart disease due to long-term exposure to road or aircraft noise. Most of them are cross-sectional. The few observational analytic investigations (case-control and cohort studies) refer to road traffic noise. For reviews see: Babisch (2000); Passchier-Vermeer and Passchier (2000); van Kempen *et al.* (2002).

With regard to *hypertension*, the relative risks found in four significantly positive studies, ranged between 1.5 and 3.3 for subjects who lived in areas with a daytime average sound pressure level in the range of 60-70 dB(A) or more. However, also significantly negative associations were found. Across all studies no consistent pattern for the relationship between traffic noise level and prevalence of hypertension can be seen. Dose-response relationships, which may support a causal interpretation of the findings, were rarely studied. When subjective ratings of noise or disturbances due to traffic noise were considered, the relative risks ranged from 0.8 to 2.3.

With regard to *ischaemic heart disease* (IHD), there is not much indication of an increased risk for subjects who live in areas with a daytime average sound pressure level of less than 60 dB(A) across the studies. For higher noise categories increases in risk were relatively consistently found amongst the studies. However, statistical significance was rarely achieved. Some studies permit reflections on dose-response relationships. These mostly prospective studies suggest an increase in risk for outdoor noise levels above 65-70 dB(A) during the daytime, the relative risks ranging from 1.1 to 1.5. Noise effects were larger when mediating factors like years in residence, room orientation and window opening habits were considered in the analyses. Regarding subjective responses of disturbances and annoyance, relative risks between 0.8 and 2.7 were found (Babisch, 2000).

Figure 3 sums up the results regarding IHD (Babisch, 2000). The entries are estimates of the relative risks with 95%-confidence intervals for the comparisons of extreme groups of noise exposure as given in the publications. The dark-shaded bars in the diagram refer to studies where the noise exposure was determined objectively (sound levels), the light-shaded bars where it was determined subjectively (annoyance). Road traffic and aircraft noise studies are viewed here together. No corresponding results are available for rail traffic studies. (Note: If different subgroups of the population were taken into account (males/females) or different final health outcomes observed, then specific studies would appear several times in the illustration. When a series of studies from a particular area under investigation were published in the same year, it is indicated by a serial number behind the year. For example, Amst77/1-mpoa means Amsterdam, 1977, Study 1, males, angina pectoris, objective exposure = sound level, aircraft noise). Bars shifted to the right for relative risks greater than '1' indicate an increase in risk in higher traffic noise-exposed subjects compared to those less exposed. All in all, a fairly consistent shift to higher risks can be seen.

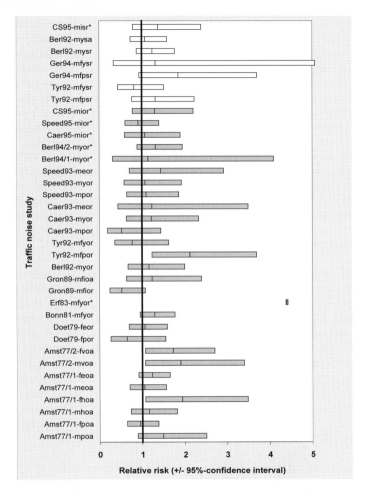

sex: f female, m male; noise measurement: o objective (sound level), dark-shaded bar; s subjective (annoyance), light-shaded bar; type of noise: a aircraft noise; r road traffic noise; ischaemic heart disease: e ECG-ischaemic symbols, h heart complaints, i ischaemic heart disease, p Angina pectoris, v cardiovascular complaints in general, y heart attack; type of study: prevalence studies; * = cohort or case-control studies.

**Figure 3**. Results of epidemiological studies on the association between traffic noise and ischaemic heart disease (extreme group comparisons) (Babisch, 2000).

In recent years, some review articles have been published trying to assess the evidence of the relationship between community noise and cardiovascular disease outcomes (Porter *et al.*, 1998; IEH, 1997; Passchier-Vermeer and Passchier, 2000; Babisch, 2000; Berglund and Lindvall, 1995; Health Council of the Netherlands, 1994; Health Council of the Netherlands, 1999). Furthermore, an attempt to summarise the findings using the approach of the meta-analyses was undertaken (van Kempen *et al.*, 2002). The evidence as concluded in the literature was characterised as follows: biochemical effects: limited evidence, hypertension: limited evidence, ischaemic heart disease: limited/sufficient (Babisch, 2002). New studies are on their way, which may improve the database. Regarding hypertension the evidence for an association tends to increase, since new studies revealed significant relative risks in exposed subjects of 1.6 (aircraft noise) and 1.9 (road traffic noise) (Rosenlund *et al.*, 2001; Maschke, 2003). With respect to myocardial infarction a new large case-control study revealed a relative risk of 1.3 in men highly exposed to traffic noise (average noise level during day >65 dB(A)), which was significant in the sub-sample of men who had been living for at least 10 years at their present address (Babisch *et al.*, 2003; Umweltbundesamt, 2004). In future, noise studies should distinguish explicitly between the noise exposure during the day and during the night because some study results suggest, that nightly noise exposure plays a fundamentally stronger role in the emergence of health disorders than noise during daytime (Maschke, 2003). Three study examples are given below.

### 3.1.    The 'Caerphilly & Speedwell heart disease studies'

Two cohorts of 2512 (Caerphilly, South Wales) and 2348 (Speedwell, England) middle-aged men (45 to 59 years) in the United Kingdom were recruited to study the predictive power of already known and of new risk factors for ischaemic heart disease (IHD). The subjects were followed-up for approx. ten years to collect new IHD cases during that period (Babisch *et al.*, 1993; Babisch *et al.*, 1998). The subjects were grouped according to 5-dB(A)-categories of the outdoor A-weighted average sound pressure level, from 6-22 hr ($L_{day}$). Due to the high correlation between day and night noise levels in the communities (correlation coefficient r = 0.94, mean difference 8 dB(A)), this noise level was used as an indicator for the overall traffic noise exposure of the streets in the study. All statistical analyses on the relationship between traffic noise and IHD incidence were controlled (model adjusted) for a number of potential confounding factors.

Figure 4 shows the relative prevalence (prevalence odds ratio and 95%-confidence intervals) of high values of biological risk factors (upper quintiles of the distributions) for the extreme group comparison of road traffic noise exposure ($L_{day}$: >65 to 70 dB(A) versus >50 to 55 dB(A)). Bars 'shifted to the right' indicate a higher prevalence in the high exposed group of men. Blood clotting factors and blood lipid factors were more prevalent in the highest exposed group compared with the lowest. However, regarding blood pressure, an opposite association was found. All in all, it was estimated that the risk for subsequent IHD was slightly higher in the exposed group (Babisch *et al.*, 1993). Although statistically not significant, the ten-year follow-up investigations revealed that the incidence in IHD was higher in subjects from the high noise exposed group (Babisch *et al.*, 1998). The estimate of the relative risk increased from 1.1 to 1.6 when effect-modifiers of exposure such as room orientations, window-opening habits and years of residence were considered. This is illustrated in Figure 5.

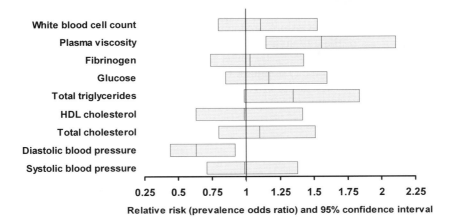

**Figure 4**. Relative prevalence of risk factors comparing men from homes with road traffic noise levels during the day of >65 to70 dB(A) with those of >50 to 55 dB(A) ['Caerphilly & Speedwell Heart Disease Studies'] (Babisch *et al.*, 1993).

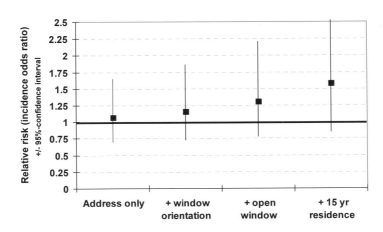

**Figure 5**. Relative risk of the incidence of ischaemic heart disease comparing men from homes with road traffic outdoor noise levels during the day of >65 to 70 dB(A) with those of >50 to 55 dB(A) - impact of the improvement of the exposure assessment ['Caerphilly & Speedwell Heart Disease Studies'] (Babisch *et al.*, 1998).

### 3.2.    The 'Berlin traffic noise studies'

In the Berlin case-control studies, interviews were conducted with 645 male patients with acute myocardial infarction (MI) collected from 17 clinics and 3390 controls taken from the population registers (Babisch *et al.*, 1994). The men were aged 41-70 years. The traffic noise exposure outdoors at the facades of the homes was taken from noise maps of the city authorities. In the subgroup of men who lived in their homes for at least fifteen years, the odds ratio (95%-confidence interval) of IHD incidence was 1.3 when the two highest categories of the noise level during the day were taken together (>70 to 80 dB(A)) as compared to the lowest noise category (≤60 dB(A)), after adjustment for potentially confounding factors. This is illustrated in Figure 6.

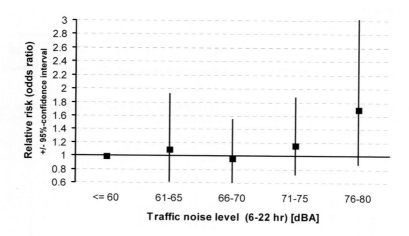

**Figure 6**. Relative risk of the incidence of ischaemic heart disease in middle aged men compared with the road traffic outdoor noise level ['Berlin Traffic Noise Studies'] (Babisch *et al.*, 1994).

### 3.3.    The 'Spandau health survey'

In the Spandau health survey, the road traffic noise exposure at the homes of 1718 subjects who took part in a follow-up investigation on the development of clinical health endpoints, was assessed using official noise maps of the area (Maschke *et al.*, 2003; Maschke 2003). The subjects, males and females, were aged 18 to 90 years. After adjustment for potentially confounding factors, cross-sectional analyses revealed a significantly higher prevalence of high blood pressure (hypertension) with increasing noise level outside the bedroom. This is shown in Figure 7. A significant increase in the risk was noted for the prevalence, if the equivalent continuous sound pressure level of the nocturnal street traffic noise exceeded 55 dB(A). The relative risk was 1.9 in comparison with locations where the equivalent continuous sound pressure level was below 50 dB(A). With regard to the noise exposure of the living rooms during the day, the results showed the same tendency of a dose-response relationship, but were statistically not significant and the magnitude of the effect was smaller. On the other hand, when only subjects were considered that slept with open windows, the effects were much larger, indicating that the noise at the sleeper's ear was of importance.

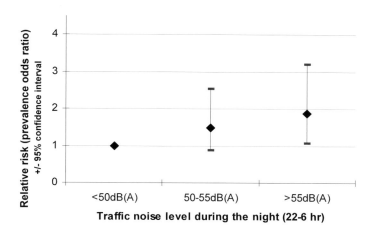

**Figure 7**. Relative risk of the prevalence of hypertension in middle aged men compared with on the road traffic noise exposure of the bedroom ['Spandau Health Survey'] (Maschke, 2003).

## 4.   Risk evaluation process

A conceptual framework for the regulation of environmental hazards was given by the US National Research Council (National Research Council, 1983; Neus and Boikat, 2000; Patton, 1993). The process of risk assessment (risk evaluation) comprises hazard identification ('Which health outcome is relevant?'), exposure assessment ('How many are affected') and dose-response assessment ('Threshold of effect?') (Babisch, 2002). This information is summarised in, and is called, 'risk characterisation' or 'health hazard characterisation' (Neus and Boikat, 2000). It involves the interpretation of the available evidence from the available data and other scientific disciplines, and is subject to the discussion of uncertainties including chance, bias and validity of studies as well as transparency, replicability and comprehensiveness of reviews. As a result of the risk evaluation process, a quantitative estimate about the likelihood that the hazard will affect exposed people will be derived. Usually attributable risk percentages will be calculated (Walter, 1998). This will serve as key information for any kind of risk management including regulatory options (Jasanoff, 1993). This is illustrated in Figure 8.

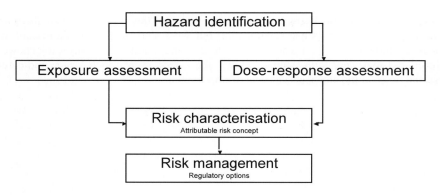

**Figure 8**. Process of risk evaluation (Neus and Boikat, 2000).


## 5.  Conclusion

Nowadays the noise and cardiovascular risk hypothesis is well established. The biological plausibility is based on clinical and experimental research and has been established for a long time on the basis of laboratory and animal experiments. For reviews see for example: Borg (1981); Kjellberg (1990); Berglund and Lindvall (1995); Babisch (2003). However, a quantitative risk assessment requires epidemiological research in this field. Only some studies of the relationship between road traffic noise and cardiovascular endpoints permit reflections on dose-response relationships. The mostly prospective studies regarding ischaemic heart diseases (IHD) suggest an increase in risk for outdoor noise levels above 65-70 dB(A) during the daytime ($L_{day}$), the relative risks (RR) ranging from 1.1 to 1.5. Approximately 13-20 per cent of the European population and 16 per cent of the German population are exposed ($P_e$) to such road traffic noise levels during the day at the facades of their homes. Assuming a 20 per cent increase in the IHD risk in these subjects (relative risk of 1.2), 17 per cent of the IHD cases would be attributable to the noise. With regard to the entire population (using the German figure of 16 per cent exposed), the population attributable risk percentage (PAR%) would be approximately 3, meaning that 3 per cent of all IHD cases would be attributable to the traffic noise (PAR% $= [P_e * (RR-1)] / [P_e * (RR-1) + 1] * 100$) (Hennekens and Buring, 1987). These figures can be relevant for decision-making in public health policy.

Since the difference between average levels of road traffic in the daytime ($L_{day}$) and at night ($L_{night}$) is normally somewhat less than 10 dB(A) in urban areas (no motorways) (Utley, 1985), and the 24-hour noise levels ($L_{24h}$) are usually 1 to 3 dB(A) lower than daytime noise levels (Rylander *et al.*, 1986), $L_{Day}$ is a good indicator for the overall traffic noise exposure from the streets. It is highly correlated with other noise metrics such as the weighted noise indicator $L_{DEN}$ according to the noise directive of the European Union (Directive 2002/49/EC, 2002). The risk assessment given above with respect to the average noise level of 65–70 dB(A) during the day corresponds with 55–60 dB(A) at night, in this respect. (Note: This approximation is only valid for urban road traffic noise).

Since at present no studies are available of data that provide a quantitative dose-response relationship between aircraft noise and cardiovascular endpoints, no other alternative exists than the approximate transfer of the quantitative risk observations derived from road traffic noise studies to aircraft noise. Contrary to road traffic noise, aircraft noise is to be heard on all sides of a building. Due to the lack of evasive possibilities from aircraft noise, and the greater annoyance reactions to aircraft noise (Miedema and Vos, 1998), the suspicion exists that the effects induced by aircraft noise could be greater than those induced by road traffic. This will have to be monitored in future studies.

# References

Babisch, W. (2000) Traffic noise and cardiovascular disease: epidemiological review and synthesis, *Noise & Health* 2 (8), 9-32.

Babisch, W. (2002) The noise/stress concept, risk assessment and research needs, *Noise & Health* 4 (16), 1-11.

Babisch, W. (2003) Stress hormones in the research on cardiovascular effects of noise, *Noise & Health* 5 (18), 1-11.

Babisch, W. (2004) Health aspects of extra-aural noise research, *Noise and Health* 6 (22), 69-81.

Babisch, W., Ising, H., Elwood, P.C., Sharp, D.S., and Bainton, D. (1993) Traffic noise and cardiovascular risk: the Caerphilly and Speedwell studies, second phase. Risk estimation, prevalence, and incidence of ischemic heart disease, *Archives of Environmental Health* 48, 406-413.

Babisch, W., Ising, H., Kruppa, B., and Wiens, D. (1994) The incidence of myocardial infarction and its relation to road traffic noise - the Berlin case-control studies, *Environment International* 20, 469-474.

Babisch, W., Ising, H., Gallacher, J.E.J., Elwood, P.C., Sweetnam, P.M., Yarnell, J.W.G., Bainton, D., and Baker, I.A. (1990) Traffic noise, work noise and cardiovascular risk factors: The Caerphilly and Speedwell Collaborative Heart Disease Studies, *Environment International* 16, 425-435.

Babisch, W., Ising, H., Gallacher, J.E.J., Sweetnam, P.M., and Elwood, P.C. (1998) Traffic noise and cardiovascular risk: The Caerphilly and Speedwell studies, third phase - 10 years follow-up, *Archives of Environmental Health* 54, 210-216.

Babisch, W., Beule, B., Schust, M., and Stark, H. (2003) Traffic noise and myocardial infarction. Results from the NaRoMI Study (Noise and Risk of Myocardial Infarction), in de Jong, R., Houtgast, T., Franssen, E.A.M., and Hofman, W. (eds) *Proceedings of the 8th International Congress on Noise as a Public Health Problem in Rotterdam*, Foundation ICBEN 2003, Schiedam, pp. 96-101.

Berglund, B., and Lindvall, T. (1995) Community Noise, Document prepared for the World Health Organization, *Archives of the Center for Sensory Research* 2 (1), Center for Sensory Research, Stockholm.

Borg, E. (1981) Physiological and pathogenic effects of sound, *Acta. Otolaryngol. Suppl.* 381, 3-67.

Cohen, S., Kessler, R.C., and Underwood, G.L. (1995) Strategies for measuring stress in studies of psychiatric and physical disorders, in Cohen, S., Kessler, R.C., and Underwood,

G.L. (eds) *A Guide for Health and Social Scientists*, Oxford University Press, New York, pp. 3-26.

Diaz, J., Ekelund, M., Gothe, R., Huber, M., Jordan, A., Kallischnigg, G., Kampet, T., Kappos, A., López Santiago, C., Hons, T.M.B.A., Maschke, C., Niemann, H., and Welker, D. (2001) *Traffic noise pollution. "Similarities and differences between European regions" - A state of the art review*, The European Commission, Directorate-General Health and Consumer Protection, Luxembourg.

Directive 2002/49/EC (2002) Directive of the European Parliament and the Council of 25 June 2002 relating to the assessment and management of environmental noise, *Official Journal of the European Communities* L 189/12, 12-25.

EEA, European Environment Agency (1999) *Indicator 4: traffic noise: exposure and annoyance*, URL: http://reports.eea.eu.int/ENVISSUENo12/en/page009.html.

Finegold, L.S., and Finegold, M.S. (2003) Development of exposure-response relationships between transportation noise and community annoyance, *J. Aviation Environ. Res.* 7 (Suppl.), 11-23.

Friedman, M., and Rosenman, R.H. (1975) *Der A-Typ und der B-Typ*, Rowohlt Verlag Gmbh, Reinbek bei Hamburg.

Health Council of the Netherlands (1994) *Noise and Health*, Report by a committee of the Health Council of the Netherlands, Health Council of the Netherlands, The Hague.

Health Council of the Netherlands (1999) *Public Health Impact of Large Airports*, Report by a committee of the Health Council of the Netherlands, Health Council of the Netherlands, The Hague.

Hennekens, C. H., and Buring, J. E. (1987) *Epidemiology in Medicine*, Little, Brown and Company, Boston/Toronto.

Henry, J.P., and Stephens, P.M. (1977) *Stress, Health, and the Social Environment, a Sociobiologic Approach to Medicine*, Springer-Verlag, New York.

IEH - Institute for Environment and Health (1997) *Workshop on Non-Auditory Health Effects of Noise*, Draft report, Institute for Environment and Health, Leicester.

Ising, H., Dienel, D., Günther, T., and Markert, B. (1980) Health effects of traffic noise, *Int. Arch. Occup. Environ. Health* 47, 179-190.

Ising, H., and Braun, C. (2000) Acute and chronic endocrine effects of noise: review of the research conducted at the Institute for Water, Soil and Air Hygiene, *Noise & Health* 2 (7), 7-24.

Jasanoff, S. (1993) Relating risk assessment and risk management. Complete separation of the two processes is a misconception, *EPA Journal* Jan/Feb/Mar 1993, 35-37.

Kjellberg, A. (1990) Subjective, behavioral and psychophysiological effects of noise, *Scand. J. Work Environ. Health* 16 (Suppl. 1), 29-38.

Lercher, P. (1996) Environmental noise and health: an integrated research perspective, *Environment International* 22, 117-128.

Lundberg, U. (1999) Coping with stress: neuroendocrine reactions and implications for health, *Noise & Health* 4, 67-74.

Maschke, C., Rupp, T., and Hecht, K. (2000) The influence of stressors on biochemical reactions - a review of present scientific findings with noise, *Int. J. Hyg. Environ. Health* 203, 45-53.

Maschke, C., Wolf, U., and Leitmann, T. (2003) *Epidemiological examinations of the influence of noise stress on the immune system and the emergence of arteriosclerosis*, Report 298 62 515 (in German, executive summary in English), WaBoLu-Hefte 01/03, Umweltbundesamt, Berlin.

Maschke, C. (2003) Epidemiological research on stress caused by traffic noise and its effects on high blood pressure and psychic disturbances, in de Jong, R., Houtgast, T., Franssen, E.A.M., and Hofman, W. (eds.) *Proceedings of the 8th International Congress on Noise as a Public Health Problem in Rotterdam*, Foundation ICBEN 2003, Schiedam, pp. 93-95.

Miedema, H.M.E., and Vos, H. (1998) Exposure-response relationships for transportation noise, *J. Acoust. Soc. Am.* 104, 3432-3445.

National Research Council (1983) *Risk Assessment in the Federal Government. Managing the Process*, National Academy Press, Washington DC.

Neus, H., and Boikat, U. (2000) Evaluation of traffic noise-related cardiovascular risk, *Noise and Health* 2 (7), 65-77.

Ortscheid, J., and Wende, H. (2002) Lärmbelästigung in Deutschland, *Zeitschrift für Lärmbekämpfung* 49, 41-45.

Ouis, D. (1999) Exposure to nocturnal road traffic noise: sleep disturbance and its after effects, *Noise & Health* 1 (4), 11-36.

Ouis, D. (2002) Annoyance caused by exposure to road traffic noise: an update, *Noise & Health* 4 (15), 69-79.

Passchier-Vermeer, W. (2003a) Relationship between environmental noise and health, *J. Aviation Environ. Res.* 7 (Suppl.), 35-44.

Passchier-Vermeer (2003b) *Night-time noise events and awakening*, TNO Inro report 2003-32, ISBN 90-5986-021-7, TNO Institute for Traffic and Transport, Delft.

Passchier-Vermeer, W., and Passchier, W.F. (2000) Noise exposure and public health, *Environmental Health Perspectives* 108 (suppl. 1), 123-131.

Patton, D.E. (1993) The ABCs of risk assessment, *EPA Journal* Jan/Feb/Mar 1993, 10-15.

Porter, N.D., Flindell, I.H., and Berry, B.F. (1998) *Health Effect-Based Noise Assessment Methods: a Review and Feasibility Study*, National Physical Laboratory, Teddington.

Rosenlund, M., Berglind, N., Pershagen, G., Järup, L., and Bluhm, G. (2001) Increased prevalence of hypertension in a population exposed to aircraft noise, *Occup. Environ. Med.* 58, 769-773.

Rylander, R., Bjorkman, M., Ahrlin, U., Arntzen, E., and Solberg, S. (1986) Dose-response relationships for traffic noise and annoyance, *Archives of Environmental Health* 41, 7-10.

Schwela, D.H. (2000) The World Health Organization guidelines for environmental health, *Noise/News International* 2000 March, 9-22.

Spreng, M. (2000) Central nervous system activation by noise, *Noise & Health* 2 (7), 49-58.

Spreng, M. (2004) Noise induced nocturnal cortisol secretion and tolerable overhead flights, *Noise & Health* 6 (22), 35-47.

Stansfeld, S., Haines, M., and Brown, B. (2000a) Noise and health in the urban environment, *Reviews on Environmental Health* 15 (1-2), 43-82.

Stansfeld, S.A., Haines, M.M., Burr, M., Berry, B., and Lercher, P. (2000b) A review of environmental noise and mental health, *Noise & Health* 2 (8), 1-8.

Suter, A.H. (1992) Noise sources and effects - a new look, *Sound and Vibration 25th anniversary issue*, 18-38.

Umweltbundesamt (2001) *Daten zur Umwelt.* Der Zustand der Umwelt in Deutschland 2000, Erich Schmidt Verlag GmbH & Co., Berlin.

Umweltbundesamt (2004) Chronic Noise as a Risik Factor for Myocardial Infarction. Results of the "NaRoMI"Study, Babisch, W. (ed), *Report 297 61 003* (in German, executive summary in English), WaBoLu-Hefte 02/04, Umweltbundesamt, Berlin.

Utley, W.A. (1985) Descriptors for ambient noise, in Bundesanstalt für Arbeitsschutz (ed.) *Proceedings of the International Conference on Noise Control Engineering in Munich*, Federal Institute for Occupational Safety Report Series, Tb Nr. 39, Wirtschaftsverlag NW, Verlag für neue Wissenschaft GmbH, Bremerhaven, pp. 1069-1073.

van Kempen, E.E.M.M., Kruize, H., Boshuizen, H.C., Ameling, C.B., Staatsen, B.A.M., and de Hollander, A.E.M. (2002) The association between noise exposure and blood pressure and ischemic heart disease: a meta-analysis, *Environ. Health Perspect.* 110, 307-317.

Walter, S.D. (1998) Attributable risk in practice, *American Journal of Epidemiology* 148, 411-413.

WHO - World Health Organisation - European Centre for Environment and Health (1995) *Concern for Europe's Tomorrow. Health and the Environment in the WHO European*

*Region.* Chapter 13: Residential Noise, Wissenschaftliche Verlagsgesellschaft mbH, Stuttgart.

WHO - World Health Organisation (2000a) *Guidelines for Community Noise*, World Health Organisation,

URL: http://www.who.int/environmental_information/Information_resources/ on_line_noise.htm, Geneva.

WHO - World Health Organisation - Regional Office for Europe (2000b) *Noise and health*, World Health Organisation, Copenhagen.

WHO - World Health Organisation - Regional Office for Europe (2003) *Environmental Health Indicators*,

URL: http://www.euro.who.int/eprise/main/who/progs/ehi/home.

# ENVIRONMENTAL NOISE, ANNOYANCE AND SLEEP DISTURBANCE

W. PASSCHIER-VERMEER[1] AND W. F. PASSCHIER[2]
*[1]TNO - Environment and Geosciences*
*Department of Environment and Health*
*P.O.BOX 6041, 2600 JA Delft, THE NETHERLANDS*
*[2]Universiteit Maastricht*
*Department of Health Risk Analysis and Toxicology*
*P.O. BOX 616, 6200 MD Maastricht, THE NETHERLANDS*

**Summary**

A rough estimate of the number of people in the EU exposed to environmental noise (from road traffic, railway traffic and aircraft) above a day-evening-night-level of 55 dB(A) is 150 million (40 per cent), including about 120 million people exposed to road traffic noise. Adverse environmental noise-induced health effects mainly are annoyance, sleep disturbance, stress-related somatic effects, effects on learning in children, and possibly hearing damage. These effects occur in a substantial part of the EU population. In this chapter the relationships between annoyance and noise exposure to various types of environmental noise are given. With respect to sleep disturbance, this chapter discusses effects of night time noise on motility (motoric unrest), self-reported sleep disturbance, and self-assessed awakening.

## 1.    Introduction

Adverse noise-induced health effects are diverse (WHO, 2000). In the chapter *Traffic, noise and health*, Babisch discusses stress-related noise-induced effects. This chapter presents an overview on annoyance and sleep disturbance. Where available, exposure-effect relationships are given. At the end of this chapter, estimates are given about the extent of environmental noise exposure in Europe and possible differences in annoyance sensitivity between populations in different parts of the EU.

*P. Nicolopoulou-Stamati et al. (eds), Environmental Health Impacts of Transport and Mobility*, 25-38.
© 2005 *Springer. Printed in the Netherlands.*

## 2.    Noise metrics used

Noise exposure is described by various metrics. The metrics have been developed in view of their correlation with certain biological effects. Here we focus on metrics for the description of noise related annoyance and sleep disturbance.

The human hearing organ is not equally sensitive to sounds of different frequencies, therefore the sound pressure level ($L$) is A-weighted, and expressed as dB(A). Noise metrics based on the A-weighted sound pressure level and used in this chapter are:

$L_{Aeq,T}$   The equivalent sound level, i.e. the (exponential) average sound level over a period of time $T$.

$L_{den}$   Day-evening-night-level. This metric is the equivalent sound level over a 24-hour period, with the sound levels increased by 5 dB(A) and 10 dB(A) during evening time (19 – 23 h) and night time (23 – 7 h) respectively.

$L_{night}$   The equivalent sound level over night time (23 – 7 h).

$L_{max}$   The maximal sound level occurring during a single noise event.

$SEL$   Sound Exposure Level. A metric also used to describe single noise events. This metric is the equivalent sound level during the event, normalised to a period of 1 second.

To evaluate an environmental noise situation, usually values of these noise metrics are assessed outdoors. Some noise-induced effects, such as awakening and hearing impairment, have been related to indoor noise metrics.

## 3.    Effects of noise exposure on health

Table 3 gives an overview of the state-of-the-art of noise effects research. In 1994, the Committee on Noise and Health, an international committee of the Health Council of the Netherlands (HCN), assessed the health effects of environmental and occupational noise exposure (HCN, 1994; Passchier-Vermeer, 1993a). Table 3 is based on the 1994 Health Council report, but has been slightly updated since (Passchier-Vermeer and Passchier, 2000). In the latter publication recent reviews have been taken into consideration (Institute for Environment and Health, 1997; Morrell, et al., 1997; HCN, 1999; Shaw, 1996; Job, 1996; Lercher et al., 1998; Carter, 1998). The table presents those effects for which sufficient scientific evidence is available to consider them to be caused by noise. Table 3 also presents the observation threshold for an effect. The observation threshold for a noise-induced effect is the lowest noise exposure at which for an average population the effect has been observed in epidemiological studies.

Section 4 and 5 of this chapter focus on aspects of annoyance and sleep disturbance, most other effects are discussed in the chapter by Babisch, with the exception of hearing impairment. ISO 1999 (1990) presents a method to estimate noise-induced hearing impairment in populations exposed to occupational noise. In Passchier-Vermeer

(1993b) it has been made plausible that the relations presented in ISO 1999 are also applicable to environmental noise exposure, if the exposure over 24-hours is taken into account. Most likely, environmental noise exposures of the Dutch and comparable populations are too low to cause detectable hearing loss, but environmental noise exposure in mega-cities appears to be sufficiently high to cause noise-induced hearing impairment (Passchier-Vermeer, 1993b; HCN, 1994; WHO, 2000).

**Table 3**. Long-term effects of exposure to environmental noise, and information on their observation threshold.

| Effect | Observation threshold | | | References |
|--------|-----------------------|---|---|------------|
| | Metric | Value in dB(A) | Indoors/ Outdoors | |
| Hearing impairment | $L_{Aeq,24h}$ | 70 | in | ISO, 1990 |
| Hypertension | $L_{den}$ | 70 | out | HCN, 1994 |
| Ischaemic heart disease | $L_{den}$ | 70 | out | HCN, 1994 |
| Annoyance | $L_{den}$ | for %HA[#] 42 | out | Miedema and Oudshoorn, 2001 |
| Performance of children | $L_{Aeq,school}$ | 70 | out | Passchier-Vermeer, 2000 |
| *Effects related to sleep disturbance* | | | | |
| Behavioural awakening | | | | Fidell *et al.*, 1995a; b; 1998; Finegold and Elias, 2002; Passchier-Vermeer, 2003 |
| Self-reported sleep disturbance | $L_{night}$ | for %HSD[§] < 45 | out at the most exposed façade | Miedema *et al.*, 2003 |
| Increased motility, Motoric unrest | $L_{night}$, SEL $L_{max}$ | | | Fidell *et al.*, 1995a; b; 1998; Ollerhead *et al.*, 1992; Griefahn *et al.*, 1999; Passchier-Vermeer *et al.*, 2002a |
| Changes in EEG parameters: sleep pattern, sleep stages EEG-awakening | *SEL* | Various values | in | HCN, 1994; 1997 |
| Heart rate Hormone levels | *SEL* | 40 | in | HCN, 1994 Babisch *et al.*, 2001 Maschke *et al.*, 2002 |
| Mood next day | $L_{Aeq,night}$ | >60 | out | HCN, 1994 |

# HA – Highly Annoyed
§ HSD – Highly Sleep Disturbed

## 4.  Annoyance

Annoyance is any feeling of resentment, displeasure, discomfort, and irritation occurring when noise intrudes into someone's thoughts and moods or interferes with activity. Noise is annoying only when it is considered not to fit with current intentions. Annoyance in populations is evaluated using questionnaires. In this section exposure-effect relationships for noise annoyance are presented (Table 4) as derived at TNO from a large international archive of noise annoyance studies with about 60,000 people exposed to environmental noise (Miedema and Oudshoorn, 2001).

Annoyance questions in different international studies do not use the same number of response categories. In order to obtain comparable annoyance measures for all different studies, all sets of response categories have been converted to a scale ranging from 0 to 100. In the definition of annoyance a cut-off point is chosen of 72 for the percentage highly annoyed (%HA), 50 for the percentage annoyed (%A) and 28 for the percentage (at least) a little annoyed (%LA).

It should be pointed out that, although the expressions in Table 4 (and the curves in Figures 9 and 10) appear to be very exact, they represent averages over large populations that were chronically exposed to noise. Especially on a local level a prediction of the fraction annoyed or highly annoyed people based on a measured (or modelled) *Lden*-value may differ considerably from the outcome of a survey.

**Table 4**. Relationships between annoyance and $L_{den}$ (day-evening-night level) for exposure to various traffic noise sources in the living environment. Exposure-effect relationships relate to adult populations (Miedema and Oudshoorn, 2001).

| Exposure-effect relationship |
| --- |
| Aircraft (little annoyed, annoyed, highly annoyed), |

%LA = -6.158 * $10^{-4}$ $(L_{den}-32)^3$ + 3.410 * $10^{-2}$ $(L_{den}-32)^2$ + 1.738 $(L_{den}-32)$,

%A  = 8.588 * $10^{-6}$ $(L_{den}-37)^3$ + 1.777 * $10^{-2}$ $(L_{den}-37)^2$ + 1.221 * $(L_{den}-37)$,

%HA = -9.199 * $10^{-5}$ $(L_{den}-42)^3$ + 3.932 * $10^{-2}$ $(L_{den}-42)^2$ + 0.2939 $(L_{den}-42)$.

Road traffic

%LA = -6.235 * $10^{-4}$ $(L_{den}-32)^3$ + 5.509 * $10^{-2}$ $(L_{den}-32)^2$ + 0.6693 $(L_{den}-32)$,

%A = 1.795 * $10^{-4}$ $(L_{den}-37)^3$ + 2.110 * $10^{-2}$ $(L_{den}-37)^2$ + 0.5353 * $(L_{den}-37)$,

%HA = 9.868 * $10^{-4}$ $(L_{den}-42)^3$ − 1.436 * $10^{-2}$ $(L_{den}-42)^2$ + 0.5118 $(L_{den}-42)$.

Railway traffic

%LA = -3.229 * $10^{-4}$ $(L_{den}-32)^3$ + 4.871 * $10^{-2}$ $(L_{den}-32)^2$ + 0.1673 $(L_{den}-32)$,

%A = 4.538 * $10^{-4}$ $(L_{den}-37)^3$ + 9.482 * $10^{-3}$ $(L_{den}-37)^2$ + 0.2129 * $(L_{den}-37)$,

%HA = 7.239 * $10^{-4}$ $(L_{den}-42)^3$ − 7.851 * $10^{-3}$ $(L_{den}-42)^2$ + 0.1695 $(L_{den}-42)$.

In Figure 9 the fraction of highly annoyed people, %HA, is given as a function of $L_{den}$ using the expressions of Table 4. Figure 10 presents the variation in degree of annoyance due to aircraft noise. Factors with a significant impact on the relationships are listed in Table 5 (Miedema and Vos, 1999), see also Guski (1999), Job (1999) and Stallen (1999).

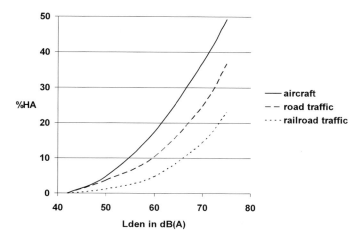

**Figure 9**. Percentage of people highly annoyed by aircraft, road traffic, and railway traffic noise as a function of day-evening-night-level (Miedema and Oudshoorn, 2001).

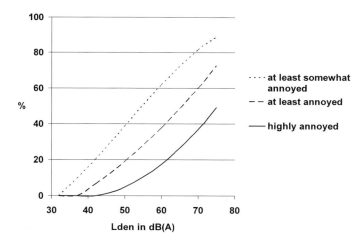

**Figure 10**. Percentage of people at least somewhat annoyed, at least annoyed, or highly annoyed by aircraft noise, as a function of day-evening-night-level (Miedema and Oudshoorn, 2001).

The data in Table 5 explain part of the differences in predicted and measured values of annoyance. Here too the data pertain to group averages of chronically exposed populations.

**Table 5**. Factors with an effect on annoyance. The difference in annoyance between the 5 per cent most and 5 per cent least annoyed subgroup, converted to a difference in $L_{den}$, and the most annoyed subgroup (Miedema and Vos, 1999).

| Factor | Difference in annoyance (converted to dB(A)) | Most annoyed subgroup |
|---|---|---|
| Gender | 0 | - |
| Age | 5 | 30 – 40 years |
| Education | 2 | university |
| Occupation | 1 | highest level |
| Size of household | 2 | two persons |
| Home ownership | 2 | owner |
| Economic dependency on source | 3 | not dependent on source |
| Use of mode of transport | 2 | if mode of transport is not used |
| Noise sensitivity | 11 | highly sensitive |
| Fear related to source | 19 | very much afraid |

## 5.   Sleep disturbance[1]

### 5.1.   Introduction

Sleep is an active physiological process, and not only the absence of waking. It has a restorative function that cannot be fulfilled by food, drink, or drug. Therefore it is plausible that a disturbed sleep negatively affects health (Carter, 1998; Passchier-Vermeer, 1993a).

Sleep disturbance can be described with physiological, hormonal and motility measures, and on the basis of self-reported observations or evaluations. This chapter is limited to motility, motoric unrest, self-registered (behavioural) awakenings, and self-reported sleep disturbance. Although several field investigations showed effects of night time noise exposure on cardiovascular and hormonal functions, exposure-effect relationships for those effect variables are not (yet) available (Passchier-Vermeer, 1993a; Ising et al., 1997; Harder et al., 1999; Babisch et al., 2001; Maschke et al., 2002).

The instantaneous physiological, hormonal, and motility effects of noise exposure during sleep can be understood as follows. Noise is a carrier of information, also with respect to possible danger in the surroundings of a person. When noise reaches a person, a signal is transmitted from the hearing organ to the cortex and interpreted. Besides, a signal is also transferred to lower parts of the central nervous system, which results, in anticipation of a physical response to a dangerous situation, in

---

[1] *In 2004 the Health Council of the Netherlands published a comprehensive review on the relationship betwee environmental noise exposure, sleep and health (Gezondheidsraad. Over de invloed van geluid op de slaap en de gezondheid [The Influence of Night-time Noise on Sleep and Health]. Den Haag: Gezondheidsraad; 2004; Publicatie nr 2004/14. In Dutch and English.)*

instantaneous changes in autonomous functions. Changes are for example an increase in stress hormones and cortisol levels in blood, an increase in heart rate and blood pressure, an increase in muscular tone which induces small movements. When the situation has been interpreted as safe, the autonomous functions return to their earlier equilibrium. These instantaneous noise-induced changes in autonomous functions have long-term effects on sleep quality, and presumably also on health (Carter, 1998; Passchier-Vermeer et al., 2002a; Born and Fehm, 2000). This process occurs during sleep at low noise levels of a noise event, if low background noise levels do not mask the presence of such a noise event.

## 5.2. Self-reported sleep disturbance

Self-reported sleep disturbance in populations is evaluated by using questionnaires. Relationships between self-reported sleep disturbance and night time noise exposure have been assessed at TNO from the large international archive of noise annoyance, including sleep disturbance, studies mentioned above. Relationships between self-reported sleep disturbance and $L_{night}$ at the most exposed façade outside dwellings are given in Miedema et al., (2003). Analogously to the definitions of %HA, %A, and %LA, the percentages of sleep disturbed persons (%HSD, %SD, and %LSD) have been specified as a function of $L_{night}$. In Figure 11 self-reported sleep disturbance has been presented for road traffic and railway traffic from Miedema et al. (2003). Since a relationship for aircraft noise could not be assessed by using the analysis technique described in Miedema et al. (2003), a preliminary relationship for aircraft noise is included in Figure 11 (HCN, 1994). In Figure 12 the variation in self-reported sleep disturbance is given.

The relationships between self-reported sleep disturbance and $L_{night}$ at the most exposed façade are given in Table 6. The curves in Figure 11 and Figure 12 demonstrate that, in terms of self-reported sleep disturbance, noise - and more specifically traffic noise exposure - clearly affects sleep, even at not uncommon outdoor levels.

**Table 6.** Relationships between self-reported sleep disturbance and the outdoors noise level $L_{night}$ for road and railway traffic (Miedema et al., 2003).

---

Only a preliminary relationship for aircraft noise

Road traffic (highly sleep disturbed, at least sleep disturbed, at least a little sleep disturbed ($L_{night}$ between 45 and 65 dB(A), at the most exposed façade),

$$\%HSD = 20.8 - 1.05L_{night} + 0.01486L_{night}^2$$
$$\%SD = 13.8 - 0.85L_{night} + 0.01670L_{night}^2$$
$$\%LSD = -8.4 + 0.16L_{night} + 0.01081L_{night}^2,$$

Railway traffic (highly sleep disturbed, at least sleep disturbed, at least a little sleep disturbed ($L_{night}$ between 45 and 65 dB(A), at the most exposed façade),

$$\%HSD = 11.3 - 0.55L_{night} + 0.00759L_{night}^2$$
$$\%SD = 12.5 - 0.66L_{night} + 0.01121L_{night}^2$$
$$\%LSD = 4.7 - 0.31L_{night} + 0.01125L_{night}^2.$$

---

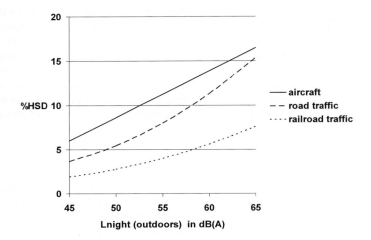

**Figure 11**. Percentage of people highly sleep disturbed, as a function of the outdoors night time equivalent sound level (Road and railway traffic: source Miedema *et al.*, 2003; Aircraft noise, preliminary relation; source HCN, 1994).

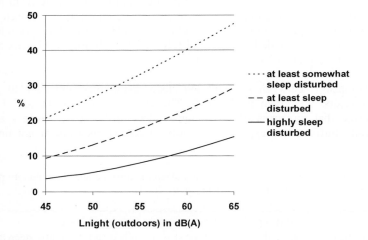

**Figure 12**. Percentage of people at least somewhat sleep disturbed, at least sleep disturbed, or highly sleep disturbed by road traffic noise, as a function of the outdoors equivalent sound level during the night (Miedema *et al.*, 2003).

### 5.3. *Behavioural awakening*

Behavioural awakening during a sleep period is usually registered by a person by pressing a button after intermittent awakening. Passchier-Vermeer (2003) reanalysed the data from eight studies (see also Fidell *et al.*, 1998; Finegold and Elias, 2002). Figure 13 shows the percentage of noise-induced reported awakenings by pressing a marker in a 5-min interval around a noise event as a function of indoor *SEL* of the noise event. The data are shown for four types of sources. The 110 data points in Figure 13 represent over 110 000 5-min intervals with noise events. A relationship between behavioural awakening and indoor *SEL* is being derived taking into account the data quality of the separate studies (Passchier-Vermeer, 2003). Table 7 gives the overall result. In all 113 157 5-min intervals with noise events, the total number of awakenings is 2058 (1.81 per cent), including 100 noise-induced awakenings.

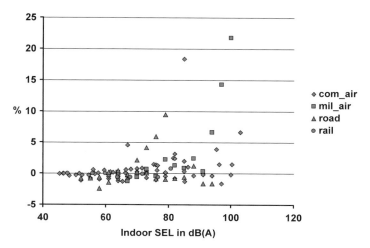

**Figure 13**. Percentage noise-induced awakenings, assessed by pressing a marker, as a function of indoor SEL, for four types of noise sources. For the explanation of the abbreviations see Table 7.

**Table 7**. Number of 5-min intervals with noise events and number of noise-induced marker pressings during these intervals, specified according to source (Passchier-Vermeer, 2003).

| Source | Number of 5-min intervals with noise events | Number of awakenings during the 5-min intervals with noise events | |
| --- | --- | --- | --- |
| | | All | Noise-induced[#] |
| Commercial aircraft | 95365 | 1691 | 41 |
| Military aircraft | 4104 | 119 | 48 |
| Road traffic | 12793 | 233 | 12 |
| Railway traffic | 895 | 15 | -1 |
| Total | 113157 | 2058 | 100 |

# Calculated as the number of all registered awakenings (third column) minus the predicted number of awakenings if no noise event would have been present in the interval

Table 7 shows a very low percentage of behavioural awakening (1.82 per cent) and of noise-induced behavioural awakening (0.09 per cent) during the 5-min intervals with noise events. This is in line with the finding in Passchier-Vermeer *et al.* (2002a) about awakenings which subjects remember after wake-up. On average 0.32 per cent extra remembered aircraft noise-induced awakenings during a 5-min interval with aircraft noise were assessed. For commercial aircraft noise a reliable relationship between percentage noise-induced behavioural awakenings and indoor SEL could be specified (Passchier-Vermeer, 2003). The probability of noise-induced behavioural awakenings increases from indoor SEL equal to 54 dB(A) upwards.

### 5.4.    Motility

#### 5.4.1.    Instantaneous Effects

Actimetry has been used in the last decade to monitor sleep disturbance in large field studies with subjects sleeping at home exposed to the usual aircraft, road traffic or railway noise (Ollerhead et al., 1992; Horne et al., 1994; Fidell et al., 1996; 1998; Griefahn et al., 1999; Möller et al., 2000; Passchier-Vermeer et al., 2001a; b; 2002a; b). In these studies, also the night-time noise exposure has been assessed and related to motility measures.

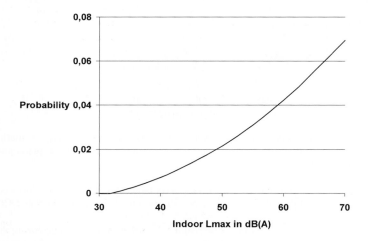

**Figure 14**. Probability of noise-induced motility during the 15-s interval with indoor $L_{max}$, as a function of indoor $L_{max}$ (Passchier-Vermeer et al., 2002a; b).

In these studies it is demonstrated that the human body does react to the night time noise event already at very low noise levels and even when the noise exposure is not remembered (see also 5.1). This is illustrated in Figure 14, in which it is shown that aircraft noise-induced increase in probability of motility during the 15-s interval with indoor $L_{max}$ of an aircraft flyover starts on average from indoor $L_{max}$ of 32 dB(A). The small movements (motility) have been measured with actimeters, worn on a wrist (Passchier-Vermeer et al., 2001a; b; 2002a; b). In the Netherlands sleep disturbance study, motility of subjects while asleep occurs in 3.3 per cent of the measurement intervals of 15-s in the absence of aircraft noise, i.e., the probability of motility during sleep in the absence of aircraft noise was 0.033. Figure 14 shows that during aircraft noise events with higher values of $L_{max}$, noise-induced increase in probability of motility is about twice the probability in the absence of aircraft noise.

*5.4.2.   Long-Term Motility Effects*

Motility during sleep is related to $L_{night}$ and to many variables of sleep and health. Average motility (motoric unrest) during sleep is higher in situations with a substantial number of noise events during sleep. This is not only the result of the instantaneous increase in motility during noise events, but increased motoric unrest also occurs during the quieter parts of sleep.

Motility during sleep is also associated with responses to questionnaires and in diaries (Passchier-Vermeer *et al.*, 2002a). Significant associations have been found between mean motility during sleep and the following variables:

-   number of behavioural awakenings during sleep period time,

-   number of awakenings remembered the following morning,

-   self-reported frequency of awakening due to aircraft noise, assessed by questionnaire,

-   sleep quality reported in a morning diary,

-   self-reported sleep quality assessed by questionnaire,

-   number of sleep complaints assessed by questionnaire,

-   number of health complaints assessed by questionnaire.

These associations show that motoric unrest is an important descriptor of sleep quality. They also suggest that chronic exposure to noise during sleep may affect health and well-being.

## 6.   Noise situation and differences in noise sensitivity in Europe

The *European Environment Agency* estimates that about 120 million people in the EU (32 per cent of the total population) are exposed to road traffic noise with $L_{den}$ over 55 dB(A), 19 per cent to $L_{den}$ between 55 and 65 dB(A), 11 per cent between 65 and 75 dB(A) and 2 per cent over 75 dB(A) (EEA, 1999). About 10 per cent of the EU population is exposed to railway noise with $L_{den}$ over 55 dB(A) (EEA, 1999). These figures are rough estimates. The extent of exposure to aircraft noise in the EU is most uncertain. Table 8 gives the number of people exposed to $L_{den}$ values over 55 dB(A) around selected airports. According to EEA (1999) 10 per cent of the EU population is highly annoyed by aircraft noise. This corresponds, according to the relationships between annoyance and $L_{den}$ given in this chapter, to about 6 per cent of the EU population being exposed to aircraft noise with $L_{den}$ over 55 dB(A).

Taking into account that a fraction of the EU population is exposed to more than one environmental noise source, a rough approximation of people in the EU exposed to environmental noise with $L_{den}$ over 55 dB(A) is 40 per cent.

**Table 8**. Estimate of number of people exposed to aircraft noise with $L_{den}$ over 55 dB(A) (M+P Raadgevende Ingenieurs, 1999).

| Airport | Number of people |
|---|---|
| Heathrow, London | 440 000 |
| Fuhlsbüttel, Hamburg | 123 000 |
| Charles de Gaulle, France | 120 000 |
| Schiphol, Amsterdam | 69 000 |
| Kastrup, Copenhagen | 54 000 |
| Barajas, Madrid | 33 000 |

In many instances, especially if road traffic noise is the predominant noise source, the differences between day time and night time noise levels are similar and $L_{night}$ is about 10 dB(A) lower than $L_{den}$. This implies that about 150 million people in the EU are exposed to night time environmental noise with $L_{night}$ over 45 dB(A). We want to underline that higher exposure during daytime and higher exposure during night-time are correlated and thus affect the same people.

Although it is interesting to estimate from these exposure data the resulting annoyance and sleep disturbance in the EU, this is too risky, since the extent of exposure below 55 dB(A) is unknown and the distribution of noise exposures above 55 dB(A) too uncertain to obtain a reliable estimate.

One might ask whether there are regional differences between the effect of (traffic) noise exposures. In 2001 the Technical University Berlin (TU, 2001) presented a state of the art review on traffic noise pollution, similarities and differences between European regions. They found that, apparently, the exposure-response relationships for annoyance do not differ systematically between different European regions.

## 7.    Conclusion

This chapter provided an overview of the relationships between annoyance, aspects of sleep disturbance, and environmental noise exposure. The exposure-effect relationships show that people start to be highly annoyed at $L_{den}$ over 55 dB(A) and report to be highly sleep disturbed at $L_{night}$ lower than 45 dB(A). Taken into account that about 150 million people in the EU are exposed to these and (much) higher noise levels, environmental noise is a serious and great threat for the well-being of people in the EU.

## References

Babisch, W., Ising, H., Gallacher, J.E.J., Sweetnam, P.M., and Elwood, P.C. (1998) The Caerphilly and Speedwell studies, 10 year follow-up, in N. Carter, and R.F.S. Job (eds), *Noise effects '98*, Proceedings of the 7th International Congress on Noise as a Public Health Problem, Sydney, Australia 1998, Vol. 1, pp. 230-235.

Babisch, W., Fromme, H., Beyer, A., and Ising, H. (2001) Increased catecholamine levels in urine in subjects exposed to road traffic noise. The role of stress hormones in noise research, *Env. Int.* 26, 474-481.

Born, J., and Fehm, H.L. (2000) The neuroendocrine recovery function of sleep, *Noise & Health* 7, 25-37.

Carter, N.L. (1998) Cardiovascular response to environmental noise during sleep, in N. Carter, and R.F.S. Job (eds), *Noise effects '98*, Proceedings of the 7th International Congress on Noise as a Public Health Problem, Sydney, Australia 1998, Vol. 2, pp. 439-444.

European Environment Agency (1999) *Environment in the European Union at the turn of the century* [Het milieu in de Europese Unie, op de drempel van een nieuwe eeuw]. Luxembourg: Official publications office of the European Communities. Internet: http://reports.eea.eu.int/92-9157-202-0/en/tab_content_RLR consulted 28-6-2003.

Fidell, S., Pearsons, K., Tabachnick B., Howe, R., Silvati, L., and Barber, D. S. (1995a) Field study of noise-induced sleep disturbance, *J. Acoust. Soc. Am.* 98 (2, 1), 1025-1033.

Fidell, S., Howe, R., Tabachnick, B., Pearsons, K., and Sneddon M. (1995b) *Noise-Induced Sleep Disturbance in Residences near Two Civil Airports*, NASA Contractor Report 198252, NASA Langley Research Centre, Hampton, VA.

Fidell, S., Howe, R., Tabachnick B., Pearsons K., Silvati L., Sneddon M., and Fletcher E. (1998) *Field Studies of Habituation to Change in Nighttime Aircraft Noise and of Sleep Motility Measurement Methods*, Report no. 8195, BBN Technologies, Canoga Park.

Fidell, S., Pearsons, K., Tabachnick B., and Howe R. (2000) Effects of sleep disturbance of changes in aircraft noise near three airports, *J. Acoust. Soc. Am.* 107 (5, 1), 2535-2547.

Finegold, L.S., and Elias, B. (2002) A predictive model of noise induced awakenings from transportation noise sources, *Proceedings of Internoise*, on CD-ROM, Filename in02_444.pdf, Dearborn, MI.

Griefahn, B., Möhler U., and Schümer R. (1999) *Vergleichende Untersuchung über die Lärmwirkung bei Strassen- und Schienenverkeh* (Hauptbericht-Textteil, Kurzfassung, Abbildungen und Tabellen, Dokumentationsanhang), München, SGS.

Guski, R. (1999) Personal and social variables as co-determinants of noise annoyance, *Noise & Health* 3, 45-56.

Harder, J., Maschke, C., and Ising, H. (1999) *Längsschnittstudie zum Verlauf von Stressreaktionen unter Einfluss von nächtlichem Fluglärm*, Umweltbundesambt, Berlin.

HCN - Health Council of the Netherlands (1994) *Noise and Health*, Publ. no. 1994/15e, Health Council of the Netherlands, The Hague.

HCN - Health Council of the Netherlands (1997) *Assessing Noise Exposure for Public Health Purposes*, Publ. no. 1997/23e, Health Council of the Netherlands, The Hague.

Horne, J.A., Pankhurst, F.L., Reyner, L.A., Hume, K., and Diamond, I.D. (1994) A field study of sleep disturbance, effects of aircraft noise and other factors on 5,742 nights of actimetrically monitored sleep in a large subject sample, *Sleep* 17 (2), 146-159.

Institute for Environment and Health (1997) *The Non-Auditory Effects of Noise*, Report R10, Leicester, UK.

ISO - International Organisation for Standardisation (1990) *ISO 1999 Acoustics - Determination of Occupational Noise Exposure and Estimation of Noise-Induced Hearing Impairment*, International Organisation for Standardisation, Geneva, Switzerland.

Ising, H., Babisch, W., Kruppa, B., Lindthammer, A., and Wiens, D. (1997) Subjective work noise, a major risk factor in myocardial infarction, *Soz. Praventivmed.* 42, 216-222.

Job, R.F.S. (1997) The influence of subjective reactions to noise on health effects of noise, *Env. Int.* 22, 93-104.

Job, R.F.S. (1999) Noise sensitivity as a factor influencing human reaction to noise, *Noise & Health* 3, 57-68.

Maschke, C., Wolf, U., and Leitman, T. (2002) *Epidemiological examinations to the influence of noise stress on the immune system and the emergence of arteriosclerosis*, Robert Koch-Institut, Berlin.

Miedema, H.M.E., and Vos, H. (1999) Demographic and attitudinal factors that modify annoyance from transportation noise, *J. Acoust. Soc. Am.* 105, 3336-3344.

Miedema, H.M.E., and Oudshoorn, C.G.M. (2001) Annoyance from transportation noise, relationships with exposure metrics DNL and DENL and their confidence intervals, *Env. Health Perspect.* 109 (4), 409-416.

Miedema, H.M.E., Passchier-Vermeer, W., and Vos, H. (2003) *Elements for a Position Paper on Night-Time Transportation Noise and Sleep Disturbance*, TNO Inro. Report 2002-59, Delft.

Möhler, U., Liepert, T.M., Schümer, R., and Griefahn, B. (2000) Differences between railway and road traffic noise, *J. Sound and Vibration* 231 (3), 853-864.

Morrell, S., Taylor, R., and Lyle, D. (1997) A review of health effects of aircraft noise, *Aust. NZ J. Publ. Health* 21, 21-36.

M+P Raadgevende Ingenieurs (1999) Present State and Future Trends in Transport Noise in Europe, Report prepared for the European Environment Agency.

Ollerhead, J.B., Jones, C.J., Cadoux, R.E., Woodley, A., Atkinson, B.J., Horne, J.A., Pankhurst, F., Reyner L., Hume, K.I., Van, F., Watson, A., Diamond, I.D., Egger, P., Holmes, D., and McKean, J. (1992) *Report of a Field Study on Aircraft Noise and Sleep Disturbance*, Civil Aviation Authority, London.

Passchier-Vermeer, W. (1993a) *Noise and Health, review*, Publication no. A93/02e, Health Council of the Netherlands, The Hague.

Passchier-Vermeer, W. (1993b) Noise-induced hearing loss from daily occupational noise exposure, extrapolations to other exposure patterns and other populations, in M. Vallet (ed.), *Proceedings 6th International Congress on Noise and Public Health*, Volume 3, INRETS, Nice, pp. 99-105.

Passchier-Vermeer, W. (2000) *Noise and Health of Children*, Report 00.042, TNO Prevention and Health, Leiden.

Passchier-Vermeer, W., and Passchier, W.F. (2000) Noise exposure and public health, *Env. Health Perspect.* 108 (1), 123-131.

Passchier-Vermeer, W., Steenbekkers, J.H.M., and Vos, H. (2001) *Sleep Disturbance and Aircraft Noise, Questionnaire, Locations and Diaries*, Report 2001.205, TNO Prevention and Health, Leiden.

Passchier-Vermeer, W., Steenbekkers, J.H.M, Waterreus, M.J.A.E, and Dam, P.J.C.M. (2001) *Sleep Disturbance and Aircraft Noise, Tables, Figures, Pictures*, Report 2001.206, TNO Prevention and Health, Leiden.

Passchier-Vermeer, W., Vos, H., van der Ploeg, F.D., and Groothuis-Oudshoorn K. (2002a) *Sleep Disturbance and Aircraft Noise, Exposure Effect Relationships*, Report 2002.027, TNO Prevention and Health, Leiden.

Passchier-Vermeer, W., Miedema, H.M.E., Vos, H., Steenbekkers, H.M.J., Houthuijs, D., and Reijneveld, S.A. (2002b) *Slaapverstoring en Vliegtuiggeluid*, TNO Inro. Rapport 2002.028, RIVM rapportnummer 441520019, Delft (in Dutch).

Passchier-Vermeer W. (2003) Night-time noise events and awakening. Delft: TNO Inro, Report 2003.023.

Shaw, H.A.W.E. (1996) Noise environments and the effect of community noise exposure, *Noise Control Eng. J.* 44, 109-119.

Stallen, P.J.M. (1999) A theoretical framework for environmental noise annoyance, *Noise & Health* 3, 69-79.

TU - Technical University Berlin (2001) *Traffic Noise Pollution. Similarities and Differences between European Regions - A State of the Art Summary*, TU-Berlin, Berlin.

WHO - World Health Organisation (2000) *Guidelines for Community Noise*, WHO, Geneva.

# PSYCHOLOGICAL AND SOCIAL ASPECTS OF "TRANSPORT AND HEALTH"

H. MOSHAMMER[1], H.-P. HUTTER[1] AND L. SCHMIDT[2]

[1]*Institute for Environmental Health, Medical University of Vienna
Kinderspitalgasse 15, 1095 Vienna, AUSTRIA*
[2]*SOMO. Social Sciences Mobility Research and Consultancy
Sonnenweg 5, 1140 Vienna, AUSTRIA*

**Summary**

Psychological well-being is often considered different from health while in fact psychological and physiological effects are part of the same continuum and often cause each other. Apart from that psychology helps us to understand and moderate attitudes and behaviour concerning transport and mobility patterns. Although psychology can contribute to the transport-and-health debate it is still in need of better quantitative data. Personal traits and coping styles as well as other personal factors obtained by questionnaires and standardised psychological tests must be considered when studying environmental health effects. Census data in many countries as well as a lot of studies provide data on annoyance, fears, and expectations as well as behavioural adaptations in connection with the physical environment. Details on the prominent example of noise annoyance are given in the chapter on "environmental noise, annoyance and sleep disturbance".

However, not only psychological effects in the individual are to be considered. The impact is also visible on the community level. While certain transport facilities are necessary for the prosperity and social status of a society, increasing transport volume will increase disparity, segregation, and isolation. Communities with higher transport noise exposure have been shown to develop fewer social contacts and are prone to more aggressive behaviour.

There is need for further research to better bridge the different disciplines that deal with the impact of our built environment. This need must not be used as an excuse to postpone policies to support more sustainable modes of transport. On the contrary those policies are urgently wanted and they should be accompanied by a close scientific evaluation and a documentation of the expected (and eventually unexpected) effects. Only a close cooperation of politics and science with a better

*P. Nicolopoulou-Stamati et al. (eds), Environmental Health Impacts of Transport and Mobility, 39-52.*

integration of social and natural sciences will in the end satisfy our needs without increasing the burdens for today and the future.


## 1.   Introduction

The importance of psychological and social factors in the cross-cutting issue of transport, environment, and health is well recognised by WHO (2000). However, a more structured approach and integration of social sciences is still needed. In a multi-national project from Austria, Switzerland, The Netherlands, France, Sweden, and Malta in the framework of the WHO-UNECE "Pan-European Program on Transport, Health, and Environment (THE PEP)" our Austrian team (sponsored by the ministries of environment and of transport) was responsible for the psychology part. We received also valuable input from teams of the other countries.

This paper sets out to collect existing concepts and ideas that bring the views of social sciences into the debate on "transport and health". Different scientific cultures hamper the dialog between social and natural sciences. For this paper medical doctors with a natural sciences background and psychologists with a social sciences background tried to work together which certainly was rather challenging. While we were not able to provide a holistic view of the topic we tried to highlight some important aspects to encourage further research and interdisciplinary debate.

The whole multi-national project was organised following the WHO (2000) booklet on Transport and Health in that each chapter of the WHO booklet was assigned one working group respectively, one responsible lead country. In the WHO booklet the chapter on "mental health and well-being" serves several goals. One is the reminder of the broader WHO approach towards health: Health is not only the absence of disease but well-being of body and soul as well. In this respect psychosocial effects are seen as sort of "soft" effects that are not as easily quantified as the "hard" effects measured in deaths, disease outcome, or physical properties (like blood pressure or airways resistance). Thus WHO discusses psychological effects that precede disease (e.g. stress from noise preceding cardiovascular disease) and psychological problems following "hard" health effects (e.g. posttraumatic stress disorder). WHO argues that these psychosocial effects are as important as the bodily health effects although more difficult to study and to quantify. One key aspect of the multi-national project was the monetary evaluation of the health effects of transport. In agreement with the WHO conclusions we found this a rather difficult task for our topic. To structure our work we decided to investigate 4 issues concerning the psychological and social aspects of transport and health:

(1)     "Well-being", an integral part of the WHO definition of health, is a key concept in this discussion. Though normally defined negatively as the absence of physical health burdens, well-being also means e.g. the possibility of taking part in social life, having social support, not suffering from violence, having no fear, being in a good temper, and having the feeling of power and self efficacy. The term implies that there is more to health than the

absence of disease or disability. It is a synonym for a good quality of life. The psychological and physical aspects of well-being are difficult to disentangle because physical damage causes mental responses (pain, anguish, distress) and psychological disturbances lead to physical ill-health.

It is sometimes believed that psychological effects are less important than "true" health effects. So for example Austrian legislation makes a distinction between "impact on health" and "annoyance". In the "Gewerbeordnung" (the Austrian business law) it is stated that an enterprise – as an example from the transport sector one could think of a garage – must not by all means endanger the health of an average human being. Annoyance on the other hand must not be unreasonable or unduly high. When the impact of an enterprise on the neighbourhood is only recognised after the launching of the enterprise only threats to health can force a cutting of emissions.

The same imbalance between physical health and psychological well-being can also be seen in the WHO 2000 Environmental Health Indicators: While lead emissions from cars are still mentioned as an important indicator (among other traffic-related air pollution indicators and accidents) noise is the only psychosocial health indicator that is related to transport. Even noise limit values cited in this indicator list are derived from "hard" health effects like cardiovascular diseases and not from psychosocial surveillance and questionnaire data although it has been argued (Fidell, 1992) that questionnaire-based assessments of environmental noise annoyance are at least as easily conducted and more relevant than sophisticated monitoring schemes of sound pressure levels.

So both national and international bodies still rely on conventional health data rather than on psychosocial impact estimates. However, in reality there is no clear cut line between medical and psychological effects. Even concerning economic impact, the psychological effects (e.g. learning ability of children impaired by noise stress, aggressive behaviour and fewer social contacts, decreased intelligence due to lead pollution) could turn out to be more important and more lasting than classical health effects.

(2)     The topic also offers the opportunity for a discussion of the more complex interactions of traffic related impact: infrastructure and design of our cities, less space for our children and for communication, perceived dangers and resulting change in mobility and social behaviour. Because these complex interacting stresses are distributed unevenly among different groups (and because the less privileged groups are less able to adjust to additional stress) psycho-social effects also raise the question of environmental justice.

While noise annoyance is mirrored by health effects (like stress-related diseases) some important social effects are totally neglected when concentrating on conventional health parameters. These include both beneficial and detrimental consequences. Modern transport infrastructure provides improved facilities for the mobility of the majority of people (beneficial effect) but a large minority (children, elderly persons, poorer families, parents with their children, handicapped persons) in certain circumstances are hindered in their mobility at the same time. Not only those people that do not own or use a car themselves but even the "motorised" persons suffer from traffic jams and lack of parking space and perceive the negative impacts of traffic on their health, freedom, mobility, and well-being.

(3)    A third issue are differences in the perception of problems and benefits of transport. Those who are involved personally often experience feelings not expected by the experts. For example in noise studies it has been shown that the measures usually applied to define noise exposure (e.g. energy equivalent average a-weighted sound pressure level) are no good predictors of the amount of annoyance. Personal factors like attitude towards the source of the noise, psychological traits like noise sensitivity or neuroticism or life history events play a role. Also several aspects of the noise itself must be considered like the number and temporal distribution of single noise events, qualitative aspects of noise like frequency distribution and information content, and way of noise transmission.

(4)    Another aspect of transport and psychology is less thoroughly covered in the WHO booklet: What can psychology tell us about the motives of people opting for one mode of mobility or another? How is knowledge about problematic behaviour transformed into action? What makes people and decision makers decide the way they do? The beneficial effects of mobility and the gains through certain mobility patterns should be stressed more because of the motivational impetus derived from it and the need for a balanced basis for future decisions.

## 2.    "Hard" and "soft" health effects

Perception of risks and the fear of accidents are important psychological factors. Reduced health as a consequence of trauma or diseases due to air pollution has relevant psychological side effects also. In a recent study (Moshammer and Neuberger, 2003) on the health effects of fine particles on school children in Linz we investigated both pulmonary function and subjective health complaints. Subjective complaints (like disturbed nights due to coughing or reduced bodily activity due to shortness of breath) in a subgroup of vulnerable (asthmatic) children are as closely linked to air pollution data as "hard" physical measures. In this study an interaction of personal and situational (environmental) variables is evident.

Psychological and social effects result from the combined impact of several sources. Susceptible subgroups are of special interest and environmental justice as well: It's the underprivileged person that cannot adapt to environmental burdens. Social deprivation and poorer environmental state of the neighbourhood often interact

(Bolte *et al.*, 2002). This is important to notice both for policy decisions and for study design when looking into causal and quantitative associations.

Physical and psychological health are always linked together. Every health related aspect of traffic (like noise, air pollution, accidents, physical activity, and climate change) has a psychosocial impact also: The literature on noise very extensively deals with annoyance (see chapter 1.1.1 on noise in this book). Some studies have been conducted (e.g. in Austria by Lercher and Kofler, 1996) on learning ability and success at school. Noise served as an indicator of traffic exposure. But certainly noise is not the only factor that for example leads to reduced tolerance against frustration. Perceived air pollution (inhaling of traffic fumes) in a study by Lercher *et al.* (1995) had a greater impact than noise and the measured air pollution on self-reported well-being. Accidents cause long-lasting post traumatic stress syndromes and even the fear of accidents influences social behaviour. Reduced physical activity has been shown to affect mood and the psycho-motor development of children (Hüttenmoser, 1995). There are increasing fears that climate change will lead to an increase in migration which would have severe effects on the social stability of whole communities. These few examples show that psychological effects can both precede and follow physical effects.

Noise annoyance or disturbed sleep is often viewed not as "bad" as death, disease, or injury. The Swiss BUWAL (2002) used an interesting approach to deal with this disparity. By defining a (small though measurable) disability adjustment per life year from disturbance of sleep and interference with communication they found that for Switzerland psychological effects of transport are as severe as cardiac and respiratory diseases from air pollution and even much more important than cardiovascular health effects of noise. The reason for this is the great number of people affected. But this approach also makes it clear that it is not so important to get very accurate estimates of each impact but to reach a socio-political agreement about the (relative) importance of the effect. The WHO concept of Disability Adjusted Life-Years (DALY: Murray, 1996) does not include Disability Weights for "minor" health effects like sleep disturbance or interference with speech. So an expert panel of medical doctors from the Swiss Workers Compensation Board was asked to rate these effects following the WHO principles. The weights for similar effects like minor forms of bronchial asthma (inducing some sleep disturbance at night) or slight hearing impairment were indicated for comparison reasons. The arithmetical means of forty-one ratings (0.033 for interference with speech, 0.055 for disturbance of sleep) were multiplied by the number of people additionally suffering from each effect under the assumption of an increase in transport volume by 1000 km (both for day and night and for passenger and freight road transport). Per 1000 km of passenger cars driven in Switzerland in the daytime a loss of 0.00013 DALYs was estimated due to interference with speech. The estimates for the night (due to sleep disturbance) amounted to 0.0027. Every 1000 km of freight transport were estimated 10 times higher (0.0013 and 0.026 DALYs respectively). In contrast overall noise effects on cardiac infarction were rated about one magnitude lower (with a greater range of uncertainty concerning the dose-response relation).

Another important psychological effect that was raised in the discussion within our expert group is that of perceived safety. This is of great concern especially for elderly people. Also for the children and their parents this is a vital issue (Körmer, 2002). Fear of accidents tends to increase individual motorised traffic because people do not dare to use a bicycle even for short distances. For a more detailed discussion of perceived and accepted risk see also Klebelsberg, (1982), Wilde (1974), and Schmidt (1994, 1995). It is important to note that scientists have a concept and understanding of risk different from the general population. While in risk management risk is defined by probability and magnitude of effect, lay people tend to rate risks in categories like familiarity, preventability, locus of control, social distribution of gains and (possible) losses, possibilities of individual counter-measures, etc. Magnitude of a possible effect generally gets more weight in the public perception than the probability of its occurrence. In general, motorists would rate the risk of accidents lower but this attitude would change in case of an actual accident in which oneself or near relatives are involved.

## 3. Measuring psychological impact

Epidemiology as a discipline of natural sciences seeks to find associations between single variables. For example an investigator might be interested in finding the association between noise exposure and blood pressure. He is very keen not to confound his results by socio-economic status, air pollution and psychosocial conditions of the population under study. But in true life noise and air pollution from the same source and the economic and social position of the exposed individuals are interlinked and act together and exert their impact. For policy action it is less important if the ultimate reason for an increased rate of high blood pressure comes from noise or another factor that is strongly connected with the source of the noise. The mere fact that one group of people is more prone to disease than another should by itself lead to preventive action.So much the better when noise is a good indicator of a detrimental situation. If it is also the causative factor this is often of theoretical interest only. In that sense there is danger from over-controlling in epidemiology. One might be interested in the contribution of a certain environmental factor (e.g. noise) to a special disease (e.g. ischaemic heart disease). We know that certain personal traits and behaviour or the socio-economic status of a person contribute to this risk. Therefore, we control these known risk factors so as to be able to calculate the contribution of noise. But what if environmental noise leads to a certain harmful behaviour or to a less supportive social environment? What if persons at risk (for several reasons) are less able to cope with noise stress or even do not have the financial means to find a less noisy living place? Actually the important risk factors are linked with each other and by looking at each risk factor separately (by controlling the others) we risk missing this important interaction!

The study of psychosocial impacts soon leads to the complicated interaction of causes and effects. A certain exposure (e.g. noise) might have psychological (e.g. annoyance) or sociological effects (e.g. aggressive, less supportive social environment). Personal traits and coping styles as well as the social environment might help to mitigate the unwanted impact of noise. Psychological effects while being true health effects themselves might result in more permanent biological effects mediated by stress hormones. Functional and morphological diseases always have consecutive psychological effects also. By our behaviour we increase or decrease the amount of exposure (our own and that of other people). Psychological and sociological causes for our behaviour (motives) have to be studied more thoroughly.

Psychological effects are traditionally measured with questionnaires. These instruments often provide binary data only (yes/no) or quite a coarse scaling. So often only non-parametric tests can be applied which make it difficult to quantify a dose-response relation. Even though the questionnaires provide a scaling, it is often difficult for medical doctors to interpret the findings. Psychologists have their own methods of validation of questionnaires. On the other hand psychologists might find it difficult to understand the clinical meaning of biological measures like lung function values or blood pressure. Many of the questionnaire studies are of a cross-sectional design. Therefore it is not always possible to discern cause and effect. This lies at the bottom of the debate on the interrelation of annoyance and sensitivity towards noise: Does repeated annoyance (under certain circumstances) introduce an increased sensitivity or is sensitivity a personal trait which makes a person more prone to feel annoyance under certain pressures? This is also interesting because noise sensitivity is correlated with certain psychiatric disorders. So while reported noise annoyance and noise sensitivity are also correlated with each other the debate is still open whether noise induces annoyance (this is quite clear) and in that way leads to sensitivity and further to disease or if a certain trait and habit would lead both to an increased risk of disease and a higher degree of annoyance.

In any case individual psychological traits, behaviour and coping strategies are important covariates to be considered when studying environmental health impacts. The negative impact of transport on annoyance, well being, sleep disturbance, impaired mood, and so on are well acknowledged in noise research (see chapter 1.1.1 on noise in this book). The more complex and indirect interactions are not so well documented.

There are several data sets that contain information on some aspects of psychological effects. Good examples are the census data on annoyance ratings (e.g. Austrian micro-census or Eurobarometer data on environment perception, spring 2002). From Austrian census data (the last census on perception of environment and environmental attitudes was performed in December 1998) it is known that the most often recognised environmental stress factor is noise followed by odour annoyance and air pollution (dust, soot). The most important source of noise is traffic (80 percent) with the majority stemming from road traffic (heavy road vehicles and private cars roughly equally responsible). Road traffic covers the whole area while other sources of noise are typically spot sources or at least confined to small corridors (e.g. railway noise). Road traffic noise therefore has the broadest impact on society. Yet, the impact of the other sources must not be neglected.

A similar approach is possible for other secondary psychological and social effects. For example there are data on health effects of air pollution. These effects have an impact on the well-being, activity and performance of the patients. There are data from some studies on children's reduced mobility and activity due to respiratory disease caused by air pollution and the rate of subjective health symptoms and reported impairment of sleep due to noise or to cough (caused by air pollution episodes).

The existing data and data gaps shall be outlined using the examples the WHO gave in the booklet on Transport & Health (2000) in the chapter on mental health and well-being:

**Effects of lead:** It is well known that lead even in small quantities has an adverse effect on the cognitive development of the unborn and new born child. Leaded petrol until recently contributed to the overall environmental load of lead. However, nowadays in most European countries only unleaded petrol is sold and it is being phased out in other countries. So leaded petrol is a threat of the past.

**Posttraumatic stress from accidents:** From several studies the amount of posttraumatic stress disorder in children after a traffic accident is known (Goldberg and Gara, 1990; Green *et al.*, 1993). Hence from the statistics on road accidents the number of children and/or adults with posttraumatic stress disorder could be estimated. It is easy to compile data on road accidents. It is not so clear how to define "posttraumatic stress" and from which point on this does constitute a "disorder" or a "psychiatric problem"? Normal psychological reactions that follow danger, pain, and injury can also be quite painful. On the other hand these reactions can lead to a more cautious and adequate behaviour. The direct impact of accidents on well-being is important but the interpretation of quantitative data should be performed with caution. The issue of posttraumatic stress is an excellent example that not the number of individual "cases" is difficult to estimate but the evaluation of each case e.g. in monetary terms.

**Aggression and nervousness:** There are some interesting studies internationally. Most of them look at noise as an exposure parameter (Novaco *et al.*, 1990, Mayer and Treat, 1977). Some findings suggest that it depends on the special regional situation if a given noise exposure is perceived as annoying. For example in an urban setting there are usually more possibilities for adaptive and protective measures than in the countryside. Even more important in that instance are the pre-existing social bonds within a community. It is nearly impossible to disentangle the causal relationships between social situation, environmental state, and behaviour of the inhabitants of a community: Transport noise reduces the perceived quality of the neighbourhood. This has a direct effect on well-being and aggressive behaviour. But it also has an indirect impact on the social background of the inhabitants through market forces: lower prices attract poorer people. These can less afford protective and adaptive measures. They even cannot fight new traffic as easily as the well-off. So to what degree is it the traffic noise that leads to psychosocial stress and to what degree is it the social background that leads to aggressiveness and to more stress?

The same as above holds true of the items "reduced social life" and "constraints on child development". These items constitute a real problem but it is not scientifically just to quantify the impact on the whole country by extrapolating from rather few and special studies.

**Mental health benefits of exercise:** This is a proven fact (see chapter 1.3.1 on physical activity in this book). But is it traffic that reduces exercise or is it a sedentary lifestyle that calls for more traffic? Certainly a "car-friendly" environment discourages travelling and cycling. Still individual decisions, habits, and motivation as well as cultural influences (advertising!) play an important part and so no linear dose-response relationship should be expected.

## 4.    Assessment of the state of the environment

There is a manifest inequity concerning living conditions, transport facilities, access to public services, opportunities of work and environmental conditions like air pollution or noise. This inequity can be summarised in terms of poor environmental quality. Such conditions can be documented and their impact on quality of life can be assessed by looking at the relation between objective equipment and services, and the perception and evaluation of the quality of life by specific populations.

Numerous researches analysed the relation between stress and somatic and/or psychic disorders (Totman, 1983). Some of this research is based on lists of "life events" (Dohrenwend and Dohrenwend, 1978) others analyse the individual's relation with "events, daily life problems, perceived as irritating, frustrating or anxiogeneous". These latter approaches seem to be better predictors of well-being and health. Among six factors (professional sphere, environmental preoccupations, family sphere, work overload and excess of responsibilities, concern about one's body, and the management of daily life), environmental concerns are ranked as the most worrying, followed by the management of daily life. Similarly, among the twenty most disturbing stress factors in our industrialised society identified by the "Hassle-scale" of Lazarus, the two factors cited by the majority of subjects are (1) health problems of a family member, and (2) air pollution. These two factors affect women more than men (women feel more often and more severely affected; Badoux-Lévy and Robin, 2002). This is true whatever the size of the municipality.

Confronted with an undesired situation, the individual can efficiently cope (protecting himself, for instance), and thus show a successful adaptation. But the individual can also engage in several coping processes, which turn out to be unsuccessful, or even find it impossible to cope, which augments the adversity of the situation. This is the characteristic of a stress situation, which leads to discomfort, annoyance, and in some cases to complaints (Moser, 1992). One can therefore ask where the limits of acceptability are in terms of environmental qualities; from where on are adaptive strategies no longer sufficient to cope with the noxious situation? What is the threshold for the expression of annoyance or complaints and in what objective life conditions are they expressed?

Furthermore, the environment is not just a figment of our imagination or a social construct; it is real. In the first place the environment has a physical manifestation in order to confer meaning; it embodies the psychology of those who live in it. It is used to confer meaning, to promote identity, to locate the person socially, culturally and economically (Moser and Uzzell, 2003). In that context, research implementing social representations is of particular importance when looking at people-environment relations. Their dynamics can only be understood through systematic social representation studies. Only through the knowledge of the social representations of well-being and quality of life of specific populations in particular environmental contexts, can we assess people's values, meanings and worldviews and therefore understand and explain their impact on their individual relation to their environment. A shift from attitudes and perception studies to the assessment of environmental values and their ideological background should provide a better understanding of an individual's behaviour in a specific situation or environment.

This methodology would lead to focus on behavioural explications at a more general level than the one classically addressed until now. Culture-specific worldviews are the necessary conditions for the understanding of individual needs and expectations. The relative lack of integrative theories in environment-behaviour matters can be seen as a consequence of not taking into account those cultural particularities. Looking at individual needs in order to identify conditions of well-being and satisfactory quality of life, permits to integrate cultural differences in a general theoretical framework. Such attempts have been successfully performed and give promising results concerning, for instance, the conditions of ecological behaviour (Poortinga *et al.*, 2002). Worldviews bring coherence in the individual and social differences of attitudes and behaviour towards the environment, help to understand individual exigencies of quality of life, and provide a framework for interventions to preserve people-environment congruity.

## 5.    The psychology of the "homo automobilicus"

Transport patterns are not the sole factor that affects psychological well-being. Also psychological traits affect the choice of transport modes. Human psychology is necessary for the understanding of the behaviour of drivers and others interacting with each other on the roads or in public transport. Expectations of the peer group, needs generated by advertising, and unquestioned habits often have a stronger influence on individual behaviour than rational considerations and cost-benefit analyses. "Speed" is not just a parameter of pure physics (distance per time) but also is associated with feelings of power, superiority, and joy. On the other hand risk and dangers might not only raise negative feelings but also those of thrill, adventure and bravery. Gender and age-specific differences in accidents and risk behaviour give ample proof to this fact.

Our civilised society leaves little room for aggressiveness in direct social interactions. While driving a motor vehicle the opportunity for interacting with each other on a basically impersonal level (usually without eye-contact although in short distance) encourages aggressive driving styles. Speed then becomes a competitive value and the highway replaces the sports arena.

Transport policies often tend to fall into the trap of misinterpreting the wants of societies for the needs of societies. But ever more room for individual motorised traffic, free choice of individual speed on all roads, and an ever increasing capacity of the road network generate even more demands. Abatement of bottlenecks in one part of the road system will thus lead to new problems in another part.

The reputation of public transport is often rather low. Part of this bad reputation is permitted but in many instances this is just a case of misperception. Even the number of regular users of public transport is often heavily underestimated. Use of public transport is often seen as caused by a lack of alternatives. But in fact public transport offers more alternatives than the private car. A sophisticated use of different means (including physical active modes of mobility, public transport, as well as semi-private (e.g. car sharing) and private motorization demands greater skills and offers more opportunities.

## 6.    Conclusions

### 6.1.    Scientific research needs

Psychological and socio-economic parameters are correlated both with exposure and with health outcome. The possible causal connections are manifold:
1.  Perceived exposure leads to psychological reactions (e.g. annoyance)
2.  Psychological stress causes bodily reactions and acts as effect mediator
3.  Socio-economic status influences both load of exposure and health
4.  Personal behaviour and community interactions are influenced by environmental stressors
5.  Ill health has psychological and social consequences

Thus psychological, social, and economic parameters should be monitored regularly in studies on the interaction of environment and health. Until now this has been done rather unsystematically because there is a lack of standardisation of the parameter-set. Even the data monitored are difficult to interpret in the epidemiological context: The complicated interactions don't allow simple assumptions on linearity and causality. It is not so clear if psychological and socio-economic parameters should be handled as confounders, co-factors, outcome variables, or indicators of exposure. Overmatching and under-controlling are both to be avoided.

Obstacles to overcome these scientific restrictions should be considered on two levels. The first concerns the methods of quantification and comparability of psychological data. The intern validity of the data can be controlled but it is not so straightforward to assure extern validity and comparability between different cultural, socio-economic, and language settings. Even in the same language context the connotation of a word or the relative meaning of a certain income or property could differ to a large extent. These differences hamper the development of a standard data set on psychological and socio-economic parameters.

The second level is of a more profound nature: There is a lack of integration of natural and social science. The dualism of body and soul is deeply rooted in our current paradigm. Even when speaking about the same phenomenon the two approaches of science use different terms or the same terms might have a different meaning. Natural and social sciences do use slightly different tools (e.g. statistical methods) to handle their problems. A sort of meta-theory is needed that overcomes the dualism both in our perception of the world and in our methodology. Only this would enable us to understand all interactions and develop a coherent theory on the impact of the entire environment (both natural and social) on our health (in its broadest sense). Only then will we be able to fully deal with the impact of transport and all its social and physicochemical implications (Kofler *et al.*, 2001).

Psychological effects result from complex interactions of different impacts. Personal traits and the given situation at the time of impact must be considered. In many respects traffic is just one stress factor among others. Some effects e.g. on mobility behaviour are quite specific. There is still a need for improvement of data on psychological and social parameters. This is true of the validity, specificity, and sensitivity of questionnaires and for the statistical methods dealing with results of questionnaires. The interconnection and correlation of several different effects and parameters make it nearly impossible to quantify the impact of one single parameter independently. Several impacts from the natural, technical, and social environment interact with personal coping styles, habits and beliefs in the production of psychological effects that themselves again have an impact of the individual's biophysical and social environment.

Methods to measure and evaluate well-being, mental health and public awareness have to take into account the process character of learning and the interactive relation between person and situation. Unfortunately, there are few studies which compare objective data and perceptions, which are of a problem-centred or narrative type, or which are especially necessary if one is interested in action-relevant data (Huber and Mandl, 1994, Lamnek, 1995). Most observations lack a gender-sensitive interpretation. The same is true of longitudinal studies and systematically documented and evaluated case studies including psychological variables relevant to well-being.

The classical methodology looking at each exposure indicator and every health effect independently (taking every mathematical trouble to stratify other "confounding" variables) cannot provide the full picture. A more holistic approach is wanted that does not separate psychological or sociological effects from "other" health impacts. Behaviour, psychological traits, and socio-economic status are influenced by environmental stressors and act on the same stressors by either changing the individual's exposure or response. Permanent health effects can be derived from chronic psychological responses but changes in health status also have an effect on the individual's socio-economic status and his ability to cope with further stressors.

A holistic approach should not only study linear dose-response interactions but bear in mind the complex connections and the mechanisms that link all the indicator variables of health, environment, and well-being with each other.

## 6.2. Consequences for decision making

Decisions on new roads and other transport infrastructure, in spatial and urban planning are among others influenced by clearly defined parameters of the physical environment (e.g. measurement and forecast of sound pressure levels and concentrations of air pollutants). The decision making process is highly sophisticated with many experts from several fields taking part in it. The population that is directly affected by these decisions usually has very little opportunity to follow or even understand the decisions and the facts and forecasts which are the basis of the whole process.

When preparing to ask for public participation one should keep in mind that it is very difficult to withdraw from that course later on. "Public" is not a homogeneous entity. Different interest groups will present conflicting proposals. Therefore integrating public opinion into decision making processes is a very tedious job which needs ample room, time and resources. Mechanisms that formalise the involvement of the public are well established and include mediation processes, public hearings, and the use of new information technologies. Where these mechanisms have been applied in a proper way (although at a higher cost of time and energy in the beginning) in the long run they have been shown to be cost effective when taking into account the higher satisfaction of all interested parties and the lower rate of clashes and need for reconstruction afterwards.

## References

Badoux-Lévy, A., and Robin, M. (2002) Analyse multidimensionnelle d'une échelle de «stresseurs de la vie quotidienne», *Les Cahiers Internationaux de Psychologie Sociale* 56, 64-73

Bolte, G., Elvers, H.-D., Cyrys, J., Schaaf, B., Berg, A. von, Borte, M., Wichmann, H.-E., and Heinrich, J. (2002) *Soziale Ungleichheit in der Belastung mit verkehrsabhängigen Luftschadstoffen: Ergebnisse der Kinderkohortenstudie LISA*, Gesundheitswesen 64, A13.

BUWAL (2002) *Schriftenreihe Umwelt Nr. 339 Lärm: Zurechnung von lärmbedingten Gesundheitsschäden auf den Strassenverkehr*, Bern

Dohrenwend, B.S., and Dohrenwend, B.P. (1978) Some issues in research on stressful life events, *J. Nerv. Ment. Dis.* 166, 7-15.

Fidell, S. (1992) Interpreting findings about community response to environmental noise exposure: what do the data say? In: *Procceedings of euro.noise* 92, Book 2, pp. 235-255, Institute of Acoustics, St. Albans.

Goldberg L., and Gara M. A. (1990) A typology of psychiatric reactions to motor vehicle accidents, *Psychopathology* 23, 15–20.

Green, M.M., McFarlane, A.C., Hunter, C.E., and Griggs, W.M. (1993) Undiagnosed post-traumatic stress disorder following motor vehicle accidents, *Medical journal of Australia* 159, 529–544.

Huber, G.L., and Mandl, H. (1994) *Verbale Daten. Eine Einführung in die Grundlagen und Methoden der Erhebung und Auswertung,* Beltz, Psychologie Verlags Union, Weinheim.

Hüttenmoser, M. (1995) Children and Their Living Surroundings: Empirical Investigations into the Significance of Living Surroundings for the Everyday Life and Development of Children, *Children's Environments* 12, 403-413.

Klebelsberg, D. (1982) *Verkehrspsychologie,* Springer Verlag, Berlin

Kofler, W., Lercher, P., and Puritscher, M. (2001) *Causally unspecific Health Risk of Environmental Incidents Part 1,2 and 3*, special version for Nobellaureate Y.T. Lee – Hopes for the Future Procedure of the papers: Kofler W., P. Lercher, M. Puritscher: The need for sufficiently taking into account unspecific effects in the understanding of health risks, Part1, 2 and 3, 12th World Clean Air and Environment Conferences, Proc., F-024a,b,c, Seoul

Körmer, C. (2002) *Implizite Verkehrserziehung von Kindern durch Eltern und Begleitpersonen,* Diplomarbeit, SOWI, Wien

Lamnek, S. (1995) *Qualitative Sozialforschung. Vol. 1. Methodologie; Vol. 2. Methoden und Techniken,* Psychologie Verlags Union, München

Lercher, P., Schmitzberger, R., and Kofler, W. (1995) Perceived traffic air pollution, associated behaviour and health in an alpine area, *Sci. Tot. Environ.* 169, 71-74.

Lercher, P., and Kofler, W. (1996) Behavioral and health responses associated with road traffic noise along alpine through-traffic routes, *Sci. Tot. Environ.* 189/190, 85-89.

Mayer, R.E., and Treat, J.R. (1977) Psychological, social and cognitive characteristics of high-risk drivers: a pilot study, *Accident analysis and prevention* 9, 1–8.

Moser, G. (1992) *Les stress urbains,* Armand Colin, Paris.

Moser, G., and Uzzell, D. (2003) Environmental psychology, in: T. Millon and M.J. Lerner (eds.), *Comprehensive Handbook of Psychology, Volume 5: Personality and Social Psychology,* John Wiley & Sons, New York, pp. 419-445.

Moshammer, H., and Neuberger, M. (2003) The active surface of suspended particles as a predictor of lung function and pulmonary symptoms in Austrian school children, *Atmospheric Environment* 37, 1737–1744.

Murray, C.J. (1996) Rethinking DALYs, in: C.J. Murray and A.D. Lopez (eds.), *The global burden of disease,* World Health Organization, Harvard School of Public Health, World Bank, Geneva.

Novaco, R.W., Stokols, D., and Milanesi, L. (1990) Objective and subjective dimension of travel impedance as determinants of commuting stress, *American journal of community psychology* 18, 231–257.

Poortinga, W., Steg, L., and Vlek, C. (2002) Myths of nature and environmental management strategies. A field study on energy reductions in traffic and transport, in G. Moser, E. Pol, Y. Bernard, M. Bonnes, J. Corraliza, V. Giuliani (eds.) *Places, People & Sustainability,* Hogrefe & Huber, Göttingen.

Schmidt, L. (1994) The significance of accepted risk and responsible action for goals and methods in psychological traffic research, in R. Trimpop and G.J.S. Wilde (eds.) *Challenges to Accident Prevention. the issue of risk compensation behaviour,* Styx Publications, Groningen, pp. 45-50.

Schmidt, L. (1995) Mobilität - gesundheitsfördernd und umweltverträglich, in A. Keul (ed.) *Wohlbefinden in der Stadt.* Ökopsychologie und Gesundheitspsychologie im Dialog, Beltz, Psychologie Verlags Union, Weinheim, pp. 112-136.

Totman, R. (1983) Psychosomatic theories, in J.R. Eiser (ed.) *Social Psychology and Behavioral Medicine,* John Wiley and sons, New York, pp. 143-75.

Wilde, G.J.S. (1974) Wirkung und Nutzen von Verkehrssicherheitskampagnen und Forderungen - ein Überblick, *Zeitschrift für Verkehrssicherheit* 20, 227-238.

World Health Organization, Dora, C., and Phillips, M. (2000) *Transport, Environment and Health,* WHO Regional Publications, European series No. 89, Geneva.

# AIR POLLUTION AND HEALTH:
## AN OVERVIEW WITH SOME CASE STUDIES

F. BALLESTER
*Public Health Department*
*'Miguel Hernández' University*
*Elche, Alicante and*
*Unit of Epidemiology and Statistics*
*Valencian School of Studies for Health - EVES*
*Calle Juan de Garay 21*
*E -46017 Valencia, SPAIN*

**Summary**

In recent years a great advance in understanding the effects of air pollution on health has been made. In this subject, the fundamental concepts on the matter are introduced, together with the main studies carried out in recent years. The main control and intervention strategies for the minimisation of its possible impact on health are overviewed.

The main short term effects of air pollution on health range from an increase in the number of deaths, hospital admissions and emergency visits, especially for respiratory and cardiovascular causes to alterations in lung performance, cardiac problems and other symptoms and discomfort.

Chronic effects related with long term exposure have also been shown. The increase in the risk of death through chronic exposure to pollution is estimated as a few times greater than risk due to acute exposure. Chronic exposure is also associated with increased cardio-respiratory morbidity and with decline of pulmonary function, both in children and adults.

There are several strategies used for reducing the levels of air pollution and therefore minimising its impact on health. Traditionally, these strategies have been based on setting limit values for each pollutant, which should not be exceeded. Today, a global approach to the problems posed by air pollution is defended, especially to those related with vehicle transport, the main source of air pollution in most European cities nowadays.

*P. Nicolopoulou-Stamati et al. (eds), Environmental Health Impacts of Transport and Mobility, 53-77.*
© 2005 *Springer. Printed in the Netherlands.*

## 1.  Introduction

In the field of public health, air pollution is a well-known phenomenon, which has been studied for long and becomes of major importance in our contemporary world stemming from a series of episodes which happened in industrial countries during the first part of the 20th century.

The cases in the Meuse Valley (Belgium) in 1930, in Donora (Pennsylvania, USA) in 1948 and, above all, the catastrophe of London, in December 1952, are probably the most outstanding and characteristic (Ware *et al.*, 1981). These exceptional situations resulted in an increase of mortality and morbidity which left no doubt to the fact that high levels of air pollution were causally associated with an increase in early deaths (Box 1).

This evidence leads to the adoption of pollution control policies, especially in western Europe and North America , with an important reduction in air pollution levels (Figure 15).

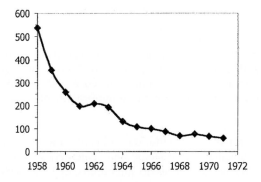

**Figure 15**. Annual average of black smoke (in µg/m³), London 1958-1971 (Schwartz and Marcus, 1990).

In recent years, a considerable number of studies carried out in different cities have found that, even below air quality levels considered safe, the increase of air pollution levels are associated with harmful effects on health. A study carried out in France, Switzerland and Austria, shows that 6 per cent of mortality and an important number of new cases of respiratory diseases in these countries may be due to air pollution. Half of this impact is caused by pollution emitted by motor vehicles (Künzli *et al.*, 2000). Additional studies have showed that vehicle-generated air pollutants are associated with a series of effects on health, especially on the respiratory system (Brunekreef *et al.*, 1997; Ciccone *et al.*, 1998; English *et al.*, 1999). On the other hand, there is growing concern on the possible risks generated by agents for which there is no satisfactory evaluation, such as exposure to polycyclic aromatic hydrocarbons (PAH), persistent organic pollutants and other air toxicants (Jedrychowski *et al.*, 2002; Turusov *et al.*, 2002). In summary, important population sectors are exposed to air pollutants with a consequent possible negative effect on their health.

Box 1. Case 1: The London Smog.

On the 5th of December 1952, a thick layer of fog developed over London at temperatures near 0°C. This fog persisted without decreasing for several days and a considerable worry generated among the population. That day and the following, there was a great demand for hospital beds and on the 8th of December, hospitals around Central London made an emergency release, declaring that they only had beds for 85 per cent of demand. That same day, newspapers revealed that some people had died because of the fog, since this fog carried dangerous chemical pollutants. The investigations carried out to evaluate the impact of air pollution on health extended for over a year.

The episode occurred because of a stationary high-pressure mass situated over Western Europe, leading to a temperature inversion in the Thames Valley. Pollutants, mostly from coal combustion in public and domestic heating systems, could not disperse and reached very high concentrations (around 10 times above current regulatory standards) (Figure 16). A very important effect on both mortality and morbidity followed such increases in air pollution during the first days. However, the mortality rates remained high for several weeks (Anderson, 1999).

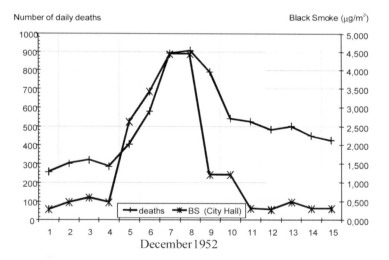

**Figure 16.** The London fog. Number of daily deaths and levels of black smoke in London, December 1-15, 1952 (Ministry of Health, 1954).

Officially, the number of exceeding deaths attributed to this episode was between 3500 and 4000 (Ministry of Health, 1954). Subsequent studies, however, have discussed such appraisal as they estimated about 12,000 exceeding deaths occurred from December 1952 through February 1953 because of acute and persisting effects of the 1952 London smog (Bell and Davis, 2001).

## 2.    Air pollutants and sources

### 2.1.    *Sources*

The air pollutants which are normally measured in urban atmosphere are mostly derived from human activities (Ritcher and Williams, 1998). There are two broad types (Figure 17).

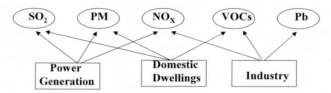

**Stationary Emission Sources**

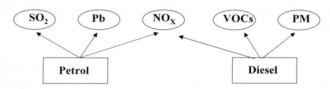

**Mobile Emission Sources (Road Traffic)**

**Figure 17**. Major emission sources of air pollutants (Ritcher and Williams, 1998).

### 2.1.1.    *Stationary sources*

They can be subdivided into: (a) Rural area sources such as agricultural production, mining and quarrying, (b) Industrial point and area sources such as manufacturing of chemicals, non-metallic mineral products, basic metal industries, power generation, and (c) Community sources, e.g. heating of homes and buildings, municipal waste and sewage sludge incinerators, fireplaces, cooking facilities, laundry services and cleaning plants. The last two are the most important.

### 2.1.2.    *Mobile sources*

They comprise any form of combustion-engine vehicles, e.g. light duty gasoline-powered cars, light and heavy-duty diesel-powered vehicles, motorcycles, aircraft, and even line sources such as fugitive dusts from vehicle traffic.

### 2.2.    *Air pollutants*

A distinction should be made between primary and secondary pollutants (WHO, 1987). Primary ones are those coming directly from the source of emission. Secondary pollutants are produced as a consequence of transformation and chemical and physical reactions suffered by primary pollutants within the atmosphere. The characteristics of the main chemical pollutants and their sources are summarised in Table 9.

**Table 9**. Summary description of the major air pollutants.

| Pollutant | Formation | Physical state | Sources |
|---|---|---|---|
| Suspended particulates(PM): $PM_{10}$, Black smoke | Primary & secondary | Solid, liquid | Vehicles (mainly diesel) Industrial processes Tobacco smoke |
| Sulphur dioxide ($SO_2$) | Primary | Gas | Industrial processes Vehicles |
| Nitrogen dioxide ($NO_2$) | Primary | Gas | Vehicles Gas heaters and cookers |
| Carbon monoxide (CO) | Primary | Gas | Vehicles (mainly gasoline) Tobacco smoke |
| Volatile organic compounds (VOCs) | Primary, secondary | Gas | Vehicles, industry, tobacco smoke |
| Lead (Pb) | Primary | Solid (fine particulates) | Vehicles , industry |
| Ozone ($O_3$) | Secondary | Gas | Vehicles (secondary to photo-oxidation of $NO_x$ and volatile organic compounds) |

*2.2.1.   Suspended particulates (PM)*

The term 'suspended particulates' refers to finely-divided, non-specific particulates, either liquid or solid, which are small enough to be suspended for hours or days, and carried through considerable distances. They represent a complex mixture of organic and non-organic substances. There are different names to describe these particles depending on the technique used for their determination and their size. Among these names we find: total suspended particulates (TPS), black smoke, breathable particulates, thoracic particles. In recent years, there is a preference in naming them after more objective features such as their diameter; particles with a diameter inferior to 10 μm ($PM_{10}$), with diameter inferior to 2,5 μm ($PM_{2,5}$) (also called fine particulates) or particulates smaller than 0,1μm (ultra fine particulates).

*2.2.2.   Sulphur compounds*

The main ones are sulphur dioxide ($SO_2$) produced by combustion in fixed sources (heating, industries) and sulphates, resulting from atmospheric oxidation of $SO_2$. Changes in the type of fuels used in Western Europe have resulted in a considerable reduction of $SO_2$ emissions, although localised high concentrations associated with occasional emissions may still occur.

### 2.2.3.   Nitrogen compounds

Their main non-natural source of emission are fossil fuels used in transport, heating and energy generation. Most combustion processes produce nitrogen monoxide (NO) which, through oxidation processes result in nitrogen dioxide ($NO_2$). Sometimes data are referred to in terms of $NO_x$, indicating a mixture of nitrogen oxides.

### 2.2.4.   Carbon oxides

They are, mainly, carbon monoxide (CO) and carbon dioxide ($CO_2$). They liberate into the atmosphere as a consequence of incomplete (CO) or complete ($CO_2$) combustion. The main source of CO are fumes coming from motor vehicle exhaust systems. $CO_2$ is also the main pollutant responsible for the greenhouse effect.

### 2.2.5.   Volatile organic compounds (VOCs)

VOCs are a varied group of compounds present in the atmosphere, which include a wide range of hydrocarbons such as alkanes, alkenes, aromatics, aldehydes, ketones, alcohols, esters, and some chlorinated compounds. Benzene is an aromatic VOC which has received much attention due to its carcinogenicity. Other aromatic compounds such as toluene are important precursors of ozone.

### 2.2.6.   Lead (Pb)

Emissions from motor vehicles are the main source of lead in air, principally present as particles with diameter less than 1 μm. In an important and increasing number of countries, lead emitted by vehicles is declining with the introduction of unleaded fuels.

### 2.2.7.   Photochemical oxides

They are pollutants resulting from chemical reactions between reactive hydrocarbons and nitrogen oxides under the effect of solar light. Ozone ($O_3$) is, from a toxicological point of view, the most important among these pollutants. Provided that vehicle-emission-generated primary pollutants react with it, it may be found in considerably large concentrations even in areas far apart from the sources of emission and its levels are frequently higher around big cities rather than inside.

A distinction is usually made between typical winter pollution ('winter smog') and typical summer pollution ('summer smog'). In winter, pollution episodes may occur due to stagnant air conditions, when pollutants coming from combustion concentrate in the atmosphere. The main pollutants are $SO_2$ and suspended particles, although these just serve as indicators of much more complex mixtures of pollutants. In summer, pollution episodes may occur in sunny hot days, when photochemical reactions of nitrogen oxides and hydrocarbons lead to the formation of ozone and other substances with toxic capacity. This pattern, however, is less clear for pollutants such as $NO_2$, CO and some particulates, which have vehicle exhaust systems as their main source.

## 3. Effects of air pollution on health

The interpretation of the reactions produced by air pollution on human health is founded on two types of studies, epidemiological or observational and toxicological or experimental. Both types of studies are considered complementary when evaluating the effects of air pollution on health.

### 3.1. Epidemiological studies

They refer to the observation of events which develop among human populations under natural conditions, circumstances which form their most significant advantage. When measuring exposure to air pollution, many studies have used data from air pollution surveillance networks, but other approaches to assses exposure are increasing, ranging from consideration of residence as an approximation to the exposure level to pollution or exposure questionnaires, to the use of personal samplers or the determination of biomarkers (Ozkaynak, 1999). Figure 18 shows the main exposure measuring methods used in the field of epidemiology. A rise in the grade of sophistication makes measurements more valid, but it also affects their cost.

- Categorical exposure (high versus low)
  ( i.e. from residence, occupation, etc.)
- Measured (or modelled) outdoor concentrations
  (usually from air pollution networks)
- Measurement of indoor and outdoor concentrations
- Estimation of individual exposure using indoor, outdoor along time-activity diaries
- Direct measurement of personal exposures
- Measurement of biomarkers of exposure

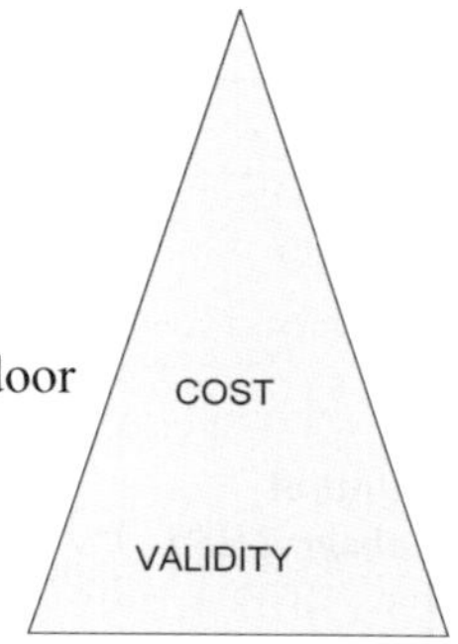

COST

VALIDITY

**Figure 18**. Exposure assessment approaches in epidemiology of air pollution (adapted from Ozkaynak, 1999).

The effects of exposure to air pollution are diverse and of different severity. Among them, its effects on the respiratory and the cardio-circulatory systems can be highlighted. This effect maintains a grading, both regarding the seriousness of its consequences and the risk population affected (Figure 19).

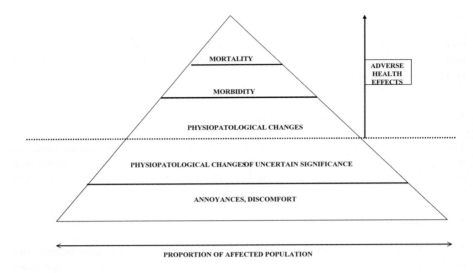

**Figure 19**. Health effects of air pollution (adapted from Andrews *et al.*, 1985).

For the study of such effects, several epidemiological designs have been used, which can be roughly divided into four categories, depending on whether the analysis is carried out at population or individual level, and whether the effects evaluated come from acute exposure or long term exposure, that is to say: chronic exposures (Figure 20).

| | | **Exposure** | |
|---|---|---|---|
| | | Acute | Chronic |
| **Unit of observation** | Aggregated | Time Series (counts): Mortality, Hospital Admissions Emergencies | Ecological (rates): Mortality Morbidity |
| | Individual | Panel studies: Symptoms/Diseases Lung Function | Cohort studies: Mortality Symptoms/Diseases Lung Function Case-control studies |

**Figure 20**. Basic study designs in air pollution epidemiology (adapted from Antó and Sunyer, 1999 and Pope and Dockery, 1999).

### 3.1.1. Time series studies

One of the most used epidemiological design type are the ecological time series studies. In them, the variations in time of exposure and the health indicator of a certain population are analysed (number of deaths, hospital admissions, etc.). One advantage of these studies is that when analysing the same population in different periods of time (day to day, generally) many of the variables which could act as confounding factors at individual level (smoking habits, age, gender, profession/occupation, etc.) are maintained stable within the same population and lose their confounding potential (Schwartz et al., 1996). In these studies possible confounding factors are those which can covariate with health indicators and pollutants along time. These factors can be summarised in four groups: geophysical, meteorological, socio-cultural (for instance, the weekly activity pattern) and correlated illnesses such as influenza with a clear seasonal behaviour (Goldsmith et al., 1996) (Figure 21). The construction of any mathematical model analysing the relationship between air pollution and health must take into account these factors. On the other hand, a series of limitations in the statistical analysis procedures have been described recently, which have raised a need for reconsidering the estimates obtained among the existing studies (Dominici et al., 2002a; Samet et al., 2003; Ramsay et al., 2003).

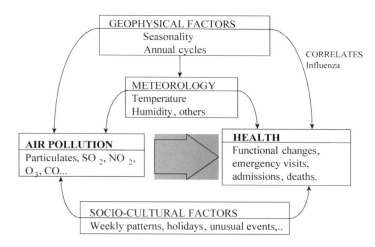

**Figure 21**. Relationship between the factors involved in the time series epidemiological studies of air pollution (adapted from Goldsmith et al., 1996).

In recent years, several multi-centric projects have been carried out, using standardised analysis criteria for the study of different aspects of the relationship between air pollution and health. In Europe, the APHEA project (Box 2) and in the United States the NMMAPS study (Samet et al., 2000a) are among those which have thrown much light upon the knowledge of acute impact of pollution on health.

In a considerable number of countries such as Canada (Burnett *et al.*, 1998), France (Quénel *et al.*, 1999), or Italy (Biggeri *et al.*, 2001), multi-centric studies have been carried out at national level. In these studies the impact of pollution has been evaluated keeping environmental, sanitary and social characteristics in mind. Also, these studies may prove useful for decision-making regarding pollution control-related measures and the improvement of information systems of each country. In Spain a study on the air pollution impact including 16 cities has been carried out (Box 3).

**Box 2**. Case 2: APHEA Project (Air Pollution Health Effects: a European Approach).

This multi-centre study started in 1993, with funding from the European Communities, with the purpose of evaluating the short-term impact of air pollution on the health of the European population. During its first phase, there were 15 European cities from 10 different countries participating, with a total study population of approximately 25 million inhabitants (Katsouyanni *et al.*, 1995).

The APHEA project has published an important number of articles, among them several meta-analysis on the short-term effect of air pollution in several European cities. These studies have been very useful in the evaluation of pollution impact in Europe and for the formulation of air quality guidelines. In the first phase of the study, in the meta-analysis for particulates a 0.4 per cent increase in death risk was obtained for 10 $\mu g/m^3$ increase in the daily levels of $PM_{10}$ and $SO_2$, and 0.3 per cent in black smoke levels (Katsouyanni *et al.*, 1997) and $NO_2$ (Touloumi *et al.*, 1997). For $PM_{10}$ and $SO_2$ lower effects were described in the eastern cities than in the western ones. A posterior sensitivity analysis controlling for long term trends and season, excluding days with very high pollution levels, reduced this heterogeneity and provided higher combined estimates (around 0.6 and 1 per cent increase in mortality for increase in 10 $\mu g/m^3$ in black smoke and $SO_2$, respectively) (Samoli *et al.*, 2001). As for ozone, a 50 $\mu g/m^3$ increase in ozone levels was associated with a 4 per cent increase in the number of hospital admissions for chronic obstructive pulmonary diseases (COPD) (Anderson *et al.*, 1997), with a 3.5 per cent in admissions for asthma in people over 15 (Sunyer *et al.*, 1997) and with 2.3 per cent in the number of deaths by all causes. (Touloumi *et al.*, 1997).

In the second phase of the project, APHEA2, 30 European cities participated, with more complete information about exposure and other variables. This has enabled the assessment of the consistency of the associations, and a better control of confounding and effect modification. For particulates, the relationship with mortality showed an increase of 0.6 per cent in the daily number of deaths for an increase of 10 $\mu g/m^3$ in black smoke, and the same for $PM_{10}$. When effect modification was investigated, cities with average levels of $NO_2$, warm climate, and low standardised mortality rates showed higher effects (Katsouyanni *et al.*, 2001). For respiratory admissions the estimates showed around 1 per cent increase fort $PM_{10}$ and smaller for black smoke (Atkinson *et al.*, 2001). On the contrary, black smoke presented a clearer association with cardiovascular admissions than $PM_{10}$, 1.1 per cent and 0.5, respectively (Le Tertre *et al.*, 2002).

*3.1.2.    Cohort studies*

Although in smaller number than time series studies, there are several cohort studies about the impact of pollution on health the first one is known as the *six cities* study. In it, a group of 8111 adults from six United States cities were studied in 1974,

(Dockery *et al.*, 1993). Results show that, once controlled for smoking habit and other risk factors, mortality indices are associated with air pollution. Death risk in the most polluted cities was 26 per cent higher compared with those with lower pollution. In another study, Pope *et al.* (1995a), evaluated the effects of air pollution by particles on mortality of participants from a follow-up study of the American Cancer Society as a part of the Cancer Prevention II study. Since 1982, data on risk factors and air pollution were collected for a total number of 500,000 adults from 151 metropolitan areas of the United States. It was found that the death risk by all causes in the most polluted areas was 15 per cent higher than in those with less pollution. The risk was higher for mortality caused by cardiovascular diseases with a 26 per cent increase in the most polluted cities compared to the least polluted ones. In March 2002 results from the follow-up of this cohort up to 1998 were published (Pope *et al.*, 2002). Fine particulates ($PM_{2.5}$) and sulphur oxide were associated with all-cause, lung cancer, and cardiopulmonary mortality. Each 10 $\mu g/m^3$ elevation in fine particulate air pollution was associated with approximately a 4, 6, and 8 per cent increased risk for all causes, cardiopulmonary, and lung cancer mortality, respectively.

**Box 3**. Case 3: EMECAS project (Spanish Multicenter Study on Health Effects of Air Pollution).

---

The EMECAS project is a collaborative effort, funded by the Spanish Ministry of Health that seeks to evaluate the short-term effect of air pollution on health in the urban Spanish population. In the first phase it undertook the assessment of the relationship between seven air pollutants and total daily mortality for all and for respiratory and cardiovascular causes in 13 Spanish cities (EMECAM, 1999).

Combined results showed a significant association between air pollution indicators and mortality. An increase of 10$\mu g/m^3$ in the levels of the average of the concurrent and one day lag for black smoke was associated with a 0.8 per cent increase in mortality (Ballester *et al.*, 2002). The estimates for TSP and $PM_{10}$ and total mortality were slightly lower. The same increase in concentrations of $SO_2$ was associated with a 0.5 per cent increase in daily deaths, and a 0.6 per cent in the case of $NO_2$. For groups of specific causes we found higher magnitude in the estimations, especially for respiratory conditions. Ozone only showed statistical significance with cardiovascular mortality (Sáez *et al.*, 2002). When two pollutant analyses were performed, estimates for all pollutants did not substantially modify, except for $SO_2$ where estimates for daily levels were strongly attenuated. On the contrary, the association for one-hour maximum levels of this pollutant did not show any change.

In the second phase of the project a total of 16 cities, accounting for more than 10 million inhabitants, are participating. More complete and updated data, both for mortality and hospital admissions, have been collected, enabling adequate control of confounding and address an exploration of potential causes *for* heterogeneity.

Combined estimates for cardiovascular admissions indicate an association with cardiovascular admissions (EMECAS, preliminary findings). An increase of 10 $\mu g/m^3$ in the $PM_{10}$ levels was associated with a 0.9 per cent increase in the number of hospital admissions for all cardiovascular diseases, and a 1.6 per cent for ischeemic heart diseases. Other pollutants also show an association with cardiovascular admissions. Results for this phase will be presented in the next few months.

Another study in the USA followed a non-smoking Seventh Day Adventists cohort of roughly 4000 adults and also found an increased risk of mortality related with the levels of pollutants in the region (Abbey et al., 1999). This study also indicated relations of particulates with chronic respiratory symptoms (Abbey et al., 1995).

The first cohort study in Europe has been published recently (Hoek et al., 2002). This has consisted of a Dutch cohort of 5000 adults followed during 7 years. Findings indicate an association between cardiopulmonary mortality and living near a major road. No clear association was found with other estimates of long term exposures.

As we can see, particulates have been among the most studied pollutants. In Table 10 a summary of results from some of the most important studies is shown.

**Table 10**. Summary of estimates of particulate matter effects (Adapted from Pope and Dockery, 1999; with addition of results from recent multicenter studies: a: Dominici et al., 2002b; b: Katsouyanni et al., 2001; c: Stieb et al., 2002; d: Samet et al., 2000a; e: Biggeri et al., 2001; f: Atkinson et al., 2001b; g: Le Tertre et al., 2002; h: Samet et al., 2000c; i: Pope et al., 2002).

| Health outcomes | Acute exposures | Chronic exposures |
|---|---|---|
| | Per cent change in health end point per increase in 10 $\mu g/m^3$ in $PM_{10}$ | Per cent change in health end point per increase in 5 $\mu g/m^3$ in $PM_{2.5}$ |
| **Increase Mortality*** | *(Population based) | *(Cohort based) |
| -   All organic causes | $0.2^a$ - $0.6^{b,c}$ - 1.0 | $2^i$ - 3 |
| -   Cardiovascular | $0.7^{c,d}$ to 1.4 | $3^i$ - 6 |
| -   Respiratory | $1.3^c$ to 3.4 | |
| -   Lung Cancer | | $4^i$ |
| | | |
| **Increased hospitalisation** | | |
| -   All respiratory | 0.8 to $2.4^e$ | |
| -   COPD | $1.0^f$ to 2.5 | |
| -   Asthma | $1.1^f$ to 1.9 | |
| -   Cardiovascular | $0.5^g$ to $1.2^h$ | |
| | | |
| **Disease:** bronchitis | | 7 |
| | | |
| **Decreased lung function (FEV$^1$)** | | |
| -   Children | 0.15 | 1 |
| -   Adults | 0.08 | 1.5 |

*3.1.3.    Studies of the effects of 'summer pollutants' on health*

The summer type pollution refers mainly to photochemical pollution coming from hydrocarbon and nitrogen oxides reactions to intense solar light. Ozone is generally considered the most toxic component of this mixture. However, not all the effects of photochemical pollution can be attributed to ozone alone. This is so for some of the discomfort caused by pollution such as ocular irritation caused by organic nitrates and aldehydes.

Recent studies have described an important number of adverse effects of ozone. Among them, increased respiratory symptoms (Thurston and Ito, 1999), declines in lung function (Galizia and Kinney, 1999; Gauderman et al., 2002), aggravation of asthma (Gauderman et al., 2002; McConnell et al., 2002), increased risk for emergency visits (Tenias et al., 2002) and hospital admissions (Anderson et al., 1997; Sunyer et al., 1997), and probably, an effect on mortality could be included (Burnett et al., 2001; Goldberg et al., 2001) . On the other hand, children and asthmatics have been described as sensitive groups to the effects of ozone.

### 3.2.    Experimental toxicological studies

These studies are carried out both with humans and animals. In them, concentration, duration and conditions of exposure are controlled by the investigator. Their main advantage is precisely the control of exposure conditions, therefore, the measurement of this exposure is more precise than in observational studies. As for its disadvantages, these are basically the fact of keeping the experiment subjects under an artificial situation -since a reduced number of pollutants are used in an ideal situation, in big doses and in special environmental circumstances. Also, the population subject to these experiments could be little representative of the general population or any susceptible groups.

It is worth highlighting that in recent years, there have been great advances in the development of studies on the effects of controlled exposure to pollutants. A substantial advance has particularly been made in the understanding of the effects and biological mechanisms related with particles and ozone.

### 3.2.1.   Particles

As far as particles are concerned, regardless of the consistency of the epidemiological studies results, about their impact on health, the lack of an explanation of their physio-pathological mechanism led to the raising of doubts on whether such association is causal . This generated an interesting debate between, on one side, some authors who considered the evidence as sufficient and pleaded for a more decisive intervention (Bates 2000; Pope et al., 1995b; Schwartz, 1994), while others raised doubts on the impact being caused by particulates (Gamble and Lewis, 1996; Moolgavkar and Luebeck, 1996; Utell and Samet, 1993). All this debate has provided an impulse for toxicological research and for more epidemiology studies which have provided firm evidence to biological coherence and robust results. Due to the access way of particles, there are fewer doubts on the mechanisms capable of causing respiratory problems, such as lung inflammation (Pope, 2000). However, the biological plausibility of the association of air pollution by particles with the cardiovascular system has raised doubts.

In recent years, several physiopathological mechanisms have been proposed (Seaton *et al.*, 1995) (Figure 22). One of the main hypotheses is that particles induce the activation of some mediators which cause increase of blood coagulation (Peters *et al.*, 1997; Schwartz, 2001). Other mechanisms studied are related with automatic cardiac control, where a significant association of particles with an increase of heart rhythm rate and a decrease of its variability has been shown (Liao *et al.*, 1999; Peters *et al.*, 1999; Pope *et al.*, 1999a; Pope *et al.*, 1999b). Lastly, in experimental studies with human beings, inflammatory changes at alveolar level and an increase of fibrinogen levels, leukocytes and platelets at alveolar level have been found (Ghio *et al.*, 2000; Salvi *et al.*, 1999). A growing number of studies support the hypothesis that the composition of ultra fine particles (Donaldson *et al.*, 2001; Frampton 2001) and their contents of transition metals (Costa and Dreher, 1997), could explain their toxic capacity in the cardiovascular system. On the other hand, an additive role, played by other gaseous pollutants, cannot be discarded.

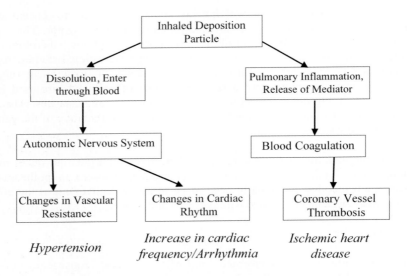

**Figure 22**. Possible physiopathological mechanisms for impact of inhaled particulates on cardiovascular system (Adapted from Muggenburg *et al.*, 2000).

### 3.2.2.    Carbon monoxide

There are few doubts about the toxicity of CO on the cardiovascular system (Allred *et al.*, 1991). It is also known that CO can produce dyspnoea and hypoxia as it causes the formation of carboxihaemoglobine (Maynard and Waller, 1999).

### 3.2.3.    Nitrogen dioxide

The mechanism by which $NO_2$ can produce harm ending in death is little known, although it has been said that it could possibly be the inflammatory response to the activation of oxidative pathways caused by this pollutant, or to its capability of damaging the alveolar macrophages function, therefore causing an increased risk of lung infection (Brunekreef and Holgate, 2002).

### 3.2.4. Ozone

Lastly, regarding toxic mechanisms for ozone, results of some studies suggest that, besides its oxidant capacity to cause lung inflammation (Bassett et al., 2001) ozone exposure can increase myocardial work and impair pulmonary gas exchange to a degree that might be clinically important in persons with significant pre-existing cardiovascular impairment, with or without concomitant lung disease (Gong et al., 1998).

With all that described above, and provided the complex mixture which forms urban air pollution, with common sources and a high correlation between pollutants, it is difficult to attribute effects to a single pollutant. Rather, particles should be considered to be a marker of pollution levels within a city. In general, knowledge about the biological mechanisms of this association is still limited and requires much additional research in different fields.

## 4. Air quality guidelines and limit values

In 1987 the World Health Organisation (WHO) published some air quality guidelines which comprised 28 pollutants (WHO, 1987). As has been mentioned above, different epidemiological studies have shown the adverse effects when they are below these levels. As a consequence of accumulating evidences that suggest so, the WHO has updated such guidelines and has increased to 38 the number of air pollutants or polluting mixtures collected in such guidelines (WHO, 1999). Although the value of these guidelines has no legal power, they are used by many countries to develop the air quality national standards. The European Union has recently published two directives with the limit values for the major air pollutants (Table 11) (EC, 1999; EC, 2000).

**Table 11**. Limit values for standard air pollutants for human health protection in the European Union (EC, [a]1992; [b]1999; [c]2000).

| Pollutant | Mean period | Limit value |
|---|---|---|
| Lead[b] | 1 year | 0.5 $\mu g/m^3$ |
| Nitrogen dioxide[b] | 1 hour | 200 $\mu g/m^3$ |
| | 1 year | 40 $\mu g/m^3$ |
| Sulphur dioxide[b] | 1 hour | 350 $\mu g/m^3$ |
| | 24 hours | 125 $\mu g/m^3$ |
| $PM_{10}$ (to be respected in 2005)[b] | 1 year | 40 $\mu g/m^3$ |
| | 24 hours | 50 $\mu g/m^3$ |
| $PM_{10}$ (to be respected in 2010)[b] | 1 year | 20 $\mu g/m^3$ |
| | 24 hours | 50 $\mu g/m^3$ |
| Carbon monoxide[c] | 8 hours | 10 $mg/m^3$ |
| Ozone[a] | 8 hours | 110 $\mu g/m^3$ |

Regarding the WHO Guidelines of 1987, there has been an important reduction of the value for $NO_2$ (from 150 $\mu g/m^3$ for a daily mean to 40 $\mu g/m^3$ for an annual mean). For suspended particles, the European Union has established $PM_{10}$ annual limit values of 40 $\mu g/m^3$ for year 2005 and of 20 $\mu g/m^3$ for year 2010. In a considerable number of European cities and regions, it will be difficult to meet these limits if decisive improvement actions are not taken to improve air quality.

## 5.    Health impact assessment of the impact of air pollution

Health impact assessment (HIA) aims at the quantification of the expected health burden due to an environmental exposure in a specific population (WHO Regional office for Europe, European Centre for Environment and Health, and Bilthoven Division, 2000). To reach this goal, HIA combines three of the four stages of the paradigm of risk assessment (Samet, 1999), that is: exposure assessment, dose-response, and risk characterisation stages.

**Box 4**. Case 4: APHEIS: Air Pollution and Health: a European Information System.

The APHEIS programme is an epidemiological surveillance system that aims to provide European, national, regional and local decision makers, environmental-health professionals and the general public with up-to-date and easy-to-use information on air pollution and public health (http://www.apheis.org).

During its first year (1999-2000), APHEIS achieved two key objectives: it defined the best indicators for epidemiological surveillance and health impact assessment (HIA) of air pollution in Europe in the form of guidelines; and it identified those entities best able to implement the surveillance system in the cities participating in the programme (APHEIS, 2001). During its second year, APHEIS implemented or adapted the organisational models designed during the first year, collected and analysed the data for HIA, prepared different health-impact scenarios and an HIA report in standardised format.

The first HIA report of APHEIS covered 26 cities in 12 Western and Eastern European countries and it assessed the acute and chronic effects of fine particles on premature mortality and hospital admissions for cardiovascular and respiratory diseases using the estimates developed by APHEA2 study and two North-American cohort studies.

The total population covered in this first HIA from APHEIS includes nearly 39 million inhabitants. $PM_{10}$ concentrations were measured in 19 cities (annual average range: 20-50 $\mu g/m^3$). Black smoke (BS) concentrations were provided by 15 cities (annual average range: 20-65 $\mu g/m^3$). The results of the HIA estimation indicates that reducing long term exposure to $PM_{10}$ levels by 5 $\mu g/m^3$ would have prevented 5,547 premature deaths annually, 800 of which were attributable to short-term exposure. If black smoke is considered, a reduction of 5 $\mu g/m^3$ in its levels would have decreased short-term deaths by over 500 per year (APHEIS, 2002).

In the context of air pollution, HIA could be useful in providing key information to public health and environmental decision makers, in the evaluation of different policies and scenarios for the reduction of air pollution levels, or in the estimation of the costs of air pollution or the benefits of preventive actions (APHEIS, 2001). The usefulness and credibility of HIA depends on the accuracy of its estimates. Because of that, an accurate procedure and a critical discussion of the strength of estimates provided by HIA are needed in each case. In Europe, the APHEIS project has developed a HIA of air pollution in 26 cities of 12 European countries (Box 4). In that way, through the integration of public health criteria in environmental decision-making, HIA may become a helpful tool for planning and evaluation of public policies.

## 6.   Intervention studies

Some studies illustrate the potential health benefits of policies and actions aimed at decreasing the exposure to air pollution. Several years ago, Pope showed that in Utah Valley, closure of an open-earth steel mill over the winter of 1987 was associated with reductions in respiratory disease and related hospital admissions among valley residents (Pope, 1996). In a later investigation, Dye *et al.* (2001) analysed composition of filters originally collected near the steel mill during the winter of 1986 (before closure), 1987 (during closure), and 1988 (after plant reopening). The authors found a higher quantity of sulphate and certain metals (i.e., copper, zinc, iron, lead, strontium, arsenic, manganese, nickel) in the 1986 and 1988 filters. They intratracheally instilled liquid extracts from filters to rats, and found that rats exposed to 1986 or 1988 extracts developed significant pulmonary injury and neutrophilic inflammation suggesting that sulphate or metals may be important determinants of the pulmonary toxicity observed (Dye *et al.*, 2001).

In the USA, Mott *et al.* (2002) evaluated the influence of national vehicle emission policies from 1968 to 1998 on deaths attributable to carbon monoxide, therefore enabling the assessment of impact of the enforcement of standards set by the 1970 Clean Air Act. They found a decline in mortality rates suggesting an important public health benefit following the decreases in CO emissions. In the same way Ostro *et al.* (1999) investigated the benefits of sulphate reductions following the 1990 Clean Air Act Amendment describing substantial health benefits.

Two recent studies have added more evidence on health benefits when reducing air pollution exposures. Clancy and colleagues evaluated the effect of air pollution control and death rates in Dublin (Clancy *et al.*, 2002). After the ban, a clear reduction (70 per cent) of black smoke concentrations was observed. Subsequently, death rates from organic causes decreased by 5.7 per cent, cardiovascular deaths by 10.3 per cent, and respiratory ones by 15.5 per cent, showing similar impact estimates than to cohort studies.

Another intervention study has focused on changes in air quality and death rates after a restriction on sulphur content in Hong Kong (Hedley *et al.*, 2002). The first year after intervention, both $SO_2$ and sulphates showed a clear reduction. Two years after the restriction, $SO_2$ maintained low levels; however, sulphate concentrations rose again and stabilised, probably, as suggested by the authors, as part of the regional pattern of sulphate pollution in southern China. Death rates showed a substantial reduction in the first 12 months, but reached a peak in the second winter after intervention. Considering years 3 to 5 after intervention, average annual trend decreased significantly in a 2.1 per cent for all causes, and 3.9 per cent for respiratory ones.

## 7.   Strategies for improvement of air quality

Some cities have established programmes intended to reduce emissions from trucks and cars. For example, in Athens a measure was introduced whereby cars were allowed to circulate on alternate days according to their number plates ending in odd or even numbers. This step is not exempt from obstacles, for it may lead to citizens not changing their old vehicles (generally more polluting than new ones) or to dispose of two vehicles with number plates ending in different numbers (odd, even). Besides that, the National Technical University of Athens has developed a traffic map that displays congestion on the city's streets updated every 15 minutes (see http://www.transport.ntua.gr/map/), this offers drivers, town planners, and public health workers the chance to monitor traffic flows, and could help to cope with plans to avoid or reduce traffic jams. Other cities, such as Cartagena in Spain, have elaborated integral operational plans for air pollution control (http://www.ayto-cartagena.es/medioam-convenios.htm). In the development of such plans all the parties involved participate (industries, city hall, environmental department, meteorology department) and certain pollution surveillance and control guidelines are defined to avoid reaching levels which would be dangerous for health. Paris is one of the cities which have most persistently fought against air pollution in recent years.

There are some international initiatives related with this issue. One major project developed by WHO and the United Nations Economic and Social Council is The Transport, Environment and Health project for Europe (http://www.the-pep.org). Besides that, an important number of cities are developing programmes to reduce the number of cars in the city and are applying important restrictions on traffic in their streets. More on that could be found at http://carfree.com. Very recently, the municipality of the city of London has implemented a system of charging a fee for cars going into central London as part of an effort to reduce traffic congestion in the centre of the city (http://www.london.gov.uk/mayor/congest/index.jsp). This congestion charge system started on 17th of February, and it is oriented towards reducing traffic, making journeys and delivery times more reliable, and raising funds to reinvest in the London transport system.

On the other hand, the energy generating and combustion processes industries and the vehicle manufacturing industry are developing new technologies which will improve air conditions. Environmental quality is considered more and more a priority among human activities. Despite this, environmental control is still necessary in order to maintain people's right to breath air under healthy conditions.

Parallel to regulation measures, and urban and transport planning activities and the development of technologies individual behaviour may contribute to the reduction of pollution. For instance, we can choose to use a less polluting vehicle, keep it in good condition to reduce emissions, avoid leaving the engine on when not moving, and driving avoiding abrupt accelerations and stops. Several people can travel in the same car, and we can use public transport more frequently, or we can use a bicycle or walk.

A very interesting strategy developed by WHO, is that which outlines an integral approach to the implications of transport policies on health (Dora, 1999), consequences due to air pollution, noises, accidents, climatic change and their capability of altering the security conditions of walking or cycling (Figure 23). From this perspective, politicians and public decision-makers need to be well-informed about the impact of the different transport options on health. Health and environment professionals play a key role in providing such information as well as in evaluating its effects on health.

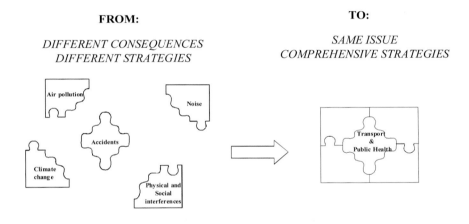

**Figure 23**. Strategies on transport, health and environment (Adapted from Dora, 1999).

## 8.   Conclusions

Today, decades after some of the most important episodes, air pollution continues to be one of the major issues of Environment and Public Health. Epidemiology studies in a considerably great number of places have established an association between increases in air pollutants and increases in symptoms, diseases, emergency visits, hospital admissions and deaths. Evidence is very consistent and shows an independent effect for particles and ozone. Other pollutants like CO, $NO_2$ or $SO_2$ could have an independent or an additive effect, and further investigation is needed to elucidate their role. Individual follow-up studies indicate a far from negligible effect of long term exposure on life expectancy. In fact their impact on mortality could be several times that of time series studies. In addition, experimental research is providing insight into the biological mechanisms of these toxicant impacts.

Road traffic is one of the major sources of exposure to air pollution in most cities of developed countries. Its association with health impact has been determined in an important number of studies.

An effort is being made to bring close different points of view, interests and needs, since policies involving air quality like those related with transport or urban planning are complex. Of special interest is the application of research in decisions, both political and individual, related to pollution and health. Health impact assessment and risk communication could provide policy makers, health, environmental and other professionals, or the public in general with clear, understandable, and good quality evidence to facilitate their making of decisions, each according to their level of responsibility.

## References

Abbey, D.E., Hwang, B.L., Burchette, R.J., Vancuren, T., and Mills, P.K. (1995) Estimated long-term ambient concentrations of PM$_{10}$ and development of respiratory symptoms in a non-smoking population, *Archives of Environmental Health* 50, 139-152.

Abbey, D.E., Nishino, N., McDonnell, W.F., Burchette, R.J., Knutsen, S.F., Lawrence, B.W., and Yang, J.X. (1999) Long-term inhalable particles and other air pollutants related to mortality in non-smokers, *American Journal of Respiratory and Critical Care Medicine* 159, 373-382.

Allred, E., Bleecker, E., Chaitman, B., Dahms, T., Gottlieb, S.O., Hackney, J., Pagano, M., Selvester, R., Walden, S., and Warren, J. (1991) Effects of carbon monoxide on myocardial ischemia, *Environ. Health Perspect.* 91, 89-132.

Anderson, H.R. (1999) Health effects of air pollution episodes, in S.T. Holgate, J.M. Samet, H. Koren, and R.L. Maynard (eds), *Air Pollution and Health*, Academic Press, San Diego, California, pp. 461-482.

Anderson, H.R., Spix, C., Medina, S., Schouten, J.P., Castellsague, J., Rossi, G., Zmirou, D., Touloumi, G., Wojtyniak, B., Pönkä, A., Bacharova, L., Schwartz, J., and Katsouyanni, K. (1997) Air pollution and daily admissions for chronic obstructive pulmonary disease in 6 European cities: results from the APHEA project, *European Respiratory Journal* 10, 1064-1071.

APHEIS - Air Pollution and Health: a European Information System (2001) *Monitoring the Effects of Air Pollution on Public Health in Europe*, Scientific report 1999-2000, Institut de Veille Sanitaire, Saint Maurice.

APHEIS - Air Pollution and Health: a European Information System (2002) *Health Impact Assessment of Air Pollution in 26 European Cities*, Second year report 2000-2001, Institut de Veille Sanitaire, Saint Maurice.

Atkinson, R.W., Ross, A.H., Sunyer, J., Ayres, J., Baccini, M., Vonk, J.M., Boumghar, A., Forastiere, F., Forsberg, B., Touloumi, G., Schwartz, J., and Katsouyanni, K. (2001) Acute effects of particulate air pollution on respiratory admissions. Results from APHEA2 project, *American Journal of Respiratory and Critical Care Medicine* 164, 1860-1866.

Ballester, F., Saez, M., Perez-Hoyos, S., Iñíguez, C., Gandarillas, A., Tobias, A., Bellido, J., Taracido, M., Arribas, F., Daponte, A., Alonso, E., Cañada, A., Guillen-Grima, F., Cirera, Ll., Perez-Boíllos, M.J., Saurina, C., Gómez, F., Tenías, J.M., and on behalf of the EMECAM group (2002) The EMECAM project: a multi-center study on air pollution and mortality in Spain, Combined results for particulates and for sulphur dioxide, *Occupational and Environmental Medicine* 59, 300-308.

Bassett, D., Elbon-Copp, C., Otterbein, S., Barraclough-Mitchell, H., Delorme, M., and Yang, H. (2001) Inflammatory cell availability affects ozone-induced lung damage, *Journal of Toxicology and Environmental Health* Part A 64, 547-465.

Bates, D.V. (2000) Lines that connect: assessing the causality inference in the case of particulate pollution, *Environ. Health Perspect.* 108, 91-92.

Bell, M.L., and Davis, D.L. (2001) Reassessment of the lethal london fog of 1952: novel indicators of acute and chronic consequences of acute exposure to air pollution, *Environ. Health Perspect.* 109 (Suppl. 3), 389-394.

Biggeri, A., Bellini, P., and Terracini, B. (eds) (2001) Meta-analysis of the Italian Studies on Short-term Effects of Air Pollution, *Epidemiologia & Prevenzione* 25 (Suppl.), 1-72.

Brunekreef, B., and Holgate, S.T. (2002) Air pollution and health, *The Lancet* 360, 1233-1242.

Brunekreef, B., Janssen, N.A., de Hartog, J., Harssema, H., Knape, M., and Van Vliet, P. (1997) Air pollution from truck traffic and lung function in children living near motorways, *Epidemiology* 8, 298-303.

Burnett, R.T., Cakmak, S., and Brook, J.R. (1998) The effect of the urban ambient air pollution mix on daily mortality rates in 11 Canadian cities, *Canadian Journal of Public Health* 89, 152-156.

Burnett, R.T., Smith-Doiron, M., Stieb, D., Raizenne, M.E., Brook, J.R., Dales, R.E., Leech, J.A., Cakmak, S., and Krewski, D. (2001) Association between ozone and hospitalization for acute respiratory diseases in children less than 2 years of age, *American Journal of Epidemiology* 153, 444-452.

Ciccone, G., Forastiere, F., Agabiti, N., Biggeri, A., Bisanti, L., Chellini, E., Corbo, G., Dell'Orco, V., Dalmasso, P., Volante, T.F., Galassi, C., Piffer, S., Renzoni, E., Rusconi, F., Sestini, P., and Viegi, G. (1998) Road traffic and adverse respiratory effects in children. SIDRIA Collaborative Group, *Occupational and Environmental Medicine* 55, 771-778.

Clancy, L., Goodman, P., and Dockery, D.W. (2002) Effect of air-pollution control on death rates in Dublin, Ireland: an intervention study, *The Lancet* 360, 1210-1214.

Costa, D.L., and Dreher, K.L. (1997) Bioavailable transition metals in particulate matter mediate cardiopulmonary injury in healthy and compromised animal models, *Environmental Health Perspectives* 105 (Suppl. 5), 1053-1060.

Dockery, D.W., Pope, C.A., Xu, X., Spengler, J.D., Ware, J.H., Fay, M.E., Ferris, B.G.J., and Speizer, F.E. (1993) An association between air pollution and mortality in six US cities, *New England Journal of Medicine* 329, 1753-1759.

Dominici, F., McDermott, A., Zeger, S.L., and Samet, J.M. (2002a) On the use of generalized additive models in time-series studies of air pollution and health, *American Journal of Epidemiology* 156, 193-203.

Dominici, F., McDermott, A., Daniels, M., Zeger, S.L., and Samet, J.M. (2002b) *A Report to the Health Effects Institute: Reanalyses of the NMMAPS Database*, Departments of Biostatistics and Epidemiology, Bloomberg school of Public Health, Baltimore, MD, USA.

Donaldson, K., Stone, V., Seaton, A., and MacNee, W. (2001) Ambient particle inhalation and the cardiovascular system: potential mechanisms, *Environ. Health Perspect.* 109 (Suppl. 4), 523-527.

Dora, C. (1999) A different route to health: implications of transport policies, *British Medical Journal* 318, 1686-1689.

Dye, J.A., Lehmann, J.R., McGee, J.K., Winsett, D.W., Ledbetter, A.D., Everitt, J.I., Ghio, A.J., and Costa, D.L. (2001) Acute pulmonary toxicity of particulate matter filter extracts in rats: coherence with epidemiologic studies in Utah Valley residents, *Environ. Health Perspect.* 109 (Suppl. 3), 395-403.

EC - European Commission (1992) Council Directive 92/72/EEC of 21 September 1992 on air pollution by ozone.

EC - European Commission (1999) Council Directive 1999/30/EC of 22 April 1999 relating to limit values for sulphur dioxide, nitrogen dioxide and oxides of nitrogen, particulate matter and lead in ambient air.

EC - European Commission (2000) Council Directive 2000/69/EC of 16 November 2000 relating to limit values for benzene, and carbon monoxide.

EMECAM (1999) El proyecto EMECAM: Estudio español sobre la relación entre la contaminación atmosférica y la mortalidad, *Revista Española de Salud Pública* 73, 165-314.

English, P., Neutra, R., Scalf, R., Sullivan, M., Waller, L., and Zhu, L. (1999) Examining associations between childhood asthma and traffic flow using a geographic information system, *Environ. Health Perspect.* 107, 761-767.

Frampton, M.W. (2001) Systemic and cardiovascular effects of airway injury and inflammation: ultrafine particle exposure in humans, *Environ. Health Perspect.* 109 (Suppl. 4), 529-532.

Galizia, A., and Kinney, P.L. (1999) Long-term residence in areas of high ozone: associations with respiratory health in a nation-wide sample of non-smoking young adults, *Environ. Health Perspect.* 107, 675-679.

Gamble, J.F., and Lewis, R.J. (1996), Health and respirable particulate ($PM_{10}$) air pollution: a causal or statistical association? *Environ. Health Perspect.* 104, 838-850.

Gauderman, W.J., Gilliland, G.F., Vora, H., Avol, E., Stram, D., McConnell, R., Thomas, D., Lurmann, F., Margolis, H.G., Rappaport, E.B., Berhane, K., and Peters, J.M. (2002) Association between air pollution and lung function growth in southern California children: results from a second cohort, *American Journal of Respiratory and Critical Care Medicine* 166, 76-84.

Ghio, A.J., Kim, C., and Devlin, R.B. (2000) Concentrated ambient air particles induce mild pulmonary inflammation in healthy human volunteers, *American Journal of Respiratory and Critical Care Medicine* 162, 981-988.

Goldberg, M.S., Burnett, R.T., Brook, J., Bailar, J.C., Valois, M.F., and Vincent, R. (2001) Associations between daily cause-specific mortality and concentrations of ground-level ozone in Montreal, Quebec, *American Journal of Epidemiology* 154, 817-826.

Goldsmith, J.R., Friger, M.D., and Abramson, M. (1996) Associations between health and air pollution in time-series analyses, *Archives of Environmental Health* 51, 359-367.

Gong, H.J., Wong, R., Sarma, R.J., Linn, W.S., Sullivan, E.D., Shamoo, D.A., Anderson, K.R., and Prasad, S.B. (1998) Cardiovascular effects of ozone exposure in human volunteers, *American Journal of Respiratory and Critical Care Medicine* 158, 538-546.

Hedley, A.J., Wong, C.M., Thach, T.Q., Ma, S., Lam, T.H., and Anderson, H.R. (2002) Cardiorespiratory and all-cause mortality after restrictions on sulphur content of fuel in Hong Kong: an intervention study, *The Lancet* 360, 1646-1652.

Ibald-Mulli, A., Stieber, J., Wichmann, H.E., Koenig, W., and Peters, A. (2001) Effects of air pollution on blood pressure: a population-based approach, *American Journal of Public Health* 91, 571-577.

Jedrychowski, W., Whyatt, R., Camann, D., Bawle, U., Peki, k., Spengler, J., Dumyahn, T., and Perera, F. (2002) Effect of prenatal PAH exposure on birth outcomes and neurocognitive development among a cohort of Polish mothers and newborns. Study design and preliminary ambient data, *European EpiMarker* 6, 1-6.

Katsouyanni, K., Zmirou, D., Spix, C., Sunyer, J., Schouten, J.P., Pönkä, A., Anderson, H.R., Le Moullec, Y., Wojtyniak, B., and Vigotti, M.A. (1995) Short-term effects of air pollution on health: a European approach using epidemiological time-series data. The APHEA project: background, objectives, design, *European Respiratory Journal* 8, 1030-1038.

Katsouyanni, K., Touloumi, G., Spix, C., Schwartz, J., Balducci, F., Medina, S., Rossi, G., Wojtyniak, B., Sunyer, J., Bacharova, L., Schouten, J.P., Pönkä, A., and Anderson, H.R. (1997) Short-term effects of ambient sulphur dioxide and particulate matter on mortality in 12 European cities: results from time series data from the APHEA project. Air pollution and health: a European approach, *British Medical Journal* 314, 1658-1663.

Katsouyanni, K., Touloumi, G., Samoli, E., Gryparis, A., Le Tertre, A., Monopolis, Y., Rossi, G., Zmirou, D., Ballester, F., Boumghar, A., Anderson, H.R., Wojtyniak, B., Paldy, A., Braunstein, R., Pekkanen, J., Schindler, C., and Schwartz, J. (2001) Confounding and effect modification in the short-term effects of ambient particles on total mortality: results from 29 European cities within the APHEA2 Project, *Epidemiology* 12, 521-531.

Künzli, N., Kaiser, J., Medina, S., Studnicka, M., Chanel, O., Filliger, P., Herry, M., Horak, F., Buybonnieux-Texier, V., Quenel, P., Schneider, J., Seethaler, R., Vergnaud, J.C., and Sommer, H. (2000) Public health impact of outdoor and traffic-related air pollution: a European assessment, *The Lancet* 356, 795-801.

Le Tertre, A., Medina, S., Samoli, E., Forsberg, B., Michelozzi, P., Boumghar, A., Vonk, J.M., Bellini, A., Atkinson, R., Ayres, J.G., Sunyer, J., Schwartz, J., and Katsouyanni, K. (2002) Short-term effects of particulate air pollution on cardiovascular diseases in eight European cities, *Journal of Epidemiology and Community Health* 56, 773-779.

Liao, D., Creason, J., Shy, C., Williams, R., Watts, R., and Zweidinger, R. (1999) Daily variation of particulate air pollution and poor cardiac autonomic control in the elderly, *Environ. Health Perspect.* 107, 521-525.

Maynard, R.L., and Waller, R.E. (1999) Carbon monoxide, in S.T. Holgate, J.M. Samet, H. Koren, and R.L. Maynard (eds), *Air Pollution and Health*, Academic Press, San Diego, California, pp. 749-796.

McConnell, R., Berhane, K., Gilliland, F., London, S.J., Islam, T., Gauderman, W.J., Avol, E., Margolis, H.G., and Peters, J.M. (2002) Asthma in exercising children exposed to ozone: a cohort study, *Lancet* 359, 386-391.

Ministry of Health (1954) *Mortality and Morbidity during the London Fog of December 1952*, Her Majesty's Stationery Office, London.

Moolgavkar, S.H., and Luebeck, E.G. (1996) A critical review of the evidence on particulate air pollution and mortality, *Epidemiology* 7, 420-428.

Mott, J.A., Wolfe, M.I., Alverson, C.J., Macdonald, S.C., Bailey, C.R., Ball, L.B., Moorman, J.E., Somers, J.H., Mannino, D.M., and Redd, S.C. (2002) National vehicle emissions policies and practices and declining US carbon monoxide-related mortality, *Journal of American Medical Association* 288, 988-995.

Ostro, B.D., Chestnut, L.G., Mills, D.M., and Watkins, A.M. (1999) in S.T. Holgate, J.M. Samet, H. Koren, and R.L. Maynard (eds), *Air Pollution and Health*, Academic Press, San Diego, California, pp. 899-915.

Peters, A., Döring, A., Wichmann, H.E., and Koenig, W. (1997) Increased plasma viscosity during an air pollution episode: a link to mortality? *The Lancet* 349, 1582-1587.

Peters, A., Perz, S., Doring, A., Stieber, J., Koening, W., and Wichmann, H.E. (1999) Increases in heart rate during an air pollution episode, *American Journal of Epidemiology* 150, 1094-1098.

Pope, C.A. (1996) Particulate pollution and health: a review of the Utah valley experience, *Journal of Exposure Analysis and Environmental Epidemiology* 6, 23-34.

Pope, C.A. (2000) Epidemiology of fine particulate air pollution and human health: biologic mechanisms and who's at risk? *Environ. Health Perspect.* 108 (Suppl. 4), 713-723.

Pope, C.A., Thun, M.J., Namboodiri, M.M., Dockery, D.W., Evans, J.S., Speizer, F. E., and Heath, C. (1995a) Particulate air pollution as a predictor of mortality in a prospective study of US adults, *American Journal of Respiratory and Critical Care Medicine* 151, 669-674.

Pope, C.A., Bates, D.V., and Raizenne, M.E. (1995b) Health effects of particulate air pollution: time for reassessment? *Environ. Health Perspect.* 103, 472-480.

Pope, C.A., Dockery, D.W., Kanner, R.E., Villegas, G.M., and Schwartz, J. (1999a) Oxigen saturation, pulse rate, and particulate air pollution, *American Journal of Respiratory and Critical Care Medicine* 159, 365-372.

Pope, C.A., Verrier, R.L., Lovett, E.G., Larson, A.C., Raizenne, M.E., Kanner, R.E., Schwartz, J., Villegas, G.M., Gold, D.R., and Dockery, D.W. (1999b) Heart rate variability associated with particulate air pollution, *American Heart Journal* 138, 890-899.

Pope, C.A., Burnett, R.T., Thun, M.J., Calle, E.E., Krewski, D., Ito, K., and Thurston, G.D. (2002) Lung cancer, cardiopulmonary mortality, and long-term exposure to fine particulate air pollution, *Journal of American Medical Association* 287, 1132-1141.

Quenel, P., Cassadou, S., Declerq, C., Eilstein, D., Filleu, L., Le Goaster, C., Le Tertre, A., Medina, S., Pascal, L., Prouvost, H., Saviuc, P., and Zeghnoun A. (1999) *Rapport Surveillance Epidémiologique 'Air & Santé'*. Surveillance des Effets sur la Santé Liés à la Pollution Atmosphérique en Milieu Urbain, Institut de Veille Sanitaire, Paris.

Ramsay, T.O., Burnett, R.T., and Krewski, D. (2003) The effect of concurvity in generalized additive models linking mortality to ambient particulate matter, *Epidemiology* 14, 18-23.

Ritcher, D.A.U., and Williams, W.P. (1998) *Assessment and Management of Urban Air Quality in Europe*, European Environmental Agency, Copenhagen.

Saez, M., Ballester, F., Barceló, M.A., Perez-Hoyos, S., Tenías, J.M., Bellido,J., Ocaña, R., Figueiras, A., Arribas, F., Aragonés, N., Tobías, A., Cirera, Ll., Cañada, A.M., and on behalf of the EMECAS group (2002) A combined analysis of the short-term effects of photochemical air pollutants on mortality within the EMECAM project, *Environ. Health Perspect.* 110, 221-228.

Salvi, S., Blomberg, A., Rudell, B., Kelly, F., Sandstrom, T., Holgate, S.T., and Frew, A. (1999) Acute inflammatory responses in the airways and peripheral blood after short-term exposure to diesel exhaust in healthy human volunteers, *American Journal of Respiratory and Critical Care Medicine* 159, 702-709.

Samet, J.M. (1999) Risk assessment and air pollution, in S.T. Holgate, J.M. Samet, H. Koren, R.L. Maynard (eds), *Air Pollution and Health*, Academic Press, San Diego, California, pp. 883-897.

Samet, J.M., Dominici, F., Curriero, F.C., Coursac, I., and Zeger, S. (2000a) Fine Particulate Air Pollution and Mortality in 20 U.S. Cities, 1987-1994, *New England Journal of Medicine* 343, 1742-1749.

Samet, J.M., Dominici, F., Zeger, S., Schwartz, J., and Dockery, D.W. (2000b) *The National Morbidity, Mortality, and Air Pollution Study. Part I: Methods and Methodologic Issues*, Health Effects Institute, Boston.

Samet, J.M., Zeger, S.L., Dominici, F., Curriero, F., Coursac, I., Dockery, D.W., Schwartz, J., and Zanobetti, A. (2000c) *The National Morbidity, Mortality, and Air Pollution Study. Part II: Morbidity and Mortality from Air pollution in the United States*, Health Effects Institute, Boston.

Samet, J.M., Dominici, F., McDermott, A., and Zeger, S.L. (2003) New problems for an old desing: time-series analyses of air pollution and health, *Epidemiology* 14 (1), 11-12.

Samoli, E., Schwartz, J., Wojtyniak, B., Touloumi, G., Spix, C., Balducci, F., Medina, S., Rossi, G., Sunyer, J., Bacharova, L., Anderson, H.R., and Katsouyanni, K. (2001) Investigating regional differences in short-term effects of air pollution on daily mortality in the APHEA project: a sensitivity analysis for controlling long-term trends and seasonality, *Environ. Health Perspect.* 109, 349-353.

Schwartz, J. (1994) What are people dying of on high air pollution days? *Environmental Research* 64, 26-35.

Schwartz, J. (2001) Air pollution and blood markers of cardiovascular risk, *Environ. Health Perspect.* 109 (Suppl. 3), 405-409.

Schwartz, J., and Marcus, A. (1990), Mortality and air pollution in London: a time series analysis, *American Journal of Epidemiology* 131, 185-194.

Schwartz, J., Spix, C., Touloumi, G., Bacharova, L., Barumandzadeh, T., Le Tertre, A., Ponce, A., Pönkä, A., Rossi, G., Sáez, M., and Schouten, J.P. (1996) Methodological issues in studies of air pollution and daily counts of deaths or hospital admissions, *Journal of Epidemiology and Community Health* 50 (Suppl. 1), S3-S11.

Seaton, A., MacNee, W., Donaldson, K., and Godden, D. (1995) Particulate air pollution and acute health effects, *The Lancet* 345, 176-178.

Stieb, D.M., Judek, S., and Burnett, R.T. (2002) Meta-analysis of time-series studies of air pollution and mortality: effects of gases and particles and the influence of cause of death, age, and season, *Journal of Air and Waste Management Association* 52, 470-484.

Sunyer, J., Spix, C., Quenel, P., Ponce, A., Barumandzadeh, T., Touloumi, G., Bacharova, L., Wojtyniak, B., Vonk, J.M., Bisanti, L., Schwartz, J., and Katsouyanni, K. (1997) Urban air pollution and emergency admissions for asthma in four European cities: the APHEA Project, *Thorax* 52, 760-765.

Tenias, J.M., Ballester, F., Perez-Hoyos, S., and Rivera, M.L. (2002) Air pollution and hospital emergency room admissions for chronic obstructive pulmonary disease in Valencia, Spain, *Archives of Environmental Health* 57, 41-47.

Thurston, G.D., and Ito, K. (1999) Epidemiological studies of ozone exposure effects, in S.T. Holgate, J.M. Samet, H. Koren, and R.L. Maynard (eds), *Air Pollution and Health*, Academic Press, San Diego, California, pp. 485-510.

Touloumi, G., Samoli, E., and Katsouyanni, K. (1996) Daily mortality and winter type air pollution in Athens, Greece - a time series analysis within the APHEA project, *Journal of Epidemiology and Community Health* 50 (Suppl. 1), s47-51.

Touloumi, G., Katsouyanni, K., Zmirou, D., Schwartz, J., Spix, C., Ponce, A., Tobías, A., Quenel, P., Rabczenko, D., Bacharova, L., Bisanti, L., Vonk, J.M., and Pönkä, A. (1997) Short-term effects of ambient oxidant exposure on mortality: a combined analysis within the APHEA project, *American Journal of Epidemiology* 146, 177-185.

Turusov, V., Rakitsky, V., and Tomatis, L. (2002) Dichlorodiphenyltrichloroethane (DDT): ubiquity, persistence, and risks, *Environ. Health Perspect.* 110, 125-128.

Utell, M.J., and Samet, J.M. (1993) Particulate air pollution and health. New evidence on an old problem, *American Review of Respiratory Disease* 147, 1334-1335.

Ware, J.H., Thibodeau, L.A., Speizer, F.E., Colome, S., and Ferris, B.G.J. (1981) Assessment of the health effects of atmospheric sulphur oxides and particulate matter: evidence from observational studies, *Environ. Health Perspect.* 41, 255-276.

WHO - World Health Organisation (1987) *Air Quality Guidelines for Europe*, WHO Regional Office for Europe, Copenhagen.

WHO - World Health Organisation (1999) *Air Quality Guidelines for Europe, 2nd Edition*, WHO Regional Office for Europe, Copenhagen.

WHO - World Health Organisation (2000) *Evaluation and Use of Epidemiological Evidence for Environmental Health Risk Assessment*, European Centre for Environment and Health, Bilthoven.

# THE RESPIRATORY EFFECTS OF AIR POLLUTION

T. ROUSSOU[1] AND P. BEHRAKIS[2]
[1] *Metropolitan Hospital*
*Ethnarhou Makariou 9 & EL. Venizelou 1*
*18547, Neo Faliro, GREECE*
[2]*Department of Experimental Physiology,*
*National and Kapodistrian University of Athens,*
*Mikras Assias 75 Goudi, GR-11527 Athens, GREECE*

## Summary

During our daily activities including working, running, talking, eating and even sleeping, cells of the body are active. A person consumes oxygen and produces carbon dioxide during such activities. Lungs are air pumping stations of our body which transfer atmospheric oxygen into our body and expel carbon dioxide in the atmosphere, through a process called respiration. Air passes through airways during respiration. The average human being consumes about 12 Kg of air each day to form carbon dioxide. It is about twelve times higher than the food we consume. Hence, even a small concentration of pollutants in the air play a significant role in the deterioration of respiratory health. Man-made vehicles consume nearly 15 Kg of air in the burning of 1 litre of fuel resulting in the emission of harmful pollutants. Combustion of fuel is one of the most significant sources of air pollution. The constantly increasing need for fuel combustion for transportation and in order to produce energy and power industry, we deplete the atmosphere of oxygen and increase the concentration of air borne pollutants. Carbon monoxide, lead and hydrocarbons are emitted in high quantities in petrol combustion. These can cause loss of visual accuracy and mental alertness. Diesel combustion emits considerably higher amounts of Nitrogen dioxide, particulate matter and sulphur dioxide. Air pollution has been associated with a variety of detrimental effects on the respiratory system, including the aggravation of bronchial asthma or the appearance of new onset asthma. Also, increased mortality from respiratory disease especially in high risk groups (elderly people, children) resulting from the exacerbation of underlying cardiopulmonary disorders and increased susceptibility due to the deleterious effects of air pollutants on immunity has a tremendous economic toll on health care systems around the world.

*P. Nicolopoulou-Stamati et al. (eds), Environmental Health Impacts of Transport and Mobility,* 79-94.
© 2005 *Springer. Printed in the Netherlands.*

## 1.   Introduction

"Air pollution" encompasses a diverse array of anthropogenic chemical emissions including gaseous combustion products, volatile chemicals, aerosols (particulate), and their atmospheric reaction products. The primary sources of chemical emissions from combustion products include primary energy production (electricity), secondary energy production (thermal and transportation) as well as industry. Outdoor air pollution harms more than 1.1 billion people, mostly in cities (Roodman, 1998) and kills an estimated 2.7 million to 3.0 million people every year—about 6% of all deaths annually (Roodman, 1998). About 9 deaths in every 10 due to air pollution take place in the developing world, where about 80% of all people live (Roodman, 1998). In the United States, the principal pollutants monitored for regulatory purposes are ozone, particles, carbon monoxide, sulphur dioxide, nitrogen dioxide, and lead. Population growth, the proliferation of roadways and a large increase in the total number of miles driven by all passenger vehicles have facilitated a glut of mobile air pollution sources (cars and trucks), resulting in substantial atmospheric pollution attributable to transportation and mobility.

Despite efforts over the past decades to reduce pollution, an alarming set of health effects attributable to air pollution have been described. Individual reactions to air pollutants depend upon the type of pollutant, how much of the pollutant is present, the degree of exposure and the types and levels of individual activity (e.g., individuals working or exercising outdoors have greater exposure). Air pollution appears to aggravate lung infections, possibly by reducing the body's ability to fight infection. In order to achieve a more precise conception of the effects of "what we breathe" on the health and functioning of our respiratory system it is important to know that for the average healthy adult, an average of ten thousand litres (10,000) of air will pass through our lungs every day of our lives through restful breathing and this may increase drastically during work or exercise. We do know that air pollution:

- is associated with chronic respiratory and cardiovascular disease,

- impairs lung immunity and enhances inflammatory processes,

- alters lung ventilation and impairs respiratory activity (reduces volume of air that can be taken in by lungs and decreases pulmonary lung function testing performance indices (FEV1, FVC, PEF, DLCO etc.),

- aggravates existing asthma and chronic obstructive pulmonary disease and perpetuates a new onset asthma,

- causes sensory irritation of eyes and nose (Abbey *et al.*, 1998; MacNee and Donaldson, 2000a; Roodman, 1998).

As a result and as is shown in recent studies, associations are made between air pollutants and decreased lung function, airway hyper-reactivity, respiratory symptoms, increased medication use and physician/emergency room visits among individuals with heart or lung disease, decreased exercise capacity, increased hospital admissions and increased mortality. Both healthy and ill individuals may be affected by certain air pollutants. Sensitive subgroups are those with asthma, individuals addicted to tobacco products, the elderly, infants, persons with coronary heart disease, and persons with chronic obstructive pulmonary disease (COPD) (Utell and Samet, 1995). These associations are of particular public health significance because infections and allergies of the respiratory tract account for a major portion of total acute illness in the general population and exact a large economic toll in terms of health care as well as time lost from school or work, visits to doctors, and admission to hospitals.

## 2. The respiratory effects of specific air pollutants

### 2.1. Ozone

Ozone is a powerful secondary pollutant formed when oxides of nitrogen and unburned volatile organic hydrocarbons, mostly from vehicle exhausts, combine with oxygen under the action of sunlight. It is a main component of smog.

Ozone has different health implications in the stratosphere and the troposphere. In the stratosphere (the "ozone layer"), 10-50 km (6-30 miles) above the earth, ozone provides a critical barrier to solar ultraviolet radiation, and protection from skin cancers, eye cataracts, and serious ecological disruption.

Ozone levels therefore tend to be highest on warm, sunny days, which are conducive to outdoor activities. In many areas, ozone concentrations peak in the mid-afternoon, when children are likely to be playing outside. It is important to distinguish between ground-level ozone air pollution and stratospheric ozone depletion by chlorofluorocarbons. These issues are unrelated (Lee et al., 1994). Excessive ozone exposure is widespread: over 70 million people lived in areas not meeting the EPA ozone standard in 1995; that number will increase markedly.

Epidemiologic studies undertaken in a variety of locations indicate a relationship between outdoor air pollution and adverse respiratory effects. The pollutants most frequently implicated in these studies have been respirable particles (notably acidic sulfates) and ozone. Examples of health outcomes found to be correlated with air pollution levels include increased prevalence of chronic cough, chest illness and bronchitis (measured by questionnaire), hospital admissions for various respiratory conditions, and decrements in lung function. The prevalence of respiratory symptoms was markedly increased among children with a history of asthma or wheezing (van Eeden et al., 2001; Official Statement of the American Thoracic Society, 1999). In a study comparing ozone ($O_3$)-induced changes in lung function and respiratory tract injury/inflammation in subjects with asthma than in normal subjects, results obtained following exposure on separate days, to $O_3$ (0.2 ppm) and filtered air for 4 h during exercise demonstrated a significant $O_3$ effect on FEV1, FVC and specific airways resistance as well as respiratory symptoms in asthmatic subjects and no significant differences in the lung function responses of the asthmatic subjects in comparison with a group of normal subjects. Findings from proximal airway lavage and bronchoalveolar lavage 18 hours following exposure in the same groups showed a significant increase in various inflammatory parameters including percentage neutrophils, total protein, lactate dehydrogenase (LDH), fobronectin, interleukin-8, granulocyte stimulating factor (GM-CSF), myeloperoxidase (MPO) and transforming growth factor-beta (TGF beta 2) concentrations whereas the asthmatic subjects showed significantly greater ($p <$ 0.05) $O_3$-induced increases in several inflammatory endpoints (percent neutrophils and total protein concentration) in BAL as compared with normal subjects who underwent bronchoscopy. These findings indicate that asthmatic persons may be at risk of developing more severe $O_3$-induced respiratory tract injury/inflammation than normal persons, and may help explain the increased asthma morbidity associated with $O_3$ pollution (Scannell *et al.*, 1996). The effects of exposures to multiple pollutants are difficult to study in humans. A few controlled investigations and field studies indicate, however, that exposures to complex mixtures of air pollutants may have synergistic acute effects on pulmonary function and, possibly, on symptoms. A recent report suggests that even brief exposure to ozone can produce allergic asthmatic responses to aeroallergens. There is, moreover, a substantial body of experimental evidence in animals indicating that ozone can lower resistance to infection, facilitate sensitization and airway responses to airborne allergens, and act synergistically with airborne acidity to damage deep lung tissues (Mac.Nee *et al.*, 2000).

Ozone is known to cause symptoms such as coughing, wheezing, and lung irritation at concentrations as low as 100 ppb (parts per billion). Symptoms often associated with exposure to ozone include a substernal chest pain, tearing and burning sensation. Relatively low amounts can cause chest pain, coughing, shortness of breath, and, throat irritation. Recovery from the harmful effects can occur following short-term exposure to low levels of ozone, but health effects may become more damaging and recovery less certain at higher levels or from longer exposures.

## 2.2.    *Particulate matter*

Particulate air pollution (PM: particulate matter) is a heterogeneous classification of liquid and solid aerosols which includes anthropogenic emissions from fuel combustion (coal, oil, biomass), transportation, and high temperature industrial processes. Smaller particles (often less than $PM_3$ (particulate matter of 3 micrometer (μm) aerodynamic diameter) include viruses and some bacteria, but mostly come from anthropogenic sources, including sulphate and nitrate aerosols and other combustion derived atmospheric reaction products; whereas larger particles ($PM_{3\ to\ 30}$) include pollen, spores, crustal dusts, and other mechanically generated dusts. Size is a critical determinant of deposition site, with larger particles (greater than $PM_{2.5}$) tending to deposit in the nasal and tracheobronchial regions), and smaller particles (less than $PM_{10}$) penetrating deeper into the lungs.

Exposure to atmospheric fine particulate matter even at low ambient concentrations has been linked to increases in mortality and morbidity. A 10 microgram increase in $PM_{10}$ has been associated with a 0.5% increase in daily respiratory mortality.

(Gavett *et al.*, 2003) Increases of $PM_{2.5}$ and $PM_{10}$ significantly correlate with increased severity of asthma attacks and increased use of asthma medication (Dominici *et al.*, 2003). In London, in December of 1952, smog coupled with unusually low temperatures and a 5 day temperature inversion resulted in a significant excess of respiratory illness and death. The major pollutants primarily result from domestic discharges of smoke with a prominence of acidic aerosols. Autopsy tissue studies demonstrated a predominance of soot and a variety of metal-bearing particle types in all lung compartments (including air space, airway, interstitium and lymph node) (Gavett *et al.*, 2003).

Similarly, pollutants originating from the destruction of the World Trade Center (WTC) on September 11th 2001 have been associated with adverse respiratory responses in rescue workers and nearby residents something that has been supported by controlled exposure studies conducted on mice exposed to samples of WTC dust primarily $PM_{2.5}$ at varying doses and length of exposure (time in hours). Results indicate that high level exposure in terms of $PM_{2.5}$ concentrations correlate with airflow obstruction and inflammation and that comparable effects in people would be consistent with a 425micrograms/$m^3$ for eight-hour exposure (Churg *et al.*, 2003).

Given a threshold dose of ambient air pollution the first effects are seen in the upper respiratory tract with symptoms including rhinitis (runny nose) the trachea and the bronchi where there is airway hyper-resposiveness, the result of inflammatory process triggering may manifest itself through symptoms including productive or not cough, shortness of breath (dyspnea), chest tightness and possibly wheezing depicting airflow through narrowed airways consequent to bronchoconstriction. The extent as well as severity of the symptoms depends on various factors including history of underlying respiratory disease, age, atopic predispostion (susceptibility to allergic reactions) as well as extent of exposure and concentrations and nature of ambient irritants. Furthermore, effects may be reversible or nonreversible in which case prolonged exposure and persistent effects may result in interference with normal daily activity of the individual, episodic respiratory illness, progressive respiratory dysfunction resulting in permanent respiratory injury and incapacitating illness (Abbey *et al.*, 1998; Albright and Goldstein, 1996; Li *et al.*, 2003; MacNee and Donaldson, 2000a).

Particulate matter ($PM_{2.5 \text{ and less}}$) increases sensitization to aeroallergens, as seen by increased airway responsiveness to metacholine challenge testing of mice exposed to industrial ambient air pollution in vitro (Dick *et al.*, 2003). It increases the frequency of exacerbations in individuals with preexisting respiratory disease (asthma, chronic obstructive pulmonary disease, cystic fibrosis etc.) and increases susceptibility to respiratory infections (Churg and Wright, 2002). Chronic inhalation studies associate particulate matter with the induction of various adverse effects including impaired lung clearance, chronic pulmonary inflammation, pulmonary fibrosis, as well as lung tumors (Oberdorster, 1996). The injurous effects of inhaled particulate matter may be seen either locally (in the lung) or systemically (Timblin *et al.*, 2002). Particle size plays a role in determinig the nature, the intensity as well as the extent of the effect. Ultrafine particles seem to have a more extensive effect primarily owing to their size (<100nm) that enables penetration deeper into the tissues. Data from in vitro and in vivo studies associate ultrafine particles with both local and systemic oxidative stress, pro-inflammatory gene regulation and altered blood coagulability (Abbey et al., 1999). In addition, in vitro studies exposing alveolar epithelial cell lines to ultrafine airborne particles (ultrafine carbon black-ufCB-a component of PM) demonstrate a dose-related expression of protooncogene and proliferation as well as apoptosis genes thereby alluding to the possible carcinogenic effects of these particles on alveolar epithelial cells (Atkinson et al., 2002).

Inflammatory effects in relation to $PM_{2.5 \text{ and less}}$ exposure locally in the lung include neutrophilia and lymphocytosis, increased concentrations of various inflammatory mediators such as increased expression of IL-8 messenger RNA in bronchial mucosa, increased levels of IL-8 in bronchial lavage fluid, up regulation of endothelial adhesion molecules, increased cytokine IL-10 in asthmatics, increased levels of TNF-∝ as well as, stimulation of NF-kappa B (Archer et al., 2004; Dick et al., 2003; Stenfors et al., 2004). On a systemic level, exposure to fine particulate matter induces a systemic inflammatory response stimulating the bone marrow to release leukocytes and platelets (Timblin et al., 2002). Studies in animals as well as humans showed that acute exposure to ambient particles accelerated the transit of polymorphonuclear (PMN) leukocytes whereas chronic exposure is associated with expansion of the PMN pool in bone marrow. Also, these polymorphonuclear leukocytes are less chemotactic and contain more damaging granules thereby strengthening their ability to destroy alveolar tissue (Timblin et al., 2002).

Finally, chronic exposure to fine particulate matter, especially high organic carbon and polycyclic aromatic hydrocarbon (PAH) ultrafine particles that localize in the mitochondria where they cause major structural damage, induces oxidative stress and propagate pulmonary inflammation (Li et al., 2003). These particles have the ability to penetrated and establish themselves within the walls of the small airways. There, through their inflammatory effects, they stimulate submucosal goblet cell proliferation and hence increased mucous production as well as NF kappaB activation, procollagen gene expression resulting in epithelial cell proliferation, increased deposition of fibrotic tissue, excess muscle tissue in the walls of the airways leading to airway remodeling (Li et al., 2003; van Eeden and Hogg, 2002). Mineral dusts are more commonly associated with the above inflammatory changes in the airways including amosite asbestos, dusts with surface complexed iron and carbonaceous aggregates of ultra fine particles.

Road traffic pollution is a significant public health problem, with particulate pollution from diesel exhaust of increasing concern. Diesel combustion products include gases, semi volatile organic substances, and particles of respirable size (particulate matter of less than 10 μm aerodynamic diameter; $PM_{10}$). Increased $PM_{10}$ concentrations are associated with acute episodes of respiratory ill health and increased mortality from both cardiovascular and respiratory causes. For example, asthma symptoms increase by 20% when daily $PM_{10}$ concentrations increase by 50 μg/m$^3$, and increases of 100 μg/m$^3$ and 200 μg/m$^3$ cause 20% increases in hospital admissions and mortality, respectively (MacNee and Donaldson, 2000b; van Eeden *et al.*, 2001).

## 3.    Transportation and traffic related pollution

As population numbers and urban expansion increase, so does the number and use of automobile vehicles. The total miles driven by all passenger vehicles in the U.S. increased 2.7 times between 1965 and 1995. In an extensive study in Japan to determine the effects of air pollution from automobiles on humans, Kagava and associates have found that respiratory symptoms were more common in people living near busy traffic roads. It was also found that the prevalence of symptoms like cough and wheezing was reduced as people moved away from busy highways. Children were more susceptible to the adverse effects of these pollutants and hence the prevalence of respiratory symptoms was higher in children than in adults. People living within 20 meters from the road were largely affected.

Vehicles, through fuel combustion, emit many pollutants including hydrocarbons (e.g., ethylene, formaldehyde, methane, benzene, phenol, 1,3-butadiene, acrolein, and polynuclear aromatic hydrocarbons), nitrogen oxides, fine particulate matter, carbon monoxide and toxic air contaminants known as hazardous air pollutants. Recently there has been increasing concern that the apparent increase in the prevalence of asthma may be due to increasing exposure to pollution, particularly from motor vehicles. Apart from asthma, toxic vehicle emissions have been associated with inflammatory exacerbations of underlying respiratory disease such as COPD, increased incidence of both upper and lower respiratory tract symptoms, increased susceptibility to infection and carcinogenesis (Nicholson and Case, 1983; Lee *et al.*, 1994; 43).

### 3.1.    *Carbon monoxide (CO)*

Carbon monoxide is formed by incomplete combustion of carbon containing fuels. Local accumulation in heavy traffic is the most important source for community ambient exposure. It is a colourless and odourless toxic gas which causes hypoxia by various mechanisms: (*a*) by the formation of carboxyhaemoglobin (COHb) with an affinity that is 200 times greater than oxygen; (*b*) by decreasing the delivery of oxygen to the tissues (the haemoglobin oxygen dissociation curve shifts to the left); (*c*) by inhibiting the action of cytochrome oxidases. Variations in uptake of CO are thought to be due to physiological variables such as lung capacity, diffusion constant of the lung, and dead space volume. Ventilation rate is also thought to affect CO uptake. In the 1980s, the concentration of COHb in the blood of city dwellers was found to be approximately double that in people living in rural traffic-free areas (Lee *et al.*, 1994). Strenuous exercise in heavy traffic for 30 minutes can increase the level of COHb 10-fold, which is the equivalent of smoking 10 cigarettes (Nicholson and Case, 1983).

Sources of Community Carbon Monoxide exposure:

- inside passenger car, commuting 5 ppm

- proximity to busy roads, intersections 15 ppm

- parking areas 4 ppm

- traffic tunnels 5 - 42 ppm

Other important contributors to CO exposure include traffic volume, traffic speed, winter season, motor vehicle density, age composition of the fleet, emissions standards for the fleet and vehicle characteristics (CO intrusion problem), community combustion of oil, gasoline, coal, wood, and use of lawn mowers, chain saws, space heaters, and charcoal.

Smokers typically have COHb levels of 5-6%. Setting aside tobacco and indoor sources, the relationship between ambient CO concentrations and blood levels is largely determined by duration of exposure and ventilation rate; the latter is roughly correlated with workload. Mean wintertime carboxyhemoglobin (COHb) (in the general population) is thought to be about 1.2%, with 3-4% of population above 2%. For example, during heavy labor in a busy traffic tunnel with CO levels of 42 ppm, the COHb would reach about 5% in 90 minutes. Risk groups include commuters, smokers, persons working in traffic. Persons with cardiac and pulmonary disease are most vulnerable; symptoms such as dyspnea and angina may develop at COHb levels of 3-4% (Abbey *et al.*, 1998; MacNee and Donaldson, 2000a; Roodman, 1998).

### 3.2.    *Diesel exhaust particles (DEPs)*

In addition to containing benzene, a known carcinogen, diesel exhaust contains high levels of fine soot, known as Small Particulate Matter, or SPM. The microscopic soot is easily inhaled deep into the lungs.

Epidemiologic and experimental studies suggest that diesel exhaust particles (DEPs) may be related to increasing respiratory mortality and morbidity. Studies have shown that DEPs augment the production of inflammatory cytokines by human airway epithelial cells *in vitro*. Benzene-extracted components showed effects mimicking DEPs on IL-8 gene expression, release of several cytokines (IL-8 (interleukin-8), granulocyte macrophage colony-stimulating factor) and nuclear factor (NF)-kappa B activation (43). DEPs have now been postulated to induce intense inflammatory reactions in the airways. Sagai and colleagues showed that intratracheal instillation of DEPs induced airway inflammatory changes associated with hyperresponsiveness and cell infiltration, such as eosinophils in mice, which mimic those found in bronchial asthma.Workers exposed to diesel exhaust report respiratory symptoms accompanied by reversible decreases in lung function. The lung function changes appear to be caused by the particulate fraction of diesel exhaust, since filtering out the particulates reduces these changes. Controlled exposure to diesel exhaust provokes symptoms, increases in airway resistance, and inflammatory changes within the lung and in peripheral blood. The studies in which the latter findings were made used whole diesel exhaust, which does not allow evaluation of the individual contributions of gases or particulates. Diesel exhaust particles (DEP) are thought to consist of a carbon core surrounded by trace metals, such as nickel, and salts into which are adsorbed organic hydrocarbons. A number of these components have inflammatory effects on the lungs of laboratory animals. For example, intratracheal instillation of ultrafine carbon particles in rats leads to neutrophil influx into the lungs, and to increases in bronchoalveolar lavage fluid (BALF) concentrations of tumor necrosis factor-(TNF)-$\alpha$. Intratracheal instillation of nickel in rats causes severe and sustained inflammation, with generation of free radicals. Inhalation of hydrocarbons also leads to lung inflammation. For example, increases in the activity of a number of enzyme markers of acute inflammation were found in rabbit lung homogenates after exposure to n-hexane. The foregoing observations indicate that diesel particles themselves can induce airway inflammation. To date, there has been no evaluation of the effects of diesel particulates alone in human volunteers (MacNee and Donaldson, 2000b).

### 3.3. *Volatile organic compounds*

The general category of VOCs consists of many chemicals, including non-methane hydrocarbons (for example, alkanes, alkenes, and aromatics), halocarbons (for example, trichloroethylene), and oxygenates (alcohols, aldehydes, and ketones). There is a preponderance of carcinogens among VOCs - for example, benzene, polyaromatic hydrocarbons, 1,3-butadiene, many of the halocarbons. Owing to the carcinogenicity of benzene and polyaromatic hydrocarbons, no safe levels are recommended by WHO.

Petroleum and gasoline consist of blends of over 250 diverse hydrocarbons. Many of these are toxic. Some, such as benzene, are carcinogenic. Hydrocarbons escape into the air during refilling, from the gasoline tank and carburetor during normal operation, and from engine exhaust. Transportation sources account for 30-50% of all hydrocarbon emissions into the atmosphere.

### 3.4. *Polycyclic aromatic hydrocarbons*

Benz[*a*]anthracene, along with a number of other polycyclic aromatic hydrocarbons, are natural products produced by the incomplete combustion of organic material. Polycyclic aromatic hydrocarbons and some of their metabolites are known to react with cellular macromolecules, including DNA, which may account for both their toxicity and carcinogenicity. The toxic effects of benz[*a*]anthracene and similar polycyclic aromatic hydrocarbons are primarily directed toward tissues that contain proliferating cells. Epithelial proliferation and cell hyperplasia in the respiratory tract have been reported following subchronic inhalation exposure (Reznik-Schuller and Mohr, 1974; Saffiotti *et al.*, 1968). The primary concern with benz[*a*]anthracene exposure is its potential carcinogenicity. There is no unequivocal, direct evidence of the carcinogenicity of the compound to humans, however, benz[*a*]anthracene and other known carcinogenic polycyclic aromatic hydrocarbons are components of coal tar, soot, coke oven emissions and tobacco smoke. There is adequate evidence of its carcinogenic properties in animals and oral exposures of mice to benz[a]anthracene have resulted in hepatomas, pulmonary adenomas and forestomach papillomas (Klein, 1963; Bock and King, 1959; U.S. EPA, 1991).

## 4.   "Acid rain"

Another powerful secondary pollutant is acid rain, formed when sulfur dioxide and oxides of nitrogen combine with water vapor and oxygen in the presence of sunlight to form a diluted "soup" of sulfuric and nitric acids. They can fall as both wet (acid rain) or dry deposition.

### 4.1.   Sulfur dioxide (SO₂)

Sulphur dioxide gas is formed during the combustion sulphur-containing fossil fuel (coal and oil), during metal smelting, paper manufacturing, food preparation, and other industrial processes. It is an important contributor to acid aerosols and "acid rain", and is typically a component of complex pollutant mixtures. Exposure to sulphur dioxide gas causes mucosal irritation of the upper respiratory tract. The level of exposure is important, as high levels of exposure may be lethal causing mucosal sloughing and alveolar hemorrhage. Odour threshold is achieved at 0.5 ppm whereas at 6-10 ppm the individual experiences irritation of the eyes, nose and throat and level as low as 0.25 ppm can provoke asthma exacerbation in exercising asthmatics.

### 4.2.   Nitrogen dioxide (NO₂)

Fossil fuel combustion generates nitrogen dioxide ($NO_2$) and nitric oxide (NO) which is rapidly oxidized to $NO_2$. $NO_2$ reacts in the presence of sunlight and VOCs (volatile organic compounds) to form ozone, Nitrogen dioxide ($NO_2$) is absorbed in both large and small airways. Very high concentrations (>200 ppm(parts per million)) are very dangerous, causing lung injury, fatal pulmonary oedema, and bronchopneumonia. Lower concentrations cause impaired mucociliary clearance, particle transport, macrophage function, and local immunity.

## 5.   High risk groups

Existing epidemiological and toxicological data indicate that exposure to ambient air pollution is associated with respiratory toxicity. The decrements in pulmonary function observed in epidemiological and experimental studies involving children exposed to ozone and other pollutants may last longer than the episodes of pollution that initiate these changes (Lee et al., 1994). In a large study to examine the association between air pollution and a number of health outcomes, the relationship between daily GP consultations for asthma and other lower respiratory diseases (LRD) and air pollution in London was investigated. The study showed that there are associations between air pollution and daily consultations for asthma and other lower respiratory disease in London. The most significant associations were observed in children and the most important pollutants were $NO_2$, CO, and $SO_2$. In adults the only consistent association was with $PM_{10}$.

### 5.1. Children

A factor that increases children's vulnerability to airborne pollution is that their airways are narrower than those of adults. Thus, irritation caused by air pollution that would produce only a slight response in an adult can result in potentially significant obstruction in the airways of a young child. Moreover, children have markedly increased needs for oxygen relative to their size. They breathe more rapidly and inhale more pollutant per pound of body weight than do adults. In addition, they often spend more time engaged in vigorous outdoor activity than adults. Experimental and epidemiological data provide grounds for concern about chronic lung damage from repeated exposures (Lee et al., 1994). Children experience reductions in peak expiratory flow measurements (PEF) and increased symptoms after increases in relatively low ambient $PM_{10}$ concentrations, and children with diagnosed asthma are more susceptible to these effects than other children (Vedal et al., 1998).

### 5.2. Elderly

The elderly are also a high risk group with respect to increased mortality stemming from prolonged exposure to ambient particulate matter. In animal studies, the use of technetium-99m diethylenetriamine penta-acetic acid (DPTA) clearance indicates that increased lung tissue permeability in older animals may be associated with an enhanced uptake of soluble components of particulate matter (American Academy of Pediatrics, 1993). This observation may provide a possible explanation for increased mortality as well as morbidity from respiratory disease in the elderly which may also be dose-related not only with respect to concentration of particulate matter in ambient air pollution (particularly $PM_{2.5 \text{ or less}}$) but also to chronicity of exposure thereby perpetuating inflammatory processes as well as respiratory symptoms with all the anticipated ensuing results including higher risk of recurrent infections and compromised respiratory functioning (Albright and Goldstein, 1996; Churg and Wright, 2002; Churg et al., 2003; Gavett et al., 2003, 11).

### 5.3. Chronic obstructive pulmonary disease

Air pollution as a trigger for exacerbations of COPD has been recognized for more than 50 years. COPD has been reported in workers exposed to particulates, and there is increasing evidence that high levels of ambient particulate pollutants may also be associated with COPD. Induction of increased mucus secretion by air pollutants such as sulphur dioxide and possibly $PM_{10}$ may contribute to the development of exacerbations of COPD, by increasing airway resistance and by the development of mucus plugging in the smaller peripheral airways, a feature commonly present in patients dying of COPD. In patients with COPD, and in cigarette smokers, there is damage to the cilia, which, together with the excess mucus produced, overwhelm the mucociliary escalator and will reduce the ability of the lungs to deal adequately with inhaled particles. Airway epithelial cells also act as a barrier to inhaled pollutants, and are an important target for the toxic and potentially inflammogenic effects of particles. On exposure to particles and other forms of air pollutants such as nitrogen dioxide, epithelial cells can release inflammatory mediators (Official Statement of the American Thoracic Society, 1999). Various studies have also investigated the hypothesis that particulates, including air pollution particles, can induce airway wall fibrosis, a process that can lead to COPD. In experiments performed on rat tracheal explants, known fibrogenic dusts such as amosite asbestos produced increased gene expression of procollagen, transforming growth factor-ß, and platelet-derived growth factor, and increased hydroxyproline in the explants indicating that mineral dusts can be directly associated with the induction of fibrosis in the airway wall and particle-induced airway wall fibrosis may lead to COPD (Li *et al.*, 2003; van Eeden and Hogg, 2002).

The dimension of the adverse effect of air pollution on respiratory function falls within a wide range. Healthy persons, for example, may sustain transient reductions in pulmonary function associated with air pollution exposure which are reversible and therefore not considered to be adverse. There is also epidemiologic evidence that air pollution may adversely affect lung growth or accelerate the age-related decline of lung function however studies are limited in their power to detect such permanent effects. Finally, air pollution exposure can evoke symptoms in persons without underlying chronic heart or lung conditions and also provoke or increase symptom rates in persons with asthma and chronic obstructive lung disease including dyspnea or shortness of breath, cough either productive or not, inspiratory and/or expiratory wheezing, chest tightness, and potentially reducing physical activity due to the above and also increasing susceptibility to respiratory tract infections.

## 6.    Biomarkers

The design of clinical studies – including controlled exposures of volunteers – has advanced and biologic specimens may be obtained after exposure, for example, using fiberoptic bronchoscopy, to identify changes in levels of markers of injury. Toxicologic studies have also improved through the incorporation of more sensitive indicators of the effect of exposure to ambient air pollution and the careful tracing of the relationship between exposure and biologically relevant doses to target sites, which may now be considered at a molecular level (Kharitinov and Barnes, 2000; MacNee and Donaldson, 2000a; WHOs 1999 Guidelines for Air Pollution Control, 2000).

Biomarkers relevant to air pollution measured in blood, exhaled air, urine, sputum, and in bronchoalveolar lavage fluids and tissue specimens collected by bronchoscopy. Also, bronchoalveolar lavage fluids are now frequently analyzed for cell numbers and types, cytokines (e.g. several interleukins and tumor necrosis factor $\alpha$), enzymes (e.g., lactate dehydrogenase and $\beta$-glucoronidase), fibronectin, protein, arachidonic acid metabolites, and reactive oxygen species all of which are markers associated with inflammatory processes of the respiratory system. Because many of the epithelial cell types of the nasopharyngeal region are similar to epithelia and responses in the trachea, bronchi, and bronchioles, responses of nasal cells have been examined as potential biomarkers for their ability to predict parallel responses in lung airways, which are more difficult to sample (Kharitinov and Barnes, 2001; MacNee and Donaldson, 2000; WHOs 1999 Guidelines for Air Pollution Control, 2000) Nitric oxide is exhaled marker indicative of ongoing inflammation particularly useful in asthmatic patients. The progressive increase in exhaled NO from asymptomatic to symptomatic asthma suggests that exhaled NO measurements may be useful in monitoring occupational asthmas, and of the environmental health effects of air pollution on inflammatory processes of the respiratory system. High levels of exhaled NO and asthmalike symptoms in subjects with occupational exposure to high levels of ozone and chlorine dioxide may indicate the presence of chronic airway inflammation (WHOs Guidelines for Air Pollution Control, 2000).

## 7. Conclusions

Ground-level ozone, acid aerosols, and particulates have created a serious health problem. There does not appear to be a "threshold level" for ozone or particles below which no hearth effects are observed. Children and the elderly, both well and those with pre-existing cardiorespiratory disease, are particularly sensitive to these air pollutants. Air pollution appears to aggravate lung infections, possibly by reducing the body's ability to fight infection. Even healthy outdoor workers show a measurable decrease in lung function when exposed to low-levels of ozone. Asthma is made substantially worse by current air concentrations of particles and ozone. Long-term exposure to air pollutants is associated with decreased lung function and increased city-specific mortality rates. Numerous studies conducted worldwide show a significant acute health consequence of exposure to particulates, and this pollutant may be responsible for between one and 10 per cent of all non-trauma mortality. Ambient air pollution is an important contributory factor compromising overall quality of life for the majority of the population inhabiting urban centres. The effects of air pollution depend on levels of exposure and susceptibility of the exposed population. Transportation and mobility seem to be significant daily perpetuators of air pollution to which every inhabitant is constantly exposed and most of all those residing in large urban centres. Air pollution also affects the workforce, is the primary cause in as many as 50 million cases of occupational chronic respiratory disease each year -a third of all occupational illnesses- and is a major burden on overall health costs. This culminates in a considerable compromise to the quality of life for millions of people of all ages throughout the world. Physicians should advise patients about the risks of smog exposure, should support more health effects research on air pollution, and should advocate the development of air pollution-related health education materials.

Efforts are being made to enhance awareness of the adverse effects of air pollution in an attempt to enable mechanisms to minimize and restrict them. Epidemiologic research designs have been refined and large sample sizes and increasingly accurate methods for exposure assessment have increased the sensitivity of epidemiologic data for detecting evidence of effects. New statistical approaches and advances in software and hardware have facilitated analyses of large databases of mortality and morbidity information.

Reducing the levels of air pollution is imperative. Legislation has facilitated the emissions from industry to be reasonably well controlled. However, given that motor vehicles are the major culprits, it will be necessary to make some hard decisions about restricting their use and/or improving their technology. De-urbanization, in order to slow the continuing growth of cities, and the development and use of public transport can constitute a potential initiation towards solving the problem of air pollution in large urban centres. Ideally, only a non-polluting renewable source of energy will solve the problem of air pollution.

# References

Abbey, D.E., Burchette, R.J., Knutsen, S., McDonnell, W., Lebowitz, D., and Enright, P.L. (1998) Long-term particulate and other air pollutants and lung function in nonsmokers, *Am. J. Respir. Crit. Care Med.*, Volume 158 (1), 289-298.

Abbey, D.E., Nishino, N., McDonnell, W.F., Burchette, R.J., Knutsen, S.F., Lawrence Beeson W., and Yang, J.X. (1999) Long-term inhalable particles and other air pollutants related to mortality in nonsmokers, *Am. J. Respir. Crit. Care Med.* Volume 159 (2), 373-382.

Air Pollution WHO's 1999 Guidelines for Air Pollution Control Fact Sheet N° 187 Revised September 2000.

Albright, J.F., and Goldstein, R.A. (1996) Airborne pollutants and the immune system, *Otolaryngol. Head Neck Surg.* 114(2), 232-238.

Ambient Air Pollution: Respiratory Hazards to Children (RE9317) American Academy of Pediatrics Policy Statement (1993) *Pediatrics* 91(6),1210-1213.

Archer, A.J., Cramton, J.L., Pfau, J.C., Colasurdo, G., and Holian, A. (2004) Airway responsiveness after acute exposure to urban particulate matter 1648 in a DO11.10 murine model, *Am. J. Physiol. Lung Cell Mol. Physiol.* 286(2), 337-343.

Atkinson, R.W., Anderson, R., Sunyer, J., Ayres, J., Baccini, M., Vonk, J.M., Boumghar, A., Forastiere, F., Forsberg, B., Touloumi, G., Schwartz, J., and Katsouyianni, K. (2001) Acute effects of particulate air pollution on respiratory admissions results from APHEA 2 Project, *Am. J. Respir. Crit. Care Med.* 164 (5), 826-830.

Brown, D.M., Donaldson, K., Borm, P.J., Schins, R.P., Dehnhardt, M., Gilmour, P., Jimenez, L.A., and Stone, V. (2004) Calcium and ROS-mediated activation of transcription factors and TNF-alpha cytokine gene expression in macrophages exposed to ultrafine particles, *Am. J. Physiol. Lung Cell Mol. Physiol.* 286(2), 344-353.

Churg, A., and Wright, J.L. (2002) Airway wall remodeling induced by occupational mineral dusts and air pollutant particles, *Chest.* 122(6) Suppl., 306S-309S

Churg, A., Brauer, M., del Carmen Avila-Casado M., Fortoul, T.I., and Wright, J.L. (2003) Chronic exposure to high levels of particulate air pollution and small airway remodelling, *Environ. Health Perspect.* 111(5), 714-718.

Dick, C.A., Singh, P., Daniels, M., Evansky, P., Becker, S., and Gilmour, M.I. (2003) Murine pulmonary inflammatory responses following instillation of size-fractionated ambient particulate matter, *J. Toxicol. Environ. Health A.* 66(23), 2193-2207.

Dominici, F., McDermott, A., Zeger, S.L., and Samet, J.M. (2003) Airborne particulate matter and mortality: timescale effects in four US cities, *Am. J. Epidemiol.* 157(12), 1055-1065.

Donaldson, K., Gilmour, M.I., and MacNee, W. (2000) Asthma and $PM_{10}$, *Respir. Res.* 1(1), 12-15.

Gavett, S.H., Haykal-Coates, N., Highfill, J.W., Ledbetter, A.D., Chen, L.C., Cohen, M.D., Harkema, J.R., Wagner, J.G., and Costa, D.L. (2003) World Trade Center fine particulate matter causes respiratory tract hyperresponsiveness in mice, *Environ. Health Perspect.* 111(7), 981-991.

Goss, C.H., Newsom, S.A., Schildcrout, J.S., Sheppard, L., and Kaufman, J.D. (2004) Effect of ambient air pollution on pulmonary exacerbations and lung function in cystic fibrosis, *Am. J. Respir. Crit. Care Med.* 169(7), 816-821.

Harrod, K.S., Jaramillo, R.J., Rosenberger, C.L., Wang, S.Z., Berger, J.A., McDonald, J.D., and Reed, M.D. (2003) Increased susceptibility to RSV infection by exposure to inhaled diesel engine emissions, *Am. J. Respir. Cell Mol. Biol.* 28(4), 451-463.

Holmes, J.R. (1995) *Respiratory disease and cancer in non-smokers increase with higher long-term air pollution exposures*, Research Division, Chief California Environmental Protection Agency, brief reports to the Scientific and Technical Community.

Hunt, A., Abraham, J.L., Judson, B., and Berry, C.L. (2003) Toxicologic and epidemiologic clues from the characterization of the 1952 London smog fine particulate matter in archival autopsy lung tissues, *Environ. Health Perspect.* 111(9),1209-1214.

Kharitonov, S.A., and Barnes, P.J. (2001) Exhaled Markers of Pulmonary Disease, *Am. J. Respir. Crit. Care Med.*, 163(7), 1693-1722.

Lambert, A.L., Trasti, F.S., Mangum, J.B., and Everitt, J.I. (2003) Effect of preexposure to ultrafine carbon black on respiratory syncytial virus infection in mice, *Toxicol. Sci.* 72(2), 331-338.

Lee, K., Yanagisawa, Y., Spengler, J.D., and Nakai, S. (1994) Carbon monoxide and nitrogen dioxide exposures in indoor ice skating rinks, *J. Sports Sci.* 12, 279–283.

Leonardi, G.S., Houthuijs, D., Steerenberg, P.A., Fletcher, T., Armstrong, B., Antova, T., Lochman, I., Lochmanova, A., Rudnai, P., Erdei, E., Musial, J., Jazwiec-Kanyion B., Niciu, E.M., Durbaca, S., Fabianova, E., Koppova, K., Lebret, E., Brunekreef, B., and van Loveren, H. (2000) Immune biomarkers in relation to exposure to particulate matter: a cross-sectional survey in 17 cities of Central Europe, *Inhal. Toxicol.* 12(4) Suppl., 1-14.

Li, N., Sioutas, C., Cho, A., Schmitz, D., Misra, C., Sempf, J., Wang, M., Oberley, T., Froines, J., and Nel, A. (2003) Ultrafine particulate pollutants induce oxidative stress and mitochondrial damage, *Environ. Health Perspect.* 111(4), 455-460.

MacNee, W., and Donaldson, K. (2000a) How can ultrafine particles be responsible for increased mortality? *Monaldi. Arch. Chest. Dis.* 55(2), 135-139.

MacNee, W., and Donaldson, K. (2000b) Exacerbations of COPD environmental mechanisms, *Chest.* 117, 390S-397S.

Nicholson, J.P., and Case, D.B. (1983) Carboxyhaemoglobin levels in New York City runners, *Physician and Sportsmedicine* 11, 135–138.

Nightingale, J.A., Maggs, R., Cullinan, P., Donelly, L.E., Rogers, D.F., Kinnersely, R., Fan Chung, K., Barnes, P.J., Ashmore, M., and Newman-Taylor, A. (2000) Airway inflammation after controlled exposure to diesel exhaust particulates, *Am. J. Respir. Crit. Care Med.* 162(1) 161-166.

Oberdorster, G. (1996) Significance of particle parameters in the evaluation of exposure-dose-response relationships of inhaled particles, *Inhal. Toxicol.* 8 Suppl., 73-78.

Official Statement of the American Thoraic Society adopted by the ATS Board of Directors, July 1999 (2000) What constitutes an adverse health effect of air pollution? *Am. J. Respir. Crit. Care Med.* 161(2), 665-673.

O'Meara, M., and Washington, D.C. (1999) *Reinventing cities for people and the planet*, Worldwatch Institute, Jun. 1999, Worldwatch Paper No. 147, 94p.

Pau-Chung Chen, Yu-Min Lai, Jung-Der Wang, Chun-Yuh Yang, Jing-Shiang Hwang, Hsien-Wen Kuo, Song-Lih Huang, and Chang-Chuan Chan (1998) Adverse Effect of Air

Pollution on Respiratory Health of Primary School Children in Taiwan, *Environmental Health Perspectives* 106 (6), 331-335.

Roodman, D.M. (1998) *The natural wealth of nations: Harnessing the market for the environment*, The Worldwatch Environmental Alert Series. W.W. Norton & Co, 303p.

Scannell, C., Chen, L., Aris, R.M., Tager, I., Christian, D., Ferrando, R., Welch, B., Kelly, T. and Balmes, J.R. (1996) Greater ozone-induced inflammatory responses in subjects with asthma, *Am. J. Respir. Crit. Care Med.* 154(1), 24-29.

Shin Kawasaki, Hajime Takizawa, Kazutaka Takami, Masashi Desaki, Hitoshi Okazaki, Tsuyoshi Kasama, Kazuo Kobayashi, Kazuhiko Yamamoto, Kazuhiko Nakahara, Mitsuru Tanaka, Masaru Sagai, and Takayuki Ohtoshi (2001), Benzene-extracted components are important for the major activity of diesel exhaust particles effect on interleukin-8 gene expression in human bronchial epithelial cells, *Am. J. Respir. Cell Mol. Biol.,* (24)4, 419-426.

Slaughter, J.C., Lumley, T., Sheppard, L., Koenig, J.Q., and Shapiro, G.G. (2003) Effects of ambient air pollution on symptom severity and medication use in children with asthma, *Ann. Allergy Asthma Immunol.* 91(4), 346-353.

Stenfors, N., Nordenhall, C., Salvi, S.S., Mudway, I., Soderberg, M., Blomberg, A., Helleday, R., Levin, J.O., Holgate, S.T., Kelly, F.J., Frew, A.J., and Sandstrom, T. (2004) Different airway inflammatory responses in asthmatic and healthy humans exposed to diesel, *Eur. Respir. J.* 23(1), 82-86.

Tankersley, C.G., Shank, J.A., Flanders, S.E., Soutiere, S.E., Rabold, R., Mitzner, W., and Wagner, E.M. (2003) Changes in lung permeability and lung mechanics accompany homeostatic instability in senescent mice, *Appl. Physiol.* 95(4), 1681-1687.

Timblin, C.R., Shukla, A., Berlanger, I., BeruBe, K.A., Churg, A., and Mossman, B.T. (2002) Ultrafine airborne particles cause increases in protooncogene expression and proliferation in alveolar epithelial cells, *Toxicol. Appl. Pharmacol.* 179(2), 98-104.

Utell, M.J., and Samet, J.M. (1995) Air pollution in the outdoor environment. In Environmental Medicine, Brooks S.M. *et al.* (Eds) *Mosby*, St. Louis Missouri - Year Book, Inc.

van Eeden, S.F., and Hogg, J.C. (2002) Systemic inflammatory response induced by particulate matter air pollution: the importance of bone-marrow stimulation, *Toxicol. Environ. Health. A.* 65(20), 1597-1613.

van Eeden, S.F., Tan, W.C., Suwa, T., Mukae, H., Terashima, T., Fujii, T., Qui, D., Vincent, R., and Hogg, J.C. (2001) Cytokines involved in the systemic inflammatory response induced by exposure to particulate matter air pollutants ($PM_{10}$), *Am. J. Respir. Crit. Care Med.* 164(5), 826-830.

Vedal, S., Petkau, J., White, R., and Blair, J. (1998) Acute effects of ambient inhalable particles in asthmatic and nonasthmatic children, *Am. J. Respir. Crit. Care Med.* 157(4),1034-1043.

# PHOTOCHEMICAL ACTIVITY OF IMPORTANCE TO HEALTH IN THE AEGEAN

C.S. ZEREFOS[1], I.S.A. ISAKSEN[2] AND K. ELEFTHERATOS[3]

[1]*Director of the National Observatory of Athens, P.O. BOX 20048, Thissio, 11810 Athens, GREECE*

*Director of the Laboratory of Atmospheric Environment, Foundation for Biomedical Research of the Academy of Athens, Soranou Efesiou 4, 11527 Athens, GREECE*

[2] *Institute of Geophysics, University of Oslo, P B 1022, Blindern, Olso 3, NORWAY*

[3] *Laboratory of Climatology & Atmospheric Environment, Department of Geology, National and Kapodistrian University of Athens, 15784 Athens, GREECE*

## Summary

The Aegean sea is the crossroads of various pollutants, as pointed out by earlier studies (e.g. Lelieveld *et al.*, 2002; Zerefos *et al.*, 2002). Ozone precursors with the high photochemical activity maintain high background values which exceed the phytotoxicity limit of 32 ppb but exceed also the human exposure limits often in the summertime. Because of this high background, ozone can be regulated only if emission abatement strategies involve several countries in Europe. The aim of the paper is to present evidence from modelling and observational results of the importance of transboundarily transported primary and secondary air pollutants in the SE Mediterranean which is experiencing the highest levels of surface ozone compared to all other European regions.

## 1.    Introduction

Atmospheric pollutants such as ozone, oxides of nitrogen and sulphur are of health concern and they may also impact the climate. Negative aspects of ozone pollution include not only health effects but also effects on crop yield and forest impacts. $NO_2$ and $SO_2$ contribute to the acid rain phenomenon, which severely impacts forest resources in some European areas. Further, ozone has a large negative respiratory impact on outdoor intensive recreational activities (such as those related to sports) and agricultural or other outdoor activities (such as tourism) and $NO_2$ and $SO_2$

*P. Nicolopoulou-Stamati et al. (eds), Environmental Health Impacts of Transport and Mobility, 95-100.*

pollution can also impact visibility (Skalkeas, 2004). Children and elderly people are believed to be especially susceptible to the negative impacts of ozone on lung function. Exposure to high levels of ozone during childhood has also been related to asthma outbreaks at a later stage of life (Katsouyanni, 2004). Because of demographic changes, the percentage of the elderly is expected to increase considerably during the next few decades, the frequency and severity of the ozone-related health impacts are also expected to increase, unless mitigating steps are taken based on sound scientific recommendations.

Therefore, several air pollutants harmful to human health, have been targeted by EU Directives, international Conventions and their subsequent protocols such as the Convention on Long-Range Transboundary Air Pollution (CLRTAP) (Geneva, 1979) and the Kyoto (1997) protocols. Although ratification of the above protocols and implementation of a large number of EU Directives have improved air quality in large areas of Europe, there are still tens of millions of people living in European countries exposed to ozone concentrations exceeding the World Health Organization's guideline for protecting human health (EEA, 1998; ETC/AQ, 1999). Similarly, several million hectares of sensitive ecosystems in Europe are still receiving acidifying and eutrophying deposits of air pollutants in excess of the critical loads (e.g. Tarrason and Schaug, 2000).

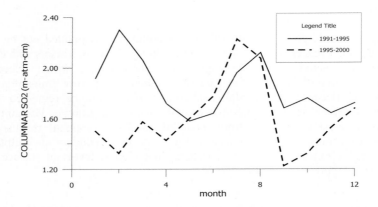

**Figure 24**. Columnar SO$_2$ amounts as measured by a Brewer monochromator at a station in SE Europe. Due to control measures, the winter peak apparent in 1991-1995 data has decreased in the last five years. The summer peak, however, mainly from LRT, remained unaffected.

A particularly suited area of concern is SE Europe. There is by now ample evidence that SE Europe is experiencing high regional values of ozone (Kouvarakis *et al.*, 2000; Kourtidis *et al.*, 2002) and transported SO$_2$ (Eisinger and Burrows, 1998; Zerefos *et al.*, 2000; Zerefos *et al.*, 2002; Lelieveld *et al.*, 2002). Figure 24 shows an example of the seasonal variation of columnar SO$_2$ amounts at Thessaloniki station downwind from major sources of SO$_2$ in the Balkan Peninsula. Due to control measures, the winter peak apparent in 1991-1995 data has decreased in the last five years. The summer peak, however, mainly from Long-Range Transport (LRT), remained unaffected. This region has also been especially targeted because the relatively high regional $SO_2$ amounts improve the accuracy of the satellite columnar $SO_2$ retrievals and in addition there is one ground-based station with continuous long-term observations of columnar $SO_2$ amounts, which can be used for additional validation of the satellite $SO_2$ observations.

Figure 25 shows the distribution of surface ozone over Europe calculated in the Oslo global chemistry model with the 2000 ozone precursor emission. It is evident (and has lately been confirmed by measurements) that SE Europe experiences relatively high ozone levels both in summer and winter. A large portion of the ozone over SE Europe is the result of LRT processes.

## 2.    Results and discussion

As mentioned in the introduction, enhanced ozone levels have been observed over the eastern Mediterranean region during the summer months (Kourtidis et al., 1997, 2000; Kouvarakis et al., 2000). Such enhancements can have several sources: Local anthropogenic emissions of ozone precursors (NOx, CO, nonmethane hydrocarbons (NMHC)), a combination of anthropogenic and natural emissions (isoprene, terpenes) or long-range transport of ozone and ozone precursors from industrial areas outside the region. A particular study has been performed using the global 3-D Oslo CTM2 to estimate the effect of local anthropogenic emissions on the ozone levels over Greece.

The Oslo Chemical Transport Model (CTM2) (Sundet, 1997) is an off-line chemical transport/tracer model (CTM) that uses precalculated transport and physical fields to simulate chemical turnover and distribution in the atmosphere. Both in the horizontal and vertical levels the model has a resolution that is determined by the input data provided. In this paper a data set for 1996 with T63 (1.9° x 1.9°) in the horizontal and 19 vertical levels from the surface up to 10 hPa was used. The model has an extensive set of chemical reactions, particularly designed to study the nonlinear ozone chemistry in the troposphere. Ozone precursor emission from anthropogenic as well as from natural sources (NOx, NMHC, CO, isoprene), atmospheric oxidation initiated by solar radiation, removal by cloud and precipitation processes and by surface deposition are included. Photodissociation is done online. The impact of local emissions is calculated by performing two model experiments:

First, the current (1996) mean monthly distribution of chemical compounds affected by the ozone chemistry is estimated by performing calculations for 1 year + spin-up time (3 months) using current emission rates (Karlsottir et al., 2000). Second, a perturbation run for 3 months is performed, starting with the calculated April distribution and giving the mean ozone distribution for the month of July. In the perturbation run, all anthropogenic emissions (NOx, CO, NMHC) over Greece (NOx, CO, NMHC) are reduced to half their values in the base case. The calculated average distributions for July in the two perturbations are compared.

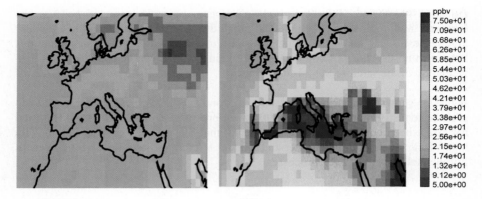

**Figure 25.** Surface ozone distribution over Europe during winter (January, left) and summer (July, right) calculated in the Oslo global chemistry model with the 2000 ozone precursor emission.

As expected, ozone precursors like NOx and CO are significantly reduced over the source region, particularly NOx, which has a short chemical lifetime. The reduction of the more long-lived precursor CO is on the order of 40 ppb (~10-15%), some reduction extending outside the source region. The perturbation in the July average distribution of surface ozone is given in Figure 26.

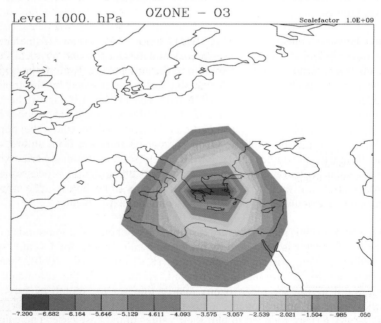

**Figure 26.** July average reduction in surface ozone distribution from a 50% reduction in ozone precursor emissions (NOx, CO, NMHC). The values are in ppbv (revised from Zerefos *et al.*, 2002).

As it appears from Figure 26 (adopted from Zerefos et al., 2002) ozone levels are reduced when precursor emissions are reduced. The reductions are largest over Greece and the Aegean Sea, with maximum values on the order of 10% (~7 ppb). Outside the region, ozone perturbations drop off rapidly. The impact of local emissions is also strongly reduced when we go above the planetary boundary layer. The calculations indicate that there are modest local contributions to the high surface levels observed in the region. On the other hand, free tropospheric ozone levels in the region seem to be dominated by long-range transport. This is in agreement with the observational results of Kourtidis et al. (2002) and Kouvarakis et al. (2002).

## 3. Conclusion

Evidence has shown that SE Europe is influenced by transboundarily transported primary and secondary air pollutants. Model calculations showed that even under strict measures which include draconian cuts in regional emissions reduce air pollutant levels (surface ozone) these are not enough for pollution abatement strategies. Therefore CLRTAP should be taken into account in an expanded version whenever political decisions are made on a European scale level, particularly in view of the EU enlargement. The Aegean sea being the crossroads of various pollutants and ozone precursors maintain high background values which exceed the phytotoxicity limit of 32 ppb but exceed also the human exposure limits often in the summertime. Because of this high background, ozone can be regulated only if emission abatement strategies involve several countries in Europe.

## References

EEA (1998) *Tropospheric Ozone - Europe's Environment: The Second Assessment (Chapter 5)*, European Environment Agency, Office for Official Publications of the European Communities.

Eisinger, M., and Burrows, J. (1998) Tropospheric sulfur dioxide observed by the ERS-2 GOME instrument, *Geophys. Res. Lett.* 25, 4177-4180.

ETC/AQ, Sluyter, R., and de Paus, T. (1999) *Ozone exceedance in EU 1997 and summer 1998: Topic Report 03/99 part II*, European Topic Centre on Air Quality.

Karlsdottir, S., Isaksen, I.S.A., Myhre, G., and Berntsen, T.K. (2000) Trend analysis of $O_3$ and CO in the period 1980-1996: A three-dimensional model study, *J. Geophys. Res.* 105, 28.907-28.934.

Katsouyanni, K. (2004) Health effects of air pollutants, in G. Skalkeas (ed.) *Proc. of the Colloquium of the Foundation for Biomedical Research of the Academy of Athens "Environment and Health"*, pp. 99-112, Athens, Greece, October 22-24, 2004.

Kourtidis, K., Ziomas, I., Zerefos, C., Balis, D., Suppan, P., Vasaras, A., Kosmidis, V., and Kentarchos, A. (1997) On the background ozone values in Greece, in B. Larsen, B. Versino and G. Angeletti (eds) *Proceedings of the 7th European Symposium on Physico-Chemical Behaviour of Atmospheric Pollutants, the Oxidizing Capacity of the Troposphere*, 387-390, European Commission, Brussels.

Kourtidis, K., Zerefos, C., Balis, D., and Kosmidis, E. (2000) Overview of concepts and results of the PAUR I and PAUR II projects, in C. Zerefos et al. (eds) *Chemistry and Radiation Changes in the Ozone Layer*, NATO ASI Ser., Ser. C 557, 55-73, Kluwer Acad., Norwell, Mass.

Kourtidis, K., Zerefos, C., Rapsomanikis, S., Simeonov, V., Balis, D.S., Kosmidis, E., Perros, P., Thompson, A., Witte, J., Calpini, B., Rappenglueck, B., Isaksen, I., Papayannis, A., Hofzumahaus, A., Gimm, H., and Drakou, R. (2002) Regional tropospheric ozone over Eastern Mediterranean, *J. Geophys. Res.*, 107, D18, 10.1029/2000JD000140.

Kouvarakis, G., Tsigaridis, K., Kanakidou, M., and Mihalopoulos, M. (2000) Temporal variations of surface regional background ozone over Crete Island in southeast Mediterranean, *J. Geophys. Res.* 105, 4399-4407.

Kouvarakis, G., Vrekousis, M., Mihalopoulos, N., Kourtidis, K., Rappengluck, B., Gerasopoulos, E., and Zerefos, C. (2002) Spatial and temporal variability of tropospheric ozone ($O_3$) in the boundary layer above the Aegean Sea (eastern Mediterranean), *J. Geophys. Res.* 107, 10.1029/2000JD000081.

Lelieveld, J., Berresheim, H., Borrmann, S., Crutzen, P.J., Dentener, F.J., Fischer, H., Feichter, J., Flatau, P.J., Heland, J., Holzinger, R., Korrmann, R., Lawrence, M.G., Levin, Z., Markowicz, K.M., Mihalopoulos, N., Minikin, A., Ramanathan, V., De Reus, M., Roelofs, G.J., Scheeren, H.A., Sciare, J., Schlager, H., Schultz, M., Siegmund, P., Steil, B., Stephanou, E.G., Stier, P., Traub, M., Warneke, C., Williams, J., and Ziereis, H. (2002) Global Air Pollution Crossroads over the Mediterranean, *Science* 298, 794-799.

Skalkeas, G. (2004) *Proc. of the Colloquium of the Foundation for Biomedical Research of the Academy of Athens "Environment and Health"*, p. 203, Athens, Greece, October 22-24.

Sundet, J.K. (1997) *Model studies with a 3-D global CTM using ECMWF data,* Ph.D. thesis, Dep. of Geophys., Univ. of Oslo, Oslo, Norway.

Tarrason, L., and Schaug, J. (2000) *Transboundary Acidification and Eutrophication in Europe*. EMEP summary report 1-2000. Oslo, Norway.

Zerefos, C., Ganev, K., Kourtidis, K., Tzortziou, M., Vasaras, A., and Syrakov, E. (2000) On the origin of $SO_2$ above Northern Greece, *Geophys. Res. Lett.* 27, 365-368.

Zerefos, C., Kourtidis, K., Melas, D., Balis, D.S., Zanis, P., Mantis, H. T., Repapis, C., Isaksen, I., Sundet, J., Herman, P.K., and Calpini, B. (2002) Photochemical Activity and Solar Ultraviolet Radiation Modulation Factors (PAUR): An overview of the project, *J. Geophys. Res.* 107, D18, 10.1029/2000JD000134.

# MONITORING OF SUSPENDED AEROSOL PARTICLES AND TROPOSPHERIC OZONE BY THE LASER REMOTE SENSING (LIDAR) TECHNIQUE:
## *A CONTRIBUTION TO DEVELOP TOOLS ASSISTING DECISION-MAKERS*

A.D. PAPAYANNIS
*National Technical University of Athens*
*School of Applied Mathematical and Physical Sciences*
*Physics Department-Lidar Group*
*Heroon Polytechniou 9*
*15780 Zografou, GREECE*

## Summary

Suspended aerosol particles and tropospheric ozone ($O_3$) play a key role in atmospheric chemistry and the climate of the earth. Recent epidemiological studies performed in the USA and Europe have shown that urban mortality rates are highly correlated with mass concentration of fine particulate matter (PM) of a diameter less than 2.5 microns ($PM_{2.5}$). Increased morbidity and mortality in association with short-term increases in exposure to PM were observed among people with respiratory and cardiovascular disease. In addition to high PM mixing ratios, high ozone concentrations can irritate the respiratory system, reduce the lung function, aggravate asthma and inflame and damage the lining of the lung. However, synergy between the effects of exposure to PM and $O_3$ is not convincingly demonstrated, yet. Therefore there is an increasing necessity for continuous monitoring of PMs and $O_3$, especially over urban areas. In this paper a brief review of the LIght Detection And Ranging (LIDAR) technique capability to monitor suspended aerosol particles ($PM_{0.3}$-$PM_{10}$) and tropospheric ozone in urban and rural sites is given. Selected examples of PM and $O_3$ measurements are presented for a photochemical smog case and a Saharan dust event, respectively. Finally, we highlight the possibilities given by the LIDAR technique to develop tools assisting decision-makers for adopting novel air pollution management and abatement strategies.

*P. Nicolopoulou-Stamati et al. (eds), Environmental Health Impacts of Transport and Mobility,* 101-113.
© 2005 *Springer. Printed in the Netherlands.*

## 1.   Introduction

Tropospheric aerosols arise from natural sources (sand storms, deserts, volcanoes, etc.), sea spray, forest fires, and from anthropogenic sources (combustion of fossil fuels, biomass burning, etc.) (Prospero and Carlson, 1972; Duce, 1995; Seinfeld and Pandis, 1998; Ramanathan et al., 2001; Prospero et al., 2002). Tropospheric ozone is mostly produced photo-chemically in the lower atmosphere, when intense solar irradiation and ozone precursors (i.e. hydrocarbons from, for example, traffic, industrial emissions and biomass burning) are available. Since the 1980's the global concentrations of aerosols (or suspended particulate matter or PM) and ozone are continuously increasing around the planet (Logan, 1994; Lelieveld et al., 1999; Ramanathan et al., 2001; Kourtidis et al., 2002).

Tropospheric ozone and suspended particulate matter contribute intensively to the radiative and the climate forcing of the earth (Houghton et al., 2001; Seinfeld et al., 2004), and can have negative effects on human health. Recent epidemiological studies performed in Europe in the frame of the APHEA project examined the short-term effects of particulate air pollution on cardiovascular diseases and total mortality over eight European cities (Touloumi et al., 1997; Katsouyanni et al., 2001; Le Tertre et al., 2002).

In addition, recent epidemiological studies performed in the USA showed that urban mortality rates are highly correlated with mass concentration of fine particulate matter (PM) with a diameter of less than or equal to 2.5 microns ($PM_{2.5}$). Increases of morbidity and mortality in association with short-term increases in exposure to PM was observed among people with respiratory or cardiovascular diseases (Dockery and Pope, 1996; Schwartz et al., 1996; Samet et al., 2000). In addition, high ozone concentrations can irritate the respiratory system, (Janic et al., 2003). However, synergy between the effects of exposure to PM and $O_3$ is not, as yet, convincingly demonstrated. Therefore, it is important to monitor PMs and $O_3$ continuously, especially in and around urban areas where the highest concentrations are recorded.

Laser remote sensing techniques offer the possibility to study a large variety of parameters and characteristics of the earth's atmosphere with high temporal and spatial resolution over periods ranging from a few hours to months (Measures, 1992; Hamonou et al., 1999; Gobbi et al., 2004). For more than thirty five years now, various types of ground-based LIDAR systems (including ceilometers) have been continuously probing the earth's atmosphere, in conjunction with several other remote sensing techniques (sun photometers, spectrophotometers, radiometers, etc), to measure ozone and aerosol properties (optical depth, spatial distribution and layering, diurnal variation, etc.), as described by Measures (1992), Davis et al., (1997) and Hamonou et al., (1999). Airborne and space-borne LIDAR systems have recently been used (e.g., LITE-GLAS/NASA) or are to be used (e.g., CALIPSO/NASA-CNES) to observe the vertically resolved variability of aerosols in the earth's atmosphere on regional and global scales (Spinhirne et al., 2004; Winker et al., 2004).

## 2.    Effects of air pollutants on human health

Among the many important air pollutants in the urban troposphere (CO, $SO_2$, $CH_4$, HCs, $CO_2$, $NO_x$) we focus here on tropospheric ozone and suspended particulate matter. Both of them have very important effects on the earth's climate change (Roelofs et al., 1997) and on human health (Touloumi et al., 1997). Aerosols can travel over long distances depending on the meteorological conditions and their size and their residence time in the troposphere ranges from a few days to a few weeks (Seinfeld and Pandis, 1998). Aerosols can be toxic, by composition or by structure (size or shape). They can cause respiratory and blood circulation problems, since they enter the human body by inhalation and affect people with respiratory or cardiovascular diseases (Dockery and Pope, 1996; Schwartz et al., 1996; Samet et al., 2000, Janic et al., 2003; Neuberger et al., 2004). Aerosols have also important effects on ecosystems and climate: the ecosystems are affected by significant mass transport of aerosols and aerosols contribute substantially to radiative forcing (Houghton et al., 2001; Seinfeld et al., 2004), through scattering of short-wavelength solar radiation and modification of the shortwave reflective properties of clouds, thereby exerting a cooling influence on the planet (Charlson et al., 1992).

High tropospheric ozone concentrations have important effects on ecosystems (phyto-toxicity) and climate (Global Warming effects, thereby exerting a large influence on the earth's thermal and radiative budget). Therefore, regarding phytotoxicity effects of ozone the respective EU legislation limit is 32 ppbv for 24 hours (EU Directive, 2002). A review of records of impacts of ambient ozone on crops in the Mediterranean concludes that there is evidence of visible injury on 24 different agricultural and horticultural crops [Fumagalli et al., 2001]. Regarding the ozone-climate interaction it plays a major role in earth's radiative forcing since it is an important greenhouse gas mainly in the upper troposphere (Roelofs et al., 1997; Houghton et al., 2001).

High tropospheric ozone concentrations have also very important effects on human health: they irritate the human respiratory system and the eyes, reduce lung function, aggravate asthma, inflame and damage the lining of the lung, aggravate chronic lung diseases (emphysema, bronchitis) and finally, make people more sensitive to allergens that cause asthma attacks (Samet et al., 2000; Ducusin et al., 2003; Oftedal et al., 2003). These important ozone effects on human health have led to the adoption of EU legislation setting an upper exposure limit of 50 ppbv for 8 hours for human health protection (EU Directive, 2002). Recently, there has been a debate over whether the body of available scientific evidence justifies the application of this limit to southern Europe (Ashmore and Fuhrer, 2000; De Santis, 2000) where ozone concentrations are generally higher due to intense solar irradiance which leads to higher ozone photochemical production.

## 3.    Instrumentation and methods

### 3.1.    *The LIDAR technique*

The LIDAR technique is an active remote sensing method that offers the possibility of acquiring the vertical profiles of air pollutants (PM, $O_3$, $SO_2$, $NO_x$, $CO_2$, Hg, benzene, toluene, etc.) in the troposphere (from ground up to 5-8 km height) with very high spatial (around 10-15 m) and temporal resolution (a few seconds up to a few minutes) (Measures, 1992). The LIDAR technique is based on the emission of laser pulses (ns or fs duration) to the atmosphere under study. According to the air pollutant or the atmospheric parameter to be studied the correct wavelength(s) have to be selected. In the case of atmospheric aerosol measurements wavelengths ranging from the ultraviolet to the infrared region (0.355 to 12 µm) can be selected. In the case of ozone measurements wavelengths in the ultraviolet region must be selected (Papayannis et al., 1990).

The backscattered laser photons by the atmospheric volume under study are collected by a receiving telescope. The wavelength selection of the LIDAR signals is performed by a set of spectral narrow-band interference filters or a high-resolution spectrometer. Photomultipliers and/or avalanche photodiodes are used to detect the backscattered photons at the respective wavelengths. Finally, the LIDAR signals are acquired and digitized in the analog and/or the photon counting mode by fast transient recorders and subsequently transferred to a personal computer for further analysis and storing (Measures, 1992).

The capability of the LIDAR technique to derive range-resolved vertical profiles of atmospheric parameters (humidity, wind velocity, temperature, etc.) and concentrations of air pollutants with very high spatial (around 10-15 m) and temporal resolution (a few seconds up to a few minutes) arises from the emission of a series of short (ns or fs duration) laser pulses with high repetition rate to the atmosphere at a given time, coupled with fast detection electronics (transient recorders). Measuring the delay time between the emitted and the received laser pulses one is able to calculate the distance of the probed atmospheric volume and thus perform range-resolved measurements of the desired air pollutants or atmospheric parameters (Measures, 1992).

In order to improve the LIDAR system's accuracy in the retrieval of molecular air pollutants profiles ($O_3$, $SO_2$, $NO_x$, $CO_2$, benzene, toluene, etc.) a set of two or three wavelengths has to be selected in order to minimize the differential scattering or absorption by the aerosol particles or other interfering molecular air pollutants (Papayannis et al., 1990; Measures, 1992). For every pair of laser wavelengths emitted a differential absorption takes place: the first wavelength ('on' wavelength) is strongly absorbed and the second ('off' wavelength) is weakly absorbed by the molecular air pollutant under study. In that case the LIDAR technique is called Differential Absorption LIDAR (DIAL) technique.

If a scanning LIDAR system is employed, three-dimensional maps of various air pollutants can be acquired, which, if combined with wind data, can provide air pollutant fluxes over selected urban areas (Schoulepnikoff, *et al.*, 1998; Fiorani *et al.*, 1998; Kambezidis *et al.*, 1998; Zwozdziak *et al.*, 2001). Additionally, if the LIDAR information is combined with wind flow data (e.g., three-dimensional air mass back trajectory analyses), large air pollution source regions can be identified Balis *et al.*, 2003). As it will be explained in Section 5, LIDAR data in conjunction with data from conventional ground air pollution (chemical analyzers, DOAS systems) and meteorological sensors can be used as input to air pollution forecasting models. Once these models are validated against real air pollution data they can be used to help decision-makers on designing and adopting the optimum strategy for air pollution reduction in large urban agglomerations (Thomasson, 2001; Vautard *et al.*, 2001). For instance, a recent publication (Hodzic *et al.*, 2004) shows the ability of the aerosol chemistry transport model CHIMERE to simulate the vertical aerosol concentration profiles at a site near the city of Paris and this is evaluated using routine elastic backscatter LIDAR and sun photometer measurements.

### 3.2. The EARLINET project

The EARLINET (European Aerosol Research Lidar Network to Establish an Aerosol Climatology) lidar Network is composed of 22 coordinated stations distributed over most of Europe (Figure 27), using advanced laser remote sensing instruments to measure directly the vertical distribution of aerosols in the Planetary Boundary Layer (PBL) and the adjacent free troposphere (FT), supported by a suite of conventional instrumentation (sun photometers, UVB radiometers, spectral photometers, meteorological stations, etc.) as described by Bösenberg *et al.*, (2001) and Bösenberg *et al.*, (2003).

The objectives of EARLINET, among others, are: (1) to establish a comprehensive and quantitative climatological data base for the horizontal and vertical distribution and variability of aerosol over Europe, (2) to provide aerosol data with unbiased sampling during Saharan dust outbreaks towards Europe, and (3) to provide the necessary data base for validating and improving numerical models describing the evolution of aerosol properties and their influence on climate and environmental conditions.

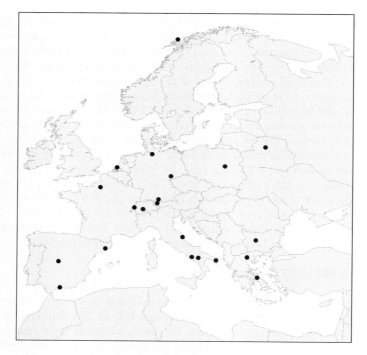

**Figure 27**. The EARLINET lidar network.

## 4.    Experimental results - examples

We present some selected experimental results from LIDAR measurements of ozone and aerosol particles in Europe. Figure 28 shows the temporal evolution (07:30 to 10:30 UT) of a Saharan dust event recorded by LIDAR over the city of Athens on 22 September 2001. In this Figure the derivative of the logarithm of the range-corrected (Measures, 1992) lidar signal (provided in the right-column grey scale) is shown which in fact delineates the atmospheric regions with low and high aerosol loads. The suspended aerosol particles over Athens are observed at around 3-4 km height asl. and correspond to the arrival of Saharan dust aerosols, as confirmed by air mass back-trajectory analysis (Papayannis *et al.*, 2004a).

The aerosols trapped at the top of the convective PBL, are observed between 1.2 and 1.7 km height asl. (in lighter colour). This information about the PBL height is extremely important for air pollution studies and forecasting models, since all emitted primary air pollutants are mostly trapped inside the PBL, thus influencing the air quality near ground. The aerosols inside the PBL (altitudes lower than 1200 m asl.) are nearly homogeneously mixed and are presented by the darker colours. No data are shown below 400 m height asl., which is the region where full overlap (Measures, 1992) between the emitted laser beam and the receiving telescope occurs, since the lidar station in Athens is located at an altitude of 220 m asl.

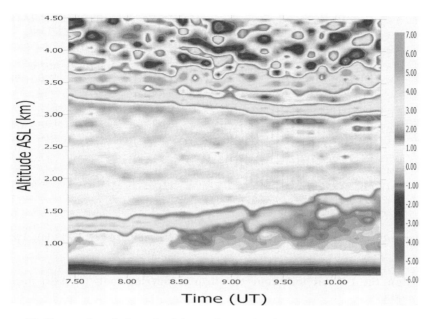

**Figure 28**. Temporal evolution of a Saharan dust outbreak over Athens, Greece (grey scale provides the derivative of the logarithm of the range-corrected lidar signal).

An example of a series of ozone vertical profile measurements performed over the city of Paris, France (Papayannis, 1989) using the differential absorption LIDAR technique is shown in Figure 29 for July 19, 1988, during a photochemical smog episode. During that day high ozone concentrations ($>10^{12}$ mol/cm$^3$) were recorded, between 0.47 and 0.65 km asl. It is interesting to note the high ozone values (around $1.5 \times 10^{12}$ mol/cm$^3$ or equivalent mixing ratios of 75 ppbv) recorded over Paris inside the PBL at altitudes below 0.6 km, between 18:30 and 18.55 hours (local time). Nevertheless, these values are much lower than the extremely high values (around 150-200 ppbv) recorded during August 2003, in conjunction with extremely high temperatures (around 40°C), causing the death of more than 14.800 persons over the French capital (Dhainaut *et al.*, 2003).

It is also very important to mention that the ozone values generally recorded during summer time over Southern Europe (Bonasoni *et al.*, 2000) and especially over the Eastern Mediterranean region (Lelevield *et al.*, 2002; Kourtidis *et al.*, 2002; Kalabokas *et al.*, 2004; Papayannis *et al.*, 2004b) are usually higher than the European Union (EU) guidelines for 8-hours and 24-hours standards (100 µg/m$^3$ or 50 ppbv and 64 µg/m$^3$ or 32 ppbv, respectively) for ambient ozone concentrations (EU Directive, 2002), due to intense solar irradiance which leads to higher ozone photochemical production.

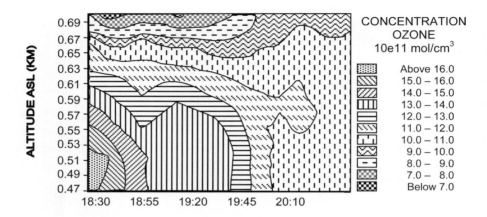

**Figure 29**. Ozone vertical profile recorded by LIDAR over Paris, France (19 July 1988).

## 5. How the LIDAR technique can help to develop tools to assist decision-makers

Decision-makers urgently need guidelines to adopt strategies to abate air pollution in large urban agglomerations all over the world (DETR, 2000). The example of more than 14.800 deaths in the city of Paris during August 2003 (caused by the synergy of very high temperatures and ozone concentrations) clearly demonstrates this need. It is well known that the ground air pollution monitoring networks are not adequate to provide data able to fully characterize and help scientists understand the mechanisms of air pollution formation and modifications processes over large urban sites. This is due to the fact that the photo-chemical reactions responsible for air pollutant formation take place in the lower troposphere inside the PBL (from ground up to several hundred meters) where the primary pollutants are emitted. In addition, wind recirculation in the lower troposphere leads to convective mixing (Millan, 1994; Soriano et al., 2001) or advection of air pollutants inside the PBL. Thus, their vertical profile may be variable.

The only technique, currently available to monitor the vertical distribution of the main air pollutants with high temporal and spatial resolution is the LIDAR technique. In addition, numerical modelling is a powerful tool for evaluating pollution abatement policies (Soriano et al., 2001). Thus, the LIDAR data in conjunction with data from conventional ground air pollution sensors (chemical analyzers, DOAS systems) and meteorological sensors/profilers can be used as input to air pollution forecasting models. Once these models are validated against real air pollution data they can be used as a tool to help decision-makers on designing and adopting the optimum strategy for air pollution reduction and management in large urban agglomerations. For example, decision-makers take steps for the reduction of emissions from primary pollutants and forecast the concentration levels of secondary air pollutants. Up to now, several air pollutants (like ozone, nitrogen oxides, aerosols, benzene, toluene, etc.) have been measured by LIDAR in urban areas in France by national agencies (e.g. INERIS and AIRPARIF) and their profiles have been compared to simulated model data, as described by Thomasson (2001), Vautard *et al.* (2001) and Hodzic *et al.* (2004).

Therefore, to design a strategy to abate and manage air pollution the following steps are recommended:

- *step 1*: identify and quantify significant sources of the individual pollutants over the area under study and produce an updated local emission inventory in combination with available meteorological data (DETR, 2000),
- *step 2:* monitor air pollution at ground level and aloft by combining LIDARS, DOAS, conventional chemical analysers and satellite data,
- *step 3*: understand the air pollution formation mechanisms and circulation patterns,
- *step 4*: use observed air pollution data (see step 1) as input to prognostic Atmospheric Chemistry Models and explore different scenarios in short- and long-range. And finally, as evidently necessary in any strategy to manage and abate air pollution,
- *step 5*: reduce air pollution (industrial and vehicle) emissions to improve local air quality (DETR, 2000). This should be the most important step for a successful air pollution abatement strategy.

Step 5 should focus on the following targets (DETR, 2000):

1. Systematic real-time monitoring of large and minor air pollution sources (industrial, power plants, domestic, traffic, etc.),
2. Application of severe penalties for polluting industries, by adoption of EU directives, using data from target 1,
3. Minimize the release of primary air pollutants of industrial or vehicle traffic origin ($SO_2$, NOx, PM, VOCs, etc.) using novel filtering technologies,
4. Improve and extend public transportation means,
5. Improve traffic circulation in urban areas and adopt *clean car* technology (green transport plans).
6. Use of renewable energy resources for production of electricity,
7. Use natural gas in transportation and industry,
8. Increase the public green and forest areas,
9. Preserve large-scale natural forests,
10. Develop infrastructure for alternative transportation (i.e. bicycles),
11. Reduce domestic wastes (use of recyclable material) and apply energy saving techniques.

## 6. Conclusions

This paper provides a short review of the effects of ozone and aerosol particles on human health and gives selected typical examples of LIDAR measurements of air pollution ($O_3$ and PMs) in Athens and Paris. Co-ordinated LIDAR measurements of aerosol particles are performed continuously since May 2000 over the European continent in the framework of the EARLINET project. In order to help decision-makers to adopt strategies to manage and reduce air pollution in large urban agglomerations, we provided guidelines toward this direction, on a 5-step procedure. One of the most important measures for a successful air pollution abatement strategy should be the one to minimize the primary air pollutant emissions from large source regions (car traffic, industrial complexes, power plants, biomass burning activities) and to increase and preserve the large green areas (forests) over the planet (such as tropical rain forests, forests in Siberia and N. America, Europe, etc.).

## Acknowledgements

EARLINET project was supported by the Environment Program of the European Union, under contract EVR1-CT1999-40003. The author would like to thank the two reviewers for their helpful comments.

## References

Ashmore, M., and Fuhrer, J. (2000) New directions: Use and abuse of the AOT40 concept, *Atmos. Environ.*, 34, 1157–1168.

Balis, D., Amiridis, V., Zerefos, C., Gerasopoulos, E., Andreae, M., Zanis, P., Kazadzidis, A., Kazazdis, S., and Papayannis, A. (2003) Raman lidar and sun photometric measurements of aerosol optical properties over Thessaloniki during a biomass burning episode, *Atmos. Environ.*, 37, 4529-4538.

Bonasoni, P., Stohl, A., Cristofanelli, P., Calzolari, F., Colombo, T., and Evangelsiti, F. (2000) Background ozone variations at the Mt. Cimone station, *Atmos. Environ.*, 34, 5183-5190.

Bösenberg, J., Ansmann, A., Baldasano, J., Balis, D., Böckmann, C., Calpini, B., Chaikovksi, A., Flamant, P., Hågård, A., Mitev, V., Papayannis, A., Pelon, J., Resendes, D., Schneider, J., Spinelli, N., Vaughan, T., Visconti, G., and Wiegner, M. (2001) EARLINET: A European Aerosol Research Lidar Network, advances in laser remote sensing, in A. Dabas, C. Loth, and J. Pelon (eds), *Proceedings of the 21st International Laser Radar Conference*, Ecole Polytechnique, Paris, pp. 155-158.

Bösenberg, J., Matthias, V., Amodeo, A., Amoiridis, V., Ansmann, A., Baldasano, J.M., Balin, I., Balis, D., Böckmann, C., Boselli, A., Carlsson, G., Chaikovsky, A., Chourdakis, G., Comeron, A., De Tomasi, F., Eixmann, R., Freudenthaler, V., Giehl, H., Grigorov, I., Hagard, A., Iarlori, M., Kirsche, A., Kolarov, G., Komguem, L., Kreipl, S., Kumpf, W., Larcheveque, G., Linné, H., Matthey, R., Mattis, I., Mekler, A., Mironova, I., Mitev, V., Mona, L., Müller, D., Music, S., Nickovic, S., Pandolfi, M., Papayannis, A., Pappalardo, G., Pelon, J., Perez, C., Perrone, R.M., Persson, R., Resendes, D.P., Rizi, V., Rocadenbosch, F., Rodrigues, J.A., Sauvage, L., Schneidenbach, L., Schumacher, R., Shcherbakov, V., Simeonov, V., Sobolewski, P., Spinelli, N., Stachlewska, I., Stoyanov, D., Trickl, T., Tsaknakis, G., Vaughan, G., Wandinger, U., Wang, X., Wiegner, M., Zavrtanik, M., and Zerefos, C. (2003) EARLINET: A European Aerosol Research Lidar Network, *Report* 348, Max-Planck Institute for Meteorology, Hamburg, pp. 200.

Charlson, R., Schwartz, S., Hales, J., Cess, R., Coakley, J., Hansen, J., and Hofmann, D. (1992) Climate forcing by anthropogenic aerosols, *Science* 256, 423-430.

Davis, A., Winker, D., Marshak, A., and Cahalan, R. (1997) Retrieval of physical and optical thicknesses from space-borne and wide-angle imaging Lidar, in A. Ansmann (ed.), *Advances in Atmospheric Monitoring with Lidar*, Springer-Verlag, Berlin, 193-196.

De Santis, F., (2000) Reply to "Use and abuse of the AOT40 concept". M. Ashmore and J. Fuhrer (eds.), *Atmos. Environ.*, 34, 1158–1169.

DETR (2000) *The air quality strategy for England, Scotland, Wales and Northern Ireland*, Department of the Environment, Transport and the Regions, 189 pp, London, UK.

Dhainaut, J.F., Claessens, Y.E., Ginsburg, C., and Riou, B. (2003) Unprecedented heat-related deaths during the 2003 heat wave in Paris: consequences on emergency departments, *Critical Care*, 8, doi:10.1186/cc2404.

Dockery, D., and Pope, A. (1996) Epidemiology of acute health effects: summary of time-series studied, in Wilson, R., and Spengler, J.D. (eds.), *Particles in our air: concentration and health effects*, Harvard University Press, Cambridge MA, USA, 123–147.

Duce, R. A. (1995) Sources, distributions, and fluxes of mineral aerosols and their relationship to climate, in Charlson, R.J. and Heintzenberg, R.J. (eds) *Dahlem Workshop on Aerosol Forcing of Climate*, pp. 43–72, John Wiley, New York.

Ducusin, R., Nishimura, M., Sarashina, T., Uzuka, Y., Tanabe, S., and Otani, M. (2003) Phagocytosis of bovine blood and milk polymorphonuclear leukocytes after ozone gas administration in vitro, *J. Vet. Med. Sci.* 65, 535-539.

EU Directive 2002/03/EC: Official Journal of the European Communities, No L67/03.03, 2002.

Fiorani, L., Calpini, B., Jaquet, L., Van den Bergh, H., and Durieux, E. (1998) A combined determination of wind velocities and ozone concentrations for a first measurement of ozone fluxes with a dial instrument during the MEDCAPHOT-TRACE campaign, *Atmos. Environ.*, 32, 2151-2159.

Fumagalli, I., Gimeno, B.S., Velissariou, D., and Mills, G. (2001) Evidence of ozone-induced adverse effects on crops in the Mediterranean region, *Atmos. Environ.*, 35, 2583– 2587.

Gobbi, G. P., Barnaba, F., and Ammannato, L. (2004) The vertical distribution of aerosols, Saharan dust and cirrus clouds in Rome (Italy) in the year 2001, *Atmos. Chem. Phys.*, 4, 351-359.

Hamonou, E., Chazette, P., Balis, D., Dulac, F., Schneider, X., Galani, E., Ancellet, G., and Papayannis, A. (1999) Characterization of the vertical structure of Saharan dust export to the Mediterranean basin, *J. Geophys. Res.* D104, 22257-22270.

Hodzic, A., Chepfer, H., Vautard, R., Chazette, P., Beekmann, M., Bessagnet, B., Chatenet, B., Cuesta, J., Drobinski, P., Goloub, P., Haeffelin, M., and Morille, Y. (2004) Comparison of aerosol chemistry transport model simulations with lidar and Sun photometer observations at a site near Paris, *J. Geophys. Res.*, 109 (D23): D23201.

Houghton, J.T., Ding, W., Griggs, D.J., Noguer, M., van der Linden, P.J., and Xiaosu, D. (2001) *Climate change* 2001: *the scientific basis,* Contribution of working group I to the Third Assessment Report of the Intergovernmental Panel on Climate Change (IPCC). Cambridge University Press, U.K.

Janic, B., Umstead, T.M., Phelps, D.S., and Floros, J. (2003) An in vitro cell model system for the study of the effects of ozone and other gaseous agents on phagocytic cells, *J. Immunol. Methods* 272, 125-134.

Kalabokas, P., and Repapis, C.C. (2004) A climatological study of rural surface ozone in central Greece, *Atmos. Chem. Phys.*, 4, 1139-1147.

Kambezidis, H., Weidauer, D., Melas, D., and Ulbricht, M. (1998) Air quality in the Athens basin during sea breeze and non-sea-breeze days using laer-remote sensing technique, *Atmos. Environ.*, 32, 2173-2182.

Katsouyanni, K., Touloumi, G., Samoli, E., Gryparis, A., Le Tertre, A., Monopolis, Y., Rossi, G., Zmirou, D., Ballester, F., Boumghar, A., Anderson, H. R., Wojtyniak, B., Paldy, A., Braunstein, R., Pekkanen, J., Schindler, C., and Schwartz, J. (2001) Confounding and effect modification in the short-term effects of ambient particles on total mortality: results from 29 European cities within the APHEA2 project, *Epidemiol.* 12, 521-531.

Kourtidis, K., Zerefos, C., Rapsomanikis, S., Simeonov, V., Balis, D., Perros, P.E., Thompson, A.M., Witte, J., Calpini, B., Sharobiem, W.M., Papayannis, A., Mihalopoulos, N., and Drakou, R. (2002) Regional levels of ozone in the troposphere over eastern Mediterranean, *J. Geophys. Res.*, 107 (D18), 8140, doi:10.1029/2000JD000140.

Lelieveld, J., Thompson, A.M., Diab, R.D., Hov, O., Kley, D., Logan, J.A., Nielsen, O.J., Stockwell, W.R., and Zhou, X. (1999) Tropospheric ozone and related processes, in Scientific Assessment of Ozone Depletion: 1998, WMO Rep. 44, 8.1–8.42, *Global Ozone Res. and Monitoring Proj.*, World Meteorol. Org., Geneva.

Leleveild, J., Berresheim, H., Borrmann, S., Crutzen, P.J., Dentener, F.J., Fischer, H., Feichter, J., Flatau, P.J., Heland, J., Holzinger, R., Korrmann, R., Lawrence, M.G., Levin, Z., Markowicz, K.M., Mihalopoulos, N., Minikin, A., Ramanathan, V., de Reus, M., Roelofs, G.J., Scheeren, H.A., Sciare, J., Schlager, H., Schultz, M., Siegmund, P., Steil, B., Stephanou, E.G., Stier, P., Traub, M., Warneke, C., Williams, J., and Ziereis, H. (2002) Global air pollution crossroads over the mediterranean, *Science* 298, 794-799.

Le Tertre, A., Medina, S., Samoli, E., Forsberg, B., Michelozzi, P., Boumghar, A., Vonk, J. M., Bellini, A., Atkinson, R., Ayres, J. G., Sunyer, J., Schwartz, J., and Katsouyanni, K. (2002) Short-term effects of particulate air pollution on cardiovascular diseases in eight European cities, *J. Epidemiol. Comm. Health* 56, 773-779.

Logan, J.A. (1994) Trends in the vertical distribution of ozone: an analysis of ozonesonde data, *J. Geophys. Res.*, 99, 25553– 25585.

Measures, R. (1992) *Laser Remote Sensing,* Krieger Publ., New York.

Millan, M. (1994) Regional processes and long range transport of air pollutants in S. Europe. The experimental evidence, *Physicochemical behaviour of atmospheric pollutants, Air Pollution Research Report* No. 50, Report EUR15609/1 EN, 447-459.

Neuberger, M., Schimek, M.G., Horak, F., Moshammer, H., Kundi, M., Frischer, T., Gomiscek, B., Puxbaum, H., Hauck, H., and AUPHEP-Team (2004) Acute effects of particulate matter on respiratory diseases, symptoms and functions: epidemiological results of the Austrian Project on Health Effects of Particulate Matter (AUPHEP), *Atmos. Environ.*, 38, 3971-3981.

Oftedal, B., Nafstad, P., Magnus, P., Bjorkly, S., and Skrondal, A. (2003) Traffic related air pollution and acute hospital admission for respiratory diseases in Drammen, Norway 1995-2000, *Eur. J. Epidemiol.* 18, 671-675.

Papayannis, A. (1989) *Etude expérimentale de la distribution verticale de l'ozone dans la troposphère par télédéction laser*, Ph.D. Thesis, University of Paris 7, Paris.

Papayannis, A., Ancellet, G., Pelon, J., and Megie, G. (1990) Multiwavelength lidar for ozone measurements in the troposphere and the lower stratosphere, *Appl. Opt.*, 29, 467-476.

Papayannis, A., Alpers, M., Balis, D., Bösenberg, J., Chaikovsky, A., de Tomasi, F., Haagaard, A., Matthias, V., Mattis, I., Mitev, V., Nickovic, S., Pappalardo, G., Pelon, J., Perez, C., Pisani, G., Puchalski, S., Stoyanov, D., Rizi, V., Sauvage, L., Simeonov, V., Trickl, T., Vaughan, G., Wiegner, M., and Castahno, A.D. (2004a) Saharan dust outbreaks towards Europe: 3 years of systematic observations by the european lidar network in the frame of the EARLINET project (2000-2003), in Pappalardo, G., and Amodeo, A. (eds), *Proc. 22nd Intern. Laser Radar Conference*, 845-848, Matera, Italy.

Papayannis, A., Tsaknakis, G., Chourdakis, G., and Georgousis, G. (2004b) Simultaneous observations of tropospheric ozone and aerosol vertical profiles over Athens using combined LIDAR/DIAL systems, vol. II, in Zerefos, C. (ed.), *Proc. Quadr. Ozone Symp.,* 901, Kos, Greece.

Prospero, J.M., and Carlson, T. (1972) Saharan air outbreaks over the tropical North Atlantic, *Pure Appl. Geophys.*, 119, 678-691.

Prospero, J.M., Ginoux, P., Torres, O., Nicholson, S.E., and Gill, T.E. (2002) Environmental characterization of global sources of atmospheric soil dust identified with the Nimbus 7 total ozone mapping spectrometer (TOMS) absorbing aerosol product, *Rev. Geophys.*, 40, 1002, doi:10.1029/2000RG000095.

Ramanathan, V., Crutzen, P.J., Kiehl, J.T., and Rosenfeld, D. (2001) Aerosols, climate, and the hydrological cycle, *Science* 294, 2119-2124.

Roelofs, G.J., Lelieveld, J., and van Dorland, R. (1997) A three-dimensional chemistry/general circulation model simulation of anthropogenically derived ozone in the troposphere and its radiative climate forcing, *J. Geophys. Res.*, 102, 23,389-23,401.

Samet, J., Zeger, S., Dominici, F., Curriero, F., Coursac, I., Dockery, D., Schwartz, J., and Zanobetti, A. (2000) National morbidity, mortality, and air pollution study. Part II. Morbidity, mortality and air pollution in the United States, *Health Effects Institute Report*, May 2000.

Schoulepnikoff, S., Van Den Bergh, H., Calpini, B., and Mitev, V. (1998) Tropospheric, air pollution monitoring, lidar, in Meyers, R.A. (ed.) *Encyclopedia of Environmental Analysis and Remediation*, 8, John Wiley, New York, 4873-4909.

Schwartz, J., Dochery, D.W., and Neas, L.M. (1996) Is daily mortality associated specifically with fine particles? *J. Air Waste Manag. Assoc.* 46, 927-939.

Seinfeld, J.H., and Pandis, S.N. (1998) *Atmospheric chemistry and physics: from air pollution to climate change*, J. Wiley and Sons Inc., NewYork, USA.

Seinfeld, J.H., Carmichael, G., Arimoto, R., Conant, W.C., Brechtel, F.J., Bates, T.S., Cahill, T.A., Clarke, A.D., Doherty, S.J., Flatau, P.J., Huebert, B.J., Kim, J., Markowicz, K.M., Quinn, P.K., Russell, L.M., Russell, P.B., Shimizu, A., Shinozuka, Y., Song, C.H., Tang, Y.H., Uno, I., Vogelmann, A.M., Weber, R.J., Woo, J.H., and Zhang, X.Y. (2004) ACE-ASIA, Regional climatic and atmospheric chemical effects of Asian dust and pollution, *Bull. Amer. Meteor. Soc.*, 85, 367-380.

Soriano, C., Baldasano, J., Buttler, W., and Moore, K. (2001) Circulatory patterns of air pollutants within Barcelona air basin in a summertime situation: lidar and numerical approach, *Bound. Layer Meteor.* 98, 33-55.

Spinhirne, J., Welton, J., Palm, S., Hart, W., Hlavka, D., Mahesh, A., and Lancaster, R. (2004) The GLAS polar orbiting lidar experiment: first year results and available data, in Pappalardo G. and Amodeo A. (eds.), *Proc. 22$^{nd}$ Intern. Laser Radar Conference,* Matera, Italy, 949-952.

Thomasson, A. (2001) *Evaluation and calibration methods for lidar/DIAL measurements in ambient air*, Ph.D. Thesis, Univ. Claude Bernard Lyon I, Lyon, France.

Touloumi, G., Katsouyanni, K., Zmirou, D., Schwartz, J., Spix, C., de Leon, AP. Tobias, A., Quennel, P., Rabczenko, D., Bacharova, L., Bisanti, L., Vonk, JM., and Ponka, A. (1997) Short-term effects of ambient oxidant exposure on mortality: a combined analysis within the APHEA project. Air pollution and health: a European approach, *Amer. J. Epidem.*, 146, 177-185.

Vautard, R., Beekmann, M., Roux, J., and Gombert, D. (2001) Validation of a hybrid forecasting system for the ozone concentrations over the Paris area, *Atmos. Environ.*, 35, 2449-2461.

Winker, D., Hostetler, C., and Hunt, W. (2004) CALIOP: the Calispo Lidar, *Proceed. 22$^{nd}$ Intern. Laser Radar Conference (ILRC 2004)*, Matera, Italy, 941-944.

Zwozdziak, J., Zwozdziak, A., Sowka, I., Ernst, K., Stacewicz, T., Szymanski, A., Chudzynski, S., Czyzewski, A., Skubiszak, W., and Stelmaszczyk, K. (2001) Some results on the ozone vertical distribution in atmospheric boundary layer from LIDAR and surface measurements over the Kamienczyk Valley, Poland, *Atmos. Res.*, 58, 55-70.

# HEALTH IMPACT AND CONTROL OF PARTICLE MATTER

L.W. OLSON AND K. BOISON
*Arizona State University*
*Technology Center*
*85212 Mesa, Arizona, USA*

## Summary

Particulate matter (PM), unlike other air pollutants such as carbon monoxide or ozone, is not a single compound. Particles can be either solids or liquids and can vary greatly in size, shape, and chemical composition. The effects of particulate aerosols on human health include damage to the cardiovascular system, aggravation of existing respiratory diseases, and increases in lung cancer mortality. Evidence of increased mortality from exposure to the types of particulate matter associated with traffic, including diesel exhaust, has been established. Children, the elderly, and those with pre-existing conditions such as asthma are particularly susceptible. Regulatory actions initially focused on larger particles, but increasingly they have targeted the smaller $PM_{10}$ and $PM_{2.5}$ fractions. Control strategies for particulate matter must be targeted to a given size fraction and take into account the various origins, both natural and man-made, for each type of particle.

## 1. Introduction

Particulate matter (PM) is part of the natural environment. It can arise from sources such as sea spray, wind blown dust, volcanic emissions, fungal spores or pollen. Anthropogenic activities also contribute to the atmospheric load of particulate matter through activities such as traffic, mining, construction, agriculture, or power plant emissions. Until fairly recently, the health effects of particulates on humans had not been studied to the same extent as other air pollutants and little attention had been paid specifically to the smaller size particulates (Commission of the European Communities, 2001). However, both toxicological and epidemiological studies have now demonstrated the negative consequences to human health of exposure to particulate matter, even at low concentrations and without apparent threshold. Inhalation of particulate matter has been linked to pulmonary inflammation and injury, cardiovascular impairment, pulmonary hypertension, and increased rates of death. Those at highest risk are the elderly, children, and those with respiratory

*P. Nicolopoulou-Stamati et al. (eds), Environmental Health Impacts of Transport and Mobility, 115-125.*

ailments or cardiovascular disease (US EPA, 1996). It is now apparent that regulating and controlling human exposure to particulate matter is essential if the health impacts of air pollution are to be minimized.

This paper will describe different types of airborne particulate matter, their sources and properties, absorbance mechanisms for particles, and evidence of adverse health effects of PM on humans from toxicological and epidemiological studies. The regulatory status for measuring and controlling particulate matter in both Europe and the United States will also be discussed.

## 2.    Classification of particulate matter

The term particulate matter seems to imply a solid, but actually PM can consist of liquid droplets, solids, or solid cores surrounded by liquid. Size is the critical factor for airborne particulate matter since the particles must be small enough to remain suspended in air for a period of days, weeks, or months. Typically, this means they must be less than about 45 μm in diameter. The shapes of the particles and their chemical composition can vary greatly, as can their source (US EPA, 2003a).

Particles can be classified in a number of ways. Primary particulate matter exists in the same chemical form as when it was generated, whereas secondary PM involves condensates of gases or liquids onto the surface of a particle, the coagulation of smaller precursor particles to form larger particles, or oxidation and subsequent reaction of molecules such as sulfur dioxide, nitrogen oxides, or volatile organic compounds (US EPA, 2003a). In Europe, suspended salt particles from evaporation of sea spray, suspended soil particles, and in Mediterranean regions Saharan dust and volcanic emissions are major natural sources of primary particulate matter (Blanchard and Woodcock, 1980; Löye-Pilot *et al.*, 1986; Nicholson, 1988; Haulet *et al.*, 1977; Malinconico, L.L., 1979).

Anthropogenic contributions tend to be more important for secondary particulate matter, especially in urban areas. Photochemical smog, for example, produces a complex soup of organic molecules, few of which were emitted directly from tailpipes. These organic chemicals can be absorbed onto what otherwise might be fairly innocuous naturally occurring dust particulates in the ambient air to create a much more toxic material. Organic compounds have been found to constitute from 10-60% of the dry mass of some particles (Turpin, 1999). Over 18,000 different organic molecules have been found adsorbed onto diesel particles (Salvi and Holgate, 1999). Thus, air pollution from traffic is a major contributor to secondary PM.

Particulate matter can also be categorized by size. Airborne PM can vary in size over five orders of magnitude, from 1 nm to 100 μm (US EPA, 2003a). Accurately determining the distribution of particulate matter over such a range poses significant technical problems and yet is essential for understanding epidemiological data and devising control strategies.

Total suspended particles (TSP) is a term used to refer to ambient PM up to 25–45 µm in size, based upon sampling conditions. The first standards in the United States for airborne PM, promulgated in 1971, were for Total Suspended Particles. In Europe during that time, there were no universally applied air quality standards and so the regulatory situation with regards to particulate matter varied widely. A number of countries in Central and Eastern Europe, Latin America, and Asia still base particulate matter standards on TSP (World Bank, 1998).

As evidence from toxicology and epidemiology studies began to accumulate that linked exposure to smaller particles to human health effects, attention began to shift to size specific fractions of particulate matter. In 1987, the U.S. EPA revised the National Ambient Air Quality Standards (NAAQS) to control $PM_{10}$ rather than TSP. $PM_{10}$ means particulate matter equal to or less than 10 µm in size. Within the European Union, a two stage process for controlling $PM_{10}$ went into effect in 1999 (EU, 1999). Member states are expected to meet 24 hour and annual limits for $PM_{10}$ by 2005 and stricter limits by 2010. There is a limited amount of reliable data on $PM_{10}$ in Europe, and no coherent overall European $PM_{10}$ data set, because $PM_{10}$ has been systematically measured in urban environments only since 1990, and then only in a few member states. Available data suggests that present $PM_{10}$ values exceed the recommended limit values in the majority of the member states (Technical Working Group, 1997).

In 1997, the U.S. EPA promulgated a proposed national $PM_{2.5}$ standard (Federal Register, 1997). After industry appeals and a tortuous path through the courts, including the Supreme Court, the standard has been finalized. State Implementation Plans are due by 2007, and compliance will be required by 2009-2014. $PM_{2.5}$ and even $PM_1$ standards are being considered in Europe as well, but such regulations have not yet been promulgated (Euro-Case, 2001). Control strategies for reducing $PM_{2.5}$ will necessarily be different from those that are effective for larger particles and must be based upon knowledge of sources and distribution patterns, as well as effective measuring and modeling technologies.

The terms coarse, fine, and ultra-fine refer to various fractions of $PM_{10}$. The $PM_{2.5}$ fraction (particles $\leq$ 2.5 µm) is called fine. The coarse fraction refers to inhalable $PM_{10}$ particles that remain when the fine fraction ($PM_{2.5}$) is removed.  In other words, coarse refers to the $PM_{2.5-10}$ fraction. Ultra-fine particles are less than 0.1 µm in size (US EPA, 2003a). There are currently no air pollution regulations that focus on ultra-fines.

Diesel vehicles produce 100-150 times more particulate matter than gasoline powered vehicles and diesel particulate matter (DPM) accounts for much of the particle load in many urban areas (Salvi and Holgate, 1999). The size distribution of DPM collected directly from a car exhaust showed three log-normal modes centered at 0.09, 0.2 and 0.7-1 µm of particle aerodynamic diameter (EAD) (Kerminen, 1997). Over 80% of the particles were found in the mode around 0.1 µm, while the largest mode contained 10% of the total particulate mass, but only 0.1% of the particles. There is evidence that DPM can coagulate as it ages and range in size up to 2.5 μm (US EPA, 2000). Thus, transportation related sources represent a major source of fine and ultra-fine particles.

Not only is there a size difference in coarse and fine PM, but the chemical composition is usually different. Mechanical processes involving break-up of larger particles or bulk material produce mainly coarse particles, with similar chemical composition to the bulk material. Thus, coarse particles can many times be traced to a given source. However, the fine fraction can contain material from many different sources because $PM_{2.5}$ is not formed by mechanical means but by combustion processes, nucleation of gas phase species, coagulation, or condensation of gases on particles (US EPA, 2003a). Particulate organic matter, especially, is poorly characterized, and can be composed of hundreds, even thousands, of different compounds (Saxena and Hildemann, 1996; Weisenberger, 1984).

Fine particles can have atmospheric lifetimes of days to weeks and can travel thousands of kilometers. They are more uniformly distributed in the atmosphere and are therefore harder to trace to their source. Coarse particles have shorter lifetimes (minutes to hours) and travel less than 10 km (US EPA, 2003a). Their effects are more localized near their source.

### 3.    Deposition and retention of particulate matter

Health effects of PM may be related to other factors than size, but that is the criteria on which current regulations are based. To understand health effects, it is not enough to simply measure the concentration and characteristics of particulate matter in the inspired air. An individual dose involves deposition of a particle in the lung which varies dramatically with size. The clearance rate of deposited particles must also be considered, and this depends upon both the initial site of deposition and physicochemical properties such as hygroscopic behavior and solubility in cellular fluids (US EPA, 2003b).

The respiratory tract is divided into three regions:
- extrathoracic (ET) consists of the nasal and oral passages through the larynx;
- tracheobronchial (TB) consists of the trachea, bronchi, and conducting bronchioles;
- alveolar (A) consists of respiratory bronchioles and alveoli.

The importance of PM size is demonstrated by varying penetration and deposition rates in the respiratory system. For particles larger than 1 μm, deposition increases with size because the process is controlled by impaction and gravitational sedimentation. Above 10 μm almost all inhaled particles are deposited in the upper respiratory tract. For particles smaller than 1 μm, deposition is controlled by diffusion processes that involve random bombardment by air molecules resulting in collisions with airway surfaces. In this region, decreasing particle size actually results in an increase in deposition (Frampton, 2002; Jacques and Kim, 2000). Thus, a significant fraction of both ultra-fine and coarse particles are deposited in the extrathoracic region. The highest deposition in the tracheobronchial region occurs for particulate matter from 5-10 μm. Deposition in the deepest regions of the lungs, the alveolar region, is greatest for the 2.5-5 μm fraction. The lowest overall respiratory tract deposition rates (as low as 10-20%) are for particles from 0.2 to 1.0 μm in size because neither sedimentation nor diffusion mechanisms are very effective for these particles (US EPA, 2003b).

Retained particles, those that are not cleared by macrophages or the mucociliary system, tend to be small (< 2.5 μm) and water insoluble. Ultra-fines can be rapidly cleared into systemic circulation and transported out of the lungs. But if the particle is deposited in the alveolar region, clearance can take months to years (US EPA, 1996; Schlesinger et al., 1997).

Airway structure and function varies with age and health. Such variations can affect deposition. Studies have shown that fractional deposition of aerosols in the ET and TB regions of the lungs is greater in children than in adults (Musante and Martonen, 1999; Oldham et al., 1997). Musante and Martonen developed a computer model that takes into account differences in airway dimensions, geometry of branching airway networks, and breathing patterns as a function of age. The model predicts a deposition rate of 73% for a 7 month old inhaling 2 μm particles during heavy breathing vs 38% for an adult (Musante and Martonen, 2000).

In a study of adults from 18 to 80 with normal lung function, Bennett et al. (1996) found that total respiratory tract deposition of 2 μm particulates was dependent on breathing pattern and airway resistance, but was independent of age. However, the elderly are more susceptible to respiratory tract diseases and these are correlated with greater total lung deposition. Increasing degrees of airway obstruction, such as in people with asthma or chronic obstructive pulmonary disease (COPD) is correlated with a marked increase in deposition (US EPA, 1996; Bennett et al., 1997; Kim and Kang, 1997). The deposition site is also affected by a disease state. For example, Brown et al. (2001) found that patients with cystic fibrosis, who have poorly ventilated lungs, deposited coarse particles (5 μm) preferentially in the TB region compared to healthy individuals. A mathematical model has been developed that successfully predicts PM deposition patterns for COPD patients by comparison with controlled breathing trials from patients with both chronic bronchitis and emphysema. The model indicates that airway obstructions are the main cause for increased depositions in the COPD lung (Segal et al., 2002).

## 4.    Toxicological effects of particulate matter

Toxicological studies on both humans and animals have been conducted to determine the effect of particulate matter on human health. Some particles have inherent toxicity, such as silica, asbestos, or coal dust. In other cases, it is the adsorbed material (e.g. sulfates, nitrates, organics, or metals) on the surface of the particulate that may represent the toxic component. In this latter situation, there is more difficulty in determining the toxicity of particulates in animal studies, since one must duplicate the exact chemical nature of the particulate found in ambient air and this is virtually impossible to do in the laboratory. Other limitations include the fact that animals have different inhalation mechanisms than humans and that typical studies have used much higher concentrations than normal ambient air exposures in order to use fewer animals (US EPA, 2003b).

The mechanism by which particulate matter causes toxic effects is not well understood. Acid aerosols have been shown to have a significantly greater effect on children with allergies or asthma (Linn *et al.*, 1997). Diesel particulate matter in particular seems to exacerbate allergic asthmatic responses (Salvi and Holgate, 1999). DPM is composed of soot, volatile hydrocarbons, and sulfates from fuel sulfur and contains more fine particulates than gasoline. The organic fraction of DPM has been linked to effects on the immune system (US EPA, 2000; US EPA, 2003b). Typically, the organic constituents of PM are poorly characterized, heterogeneous complex mixtures and it is not yet known whether particles from other combustion sources have effects similar to DPM.

Combustion particles that have a high content of soluble metals have been shown to cause lung injury with the soluble metal portion appearing to be a primary determinant. (Costa and Dreher, 1997). How much these effects depend upon the specific identity of the transition metal has yet to be completely elaborated.

Studies with ultrafine particulate matter of varying composition have shown significantly greater inflammatory response at the same dose and chemical composition when compared to fine PM. This is possibly due to greater surface area per unit mass in the ultrafine particles (Oberdorster *et al.*, 1992; Li *et al.*, 1996, 1997, 1999). Low grade inflammation caused by deposition of ultra-fine particles in the alveoli that leads to increased coagulatability may be part of the pathological mechanism linking cardiovascular mortality and morbidity (Seaton *et al.*, 1995). However, there is still comparatively limited toxicological data on which to differentiate the effects of ultra-fines from those of other size fractions.

## 5.    Epidemiological studies on health effects of particulate matter

Many recent studies have suggested that exposure to airborne particulate matter is associated with life shortening (Brunekreef and Holgate, 2002). Both U.S. cohort studies (Dockery *et al.*, 1993; Pope *et al.*, 1995, 2002) and European APHEA (Air Pollution and Health: a European Approach) studies (Katsouyanni, 2001) have shown an increase in daily mortality rates for all-cause mortality, as well as cardiopulmonary and lung cancer mortality, associated with increases in particulate air pollution. In the APHEA2 study from the early to mid 1990s, data from 21 cities showed that all-cause daily mortality increased by 0.6% (95% CI 0.4-0.8) for each 10 $\mu g/m^3$ increase in $PM_{10}$. This compared favorably to data collected from the NMMAPS (National Mortality, Morbidity and Air Pollution Studies) in the U.S.

where the all-cause mortality for each 10 μg/m$^3$ of PM$_{10}$ was found to be 0.5% (0.1-0.9) (Brunekreef and Holgate, 2002). No evidence was found in the NMMAPS data for a PM$_{10}$ threshold for all-cause or cardiopulmonary mortality (Daniels, *et al.*, 2000).

Fine particulate air pollution has been correlated with increased mortality rates. Analysis of the Harvard Six Cities study showed PM$_{2.5}$ to be significantly associated with excess mortality across all six cities, but not PM$_{10-2.5}$ (Klemm *et al.*, 2000). Other studies have not been able to distinguish between the effects of PM$_{2.5}$ and PM$_{10-2.5}$ (Lipfert *et al.*, 2000; Mar *et al.*, 2000). In a national study involving all 50 states, Pope found an increase of approximately 15 to 17% in adjusted all-cause mortality rates between the most polluted and least polluted cities, based on sulfate and PM$_{2.5}$ concentrations (Pope *et al.*, 1995). Fine particles have been associated specifically with increased risks of cardiopulmonary and lung cancer mortality, but not with other causes of death considered together (Dockery *et al.*, 1993). Each 10μg/m$^3$ increase in long-term average PM$_{2.5}$ ambient concentration was found to increase the all-cause mortality by 4%, cardiopulmonary mortality by 6%, and lung cancer mortality by 8%. More coarse particles and other gaseous pollutants, except for sulfur dioxide, were not significantly associated with increased mortality risks (Pope *et al.*, 2002).

Most cohort studies of the effects of air pollution have used area measures of particulate matter concentrations to determine individual exposures. In a specific effort to investigate the effect of traffic-related air pollution, Hoek and co-workers have used an on-going NLCS (Netherlands Cohort Study on Diet and Cancer) to determine the effects of black smoke and nitrogen dioxide on mortality rates (Hoek *et al.*, 2002). Using a geographic information system, individuals living within 100 m of a major freeway or 50 m of a major urban road were identified as being exposed. No association was seen between any exposures and non-cardiopulmonary or non-lung cancer mortality rates. But cardiopulmonary mortality for those living near a major road increased (relative risk 1.95; 95% CI 1.09-3.52). The highest effect was seen for individuals living for ten or more years near such roads.

## 6.    Control strategies

Developing effective control strategies for particulate matter requires an understanding of the type and sources of PM to be targeted. TSP or the coarse fractions of PM$_{10}$ can originate from a variety of natural sources, providing a high background level of particles in certain areas. Anthropogenic contributions to this fraction come from agriculture, mining, construction, road dust, residential wood burning and other such sources. Regulatory control of particulates from these activities has typically lagged behind industrial source control. However, best practices in agriculture and construction do exist and can be very effective in reducing coarse fractions of ambient PM. In some parts of the U.S. that are non-attainment for PM$_{10}$, very stringent air pollution regulations are now in place that require PM$_{10}$ road sweepers, pavement of virtually all roads and alleys, strict control of parking in non-paved lots, dust control measures for all construction projects, and track-out dust control for construction vehicles. Agriculture has largely escaped regulatory control until now, but is increasingly under pressure to conform to dust control measures.

Fine particles, $PM_{2.5}$, are more likely to originate from combustion sources. Effective control strategies can include cleaner burning fuels, such as replacing coal with natural gas or using newer formulations for diesel with reduced sulfur content. Industrial point sources can also be targeted, for example with bag houses or electrostatic precipitators as mandated control devices. Coordinated traffic signal systems, reduction in the use of highly polluting two stroke engines, and use of alternative fuel vehicles can also contribute to reducing anthropogenic contributions.

An example of the effects of such strategies is seen in the decision to ban the sale of bituminous coal within the city of Dublin, Ireland in 1990. Clancy and co-workers compared air pollution, weather, and deaths for 72 months before and after the ban. Black smoke and sulfur dioxide concentrations were measured from six residential monitoring stations. Overall, mean black smoke concentrations dropped by about two-thirds after the ban and sulfur dioxide dropped by one-third. Age standardized, non-trauma death rates in Dublin decreased by about 5.7% in the 72 months after the ban compared with the 72 months before the ban after adjustments for population changes, weather, respiratory epidemics, and secular changes in death rates in the rest of Ireland. The largest decrease was in respiratory death rates (15.5%) and average annual cardiovascular death rates decreased by 10.3%. In this case study, control of particulate air pollution lead to an immediate reduction in cardiovascular and respiratory deaths (Clancy, 2002).

## 7.   Conclusions

Airborne particulate matter can be divided by size into coarse, fine, and ultra-fine fractions. Each category can have both natural and man-made sources, but the fine fraction of PM is predominately from combustion sources and so the extraordinary growth in traffic and industrialization in many parts of the world has greatly increased levels of this type of particulate matter. Deposition of PM in the respiratory system depends upon its size and the health of the target individual. Children and those with respiratory diseases have a higher rate of deposition. The toxicology of exposure to particulate matter has only recently begun to be examined, but both cardiovascular and respiratory effects have been identified. Prospective cohort studies have established a clear link between exposure to airborne particles, especially the fine fraction, and increases in cardiopulmonary mortality and lung cancer mortality. In some cases, other common air pollutants have not been associated with these increased risks. The regulatory status of controlling airborne PM is actively evolving and is being driven, in part, by increasing evidence from epidemiological and toxicological studies of adverse health effects in humans from exposure to particulates.

## References

Bennett, W.D., Zeman, K.L., and Kim, C. (1996) Variability of fine particle deposition in healthy adults: effect of age and gender, *American Journal Respiratory Critical Care Medicine* 153, 1641-1647.

Bennett, W.D., Zeman, K.L., Kim, C., and Mascarella, J. (1997) Enhanced deposition of fine particles in COPD patients spontaneously breathing at rest, *Inhalation Toxicology* 9, 1-14.

Blanchard, D.C., and Woodcock, A.H. (1980) The production, concentration and vertical distribution of the sea-salt aerosol, *Annuals New York Academy Sciences* 338, 330-347.

Brown, J.S., Kirby, Z.L., and Bennett, W.D. (2001) Regional deposition of coarse particles and ventilation distribution in healthy subjects and patients with cystic fibrosis, *Journal Aerosol Medicine* 14, 443-454.

Brunekreef, B. and Holgate, S.T (2002) Air pollution and health, *Lancet* 360, 1233-1242.

Clancy, L., Goodman, P., Sinclair, H. and Dockery, D.W. (2002) Effect of air-pollution control on death rates in Dublin, Ireland: an intervention study, *Lancet* 360 1210-1214.

Commission of the European Communities (2001) The Clean Air for Europe (CAFÉ) programme: towards a thematic strategy for air quality, COM(2001) 245 final.

Costa, D.L. and Dreher, K.L. (1997) Bioavailable transition metals in particulate matter mediate cardiopulmonary injury in healthy and compromised animal models, in Driscoll, K.E., and Oberdörster, G. (eds.), *Proceedings of the sixth international meeting on the toxicology of natural and man-made fibrous and non-fibrous particles*, September 1996, Lake Placid, NY, *Environmental Health Perspectives Supplementl* 105(5), 1053-1060.

Daniels, M.J., Dominici, F., Samet, J.M., and Zeger, S.L. (2000) Estimating particulate matter mortality dose-response curves and threshold levels: an analysis of daily time-series for the 20 largest U.S. cities, *American Journal of Epidemiology* 152, 397-406.

Dockery, D.W., Pope, C.A., Xiping, X., Spengler, J.D., Ware, J.H., Fay, M.E., Ferris, B.G., and Speizer, F.E. (1993) An association between air pollution and mortality in six U.S. cities, *New England Journal of Medicine* 329, 1753-1759.

EU (1999) EU Daughter Directive 1999/30/EC.

Euro-Case (2001) European Council of Applied Sciences and Engineering, The health impact of urban pollution: present needs and future challenges,
Downloaded 15-08-03 from: http://www.euro-case.org/Activities/
workshop_Air_010426/Final%20report.pdf

Federal Register (1997) National ambient air quality standards for particulate matter, final rule, F. R. (July 18) 62: 38,652-38,752.

Frampton, M.W., Utell, M.J., Beckett, W., Oberdörster, G., Morrow, P., Zareba, W., and Cox, C. (2002) Ultrafine particles: characterization, health effects, and pathophysiological mechanisms, University of Rochester School of Medicine & Dentistry EPA Particulate Matter Center, Downloaded 15-08-03 from:
http://www2.envmed.rochester.edu/envmed/PMC/anrep02/rep02_C3.html

Haulet, R., Zettwoog, P., and Sabroux, J.C. (1977) Sulphur dioxide discharge from Mount Etna, *Nature* 268, 715-717.

Hoek, G., Brunekreef, B., Goldbohm, S., Fischer, P., and van den Brandt, P.A. (2002) Association between mortality and indicators of traffic-related air pollution in the Netherlands: a cohort study, *Lancet* 360, 1203-1209.

Jaques, P.A., and Kim, C.S., (2000) Measurement of total lung deposition of inhaled ultrafine particles in healthy men and women, *Inhalation Toxicology* 12, 715-731.

Katsouyanni, K., Touloumi G., Gryparis, A., Le Tertre, A., Monopolis, Y., Rossi, G., Zmirou D., Ballester, F., Boumghar, A., Anderson, H.R., Wojtyniak, B., Paldy, A., Braunstein, R., Pekkanen, J., Schindler, C., and Schwartz, J. (2001) Confounding and effect modification in the short-term effects of ambient particles on total mortality: results from 29 European cities within the APHEA2 project, *Epidemiology*, 12, 521-531.

Kerminen, V.M., Mäkelä, T.E., Ojansen, C.H., and Hillamo, R.E. (1997) Characterization of the particulate phase in the exhaust from a diesel car, *Environ. Sci. Technol.* 31, 1883-1889.

Kim, C.S. and Kang, T.C. (1997) Comparative measurement of lung deposition of inhaled fine particles in normal subjects and patients with obstructive airway disease, *American Journal Respiratory Critical Care Medicine* 155, 899-905.

Klemm, R.J., Mason, R.M., Heilig, C.M., Neas, L.M., and Dockery, D.W. (2000) Is daily mortality associated specifically with fine particles? Data reconstruction and replication of analysis, *Journal of Air and Waste Management* 50, 1215-1222.

Li, X.Y., Gilmour, P.S., Donaldson, K., and MacNee, W. (1996) Free radical activity and pro-inflammatory effects of particulate air pollution ($PM_{10}$) *in vivo* and *in vitro*, *Thorax* 51, 1216-1222.

Li, X.Y., Gilmour, P.S., Donaldson, K., and MacNee, W. (1997) *In vivo* and *in vitro* proinflammatory effects of particulate air pollution ($PM_{10}$), in K.E. Driscoll and G. Oberdörster (eds.), *Proceedings of the sixth international meeting on the toxicology of natural and man-made fibrous and non-fibrous particles*, September 1996, Lake Placid, NY, *Environmental Health Perspectives Supplement* 105(5), 1279-1283.

Li, X.Y., Brown, D., Smith, S., MacNee, W., and Donaldson, K. (1999) Short-term inflammatory responses following intratracheal instillation of fine and ultrafine carbon black in rats, *Inhalation Toxicology* 11, 709-731.

Linn, W.S., Gong, H., Jr., Shamoo, D.A., Anderson, K.R., and Avol, E.L. (1997) Chamber exposures of children to mixed ozone, sulfur dioxide, and sulfuric acid, *Archives Environmental Health* 52, 179-187.

Lipfert, F.W, Morris, S.C., and Wyzga, R.E. (2000) Daily mortality in the Philadelphia metropolitan area and size-classified particulate matter, *Journal of the Air and Waste Management Association* 50(8), 1501-1513.

Löye-Pilot, M.D., Martin, J.M., and Morelli, J. (1986) Influence of Saharan dust on the rain acidity and atmospheric input to the Mediterranean, *Nature* 321, 427-428.

Malinconico, L.L. (1979) Fluctuations in $SO_2$ emission during recent eruptions of Etna, *Nature* 278, 43-45.

Mar, T.F., Norris, G.A., Koenig, J.Q., and Larson, T.V. (2000) Associations between air pollution and mortality in Phoenix, 1995-1997, *Environmental Health Perspectives*, 108, 347-353.

Musante, C.J., and Martonen, T.B. (1999) Predicted deposition patterns of ambient particulate air pollutants in children's lungs under resting conditions, in Proceedings of the third colloquium on particulate air pollution and human health; June; Durham, NC. Irvine, CA: University of California, Air Pollution Health Effects Laboratory, 7-15 – 7-20.

Musante, C.J., and Martonen, T.B. (2000) Computer simulations of particle deposition in the developing human lung, *Journal Air Waste Management Association* 50, 1426-1432.

Nicholson, K.W. (1988) A review of particle resuspension, *Atmospheric Environment* 22, 2639-2651.

Oberdörster, G., Ferin, J., Gelein, R., Soderholm, S.C., and Finkelstein, J. (1992) Role of the alveolar macrophage in lung injury: studies with ultrafine particles, *Environmental Health Perspectives* 97, 193-199.

Oldham, M. J., Mannix, R. C., and Phalen, R. F. (1997) Deposition of monodisperse particles in hollow models representing adult and child-size tracheobronchial airways, *Health Physics* 72, 827-834.

Pope, C.A., Thun, M.J., Namboodiri, M.M., Dockery, D.W., Evans, J.S., Speizer, F.E., and Health, C.W. (1995) Particulate air pollution as a predictor of mortality in a prospective study of U.S. adults, *American Journal of Respiratory and Critical Care Medicine* 151, 669-674.

Pope, C.A., Burnett, R.T., Thun, M.J., Calle, E.E., Krewski, D., Ito, K., and Thurston, G.D. (2002) Lung cancer, cardiopulmonary mortality, and long-term exposure to fine particulate air pollution, *Journal of the American Medical Association* 287, 1132-1141.

Salvi. S., and Holgate, S.T. (1999) Mechanisms of particulate matter toxicity, *Clinical and Experimental Allergy* 20, 1187-1194.

Saxena, P., and Hildemann, L. M. (1996) Water-soluble organics in atmospheric particles: a critical review of the literature and applications of thermodynamics to identify candidate compounds, *Journal Atmospheric Chemistry* 24, 57-109.

Schlesinger, R.B., Ben-Jebria, A., Dahl, A.R., Snipes, M.B., and Ultman, J. (1997) Disposition of inhaled toxicants, in: E.J. Massaro (ed.), *Handbook of Human Toxicology* $2^{nd}$ *ed.,* Taylor & Francis, Washington, DC, 191-224.

Seaton, A., MacNee, W., Donaldson, K., and Godden, D. (1995) Particulate air pollution and acute health effects, *Lancet* 349, 1582-1587.

Segal, R. A., Martonen, T. B., Kim, C. S., and Shearer, M. (2002) Computer simulations of particle deposition in the lungs of chronic obstructive pulmonary disease patients, *Inhalation Toxicology* 14, 705-720.

Technical Working Group on Particles (1997) Ambient air pollution by particulate matter: Position Paper,
Downloaded 15-08-03 from: http://europa.eu.int/comm/environment/air/pdf/ pp_pm.pdf

Turpin, B. J. (1999) Options for characterizing organic particulate matter, *Environmental Science Technology* 33, 76A-79A.

U.S. EPA (1996) Air quality criteria for particulate matter, Research Triangle Park, NC: National Center for Environmental Assessment-RTP Office, report no. EPA/600/P-95/001aF-cF.3v.

U.S. EPA (2000) Health assessment document for diesel exhaust [SAB review draft], report no. EPA/600/8-90/057E, Office of Research and Development, National Center for Environmental Assessment-Washington Office, Washington DC,
Downloaded 15/08/03: http://cfpub.epa.gov/ncea/cfm/nceahome.cfm.

US EPA (2003a) Fourth external review draft of air quality criteria for particulate matter, Vol 1, EPA/600/P-99/002-aD, Downloaded 26/1/05 from:
http://cfpub.epa.gov/ncea/cfm/recordisplay.cfm?deid=58003

US EPA (2003b) Fourth external review draft of air quality criteria for particulate matter, Vol 2, EPA/600/P-99/002-aD, Downloaded 26/1/05 from:
http://cfpub.epa.gov/ncea/cfm/recordisplay.cfm?deid=58003

Weisenberger, B. (1984) Health effects of diesel emissions. An update, *J. Soc. Occup. Med.* 47, 90-92.

World Bank (1998) Airborne particulate matter, *Pollution Prevention and Abatement Handbook*, effective July 1998.

# ENDOCRINE PROBLEMS RELATED TO MOBILITY

S. LIVADAS, D. KALTSAS AND G. TOLIS
*Endocrine Unit, Hippocrateion Hospital*
*Vas Sofias 108, Athens, GREECE*

**Summary**

Human mobility is increasing day by day and maybe has several adverse effects on health, through energetic or passive mechanisms. As mobility we can define the movement from one place to another and in this paper we will discuss health impacts, especially on the endocrine axis, related to mobility.

A very complex network, whose main elements are the neural and endocrine system, govern the human body. Elegant interaction mediated through centres in hypothalamus, cortex, amygdala, hippocampus, basal ganglia and endocrine glands, mainly pituitary, regulates response of the body to any stimuli. We should keep in mind that all stimuli, including the arts, have an impact on human physiology and on health outcomes ranging from illness to health, from injury to peak performance.

Mobility produces changes to several axes and rhythms of the human functions and man needs time to adapt to the new situations and reorganize the set points. The most characteristic paradigm of such an alteration is jetlag and the well-known consequences on the human body. These changes are much more evident and serious, if mobilization is happening against a man's will, i.e. displacement due to a war. In addition we should consider alterations of homeostasis, as a result of mobility or perhaps an epiphenomenon. For example, exposure everyday to traffic or industrial noise affects many aspects of human health. We are going to discuss endocrine aspects of mobility based on the following three patterns: jetlag, violent displacement and exposure to noise.

## 1. Introduction

In the beginning before analyzing the special effects of violent mobilization to hormone homeostasis, a definition of stress and the observed metabolic-endocrine functions has to be given. Hans Selye defines stress as "the nonspecific response of the body to any demand." A nonspecific response is the body's total biological

*P. Nicolopoulou-Stamati et al. (eds), Environmental Health Impacts of Transport and Mobility*, 127-134.
© 2005 *Springer. Printed in the Netherlands.*

reaction (such as sweating or increased heart rate), which may vary in intensity but not in type. A stressor is any physical or psychological event or condition that triggers stress and stress response consists of the emotional and physical reactions. So, as stress we consider the state that accompanies the stress response. An individual's response at a stressful event depends on both genetic and developmental factors.

A person's adaptation to stress occurs in three stages: alarm, resistance and exhaustion. During the phase of alarm response is regulated by the autonomic nervous system and the physical changes are known as the fight-or-flight reaction (Tolis *et al.*, 1973). The release of cortisol, epinephrine and norepinephrine from the adrenals result in increased heart rate, breathing and blood pressure, release of extra sugar from the liver for energy, halted digestion, increased perspiration to cool the skin, release of endorphins to relieve pain and increased blood cell production (Chrousos, 2000). These processes are initiated through intracellular receptors for steroid hormones, plasma-membrane receptors, and second-messenger systems for catecholamines. Cross talk between catecholamines and glucocorticoid-receptor signalling systems can occur (McEwen, 1998).

At the phase of resistance the body resists dramatic changes and once a stressor recedes, the parasympathetic branch of the autonomic nervous system stops the alarm reaction and restores the body to its normal state. Due to a great loss of energy during these two phases the body proceeds to the third phase, exhaustion. Exhaustion is a life-threatening form of physiological depletion characterized by distorted perceptions and disorganized thinking. A crucial observation is the close relationship between stressful life events and the onset of psychological and physical illnesses (Breslau, 2001). Data gathered from 300 studies showed that exposure to brief, contrived stressors was typically associated with adaptive up-regulation of immune function, in particular for natural killer cell activity. Only chronic stressors were associated with suppression of both cellular and humoral measures of the immune system (Segerstrom and Miller, 2004). These findings suggest that time-limited stressor exposure may actually serve to strengthen immune function and entail a vaccination-type effect.

There is considerable evidence in humans, derived largely from studies of adopted children, and identical and fraternal twins brought up together or apart, that a tendency toward anxiety and fear is a heritable trait. The tendency to develop fearful or anxious responses to the environment in general has a clear genetic component. It appears that a major portion of the genetic contribution to human fear and anxiety involves neurotransmitters and their receptors. Moreover, substances such as GABA, serotonin and their receptors play a key role. For example, variability in the receptors responsible for clearing serotonin from the synaptic space between two communicating neurons correlates quite well with variation in anxiety among different individuals. Even more, intriguing data arise from experiments on mice. Mice that lack a receptor in the brain for glucocorticoids are much less anxious than control mice. Other studies reported that genes related to biological clocks (a possible key role of melatonin secretion) are associated with fearfulness (Stam *et al.*,

1999). These observations hint at the existence of a close loop between stressors, body rhythms and response.

## 2. Jet-lag

At base line, hormones are secreted on circadian rhythm, according to the demands of life. For example peak cortisol levels are observed early in the morning, before waking up, in normal people and their nadir is observed during deep sleep at night, where the needs are minimal. The rhythm of secretion is totally different in people who work late at night, where maximal cortisol concentrations are observed at night (Sedgwick, 1998).

Biological oscillations with an endogenous period of near 24h (circadian rhythms) are generated by the master circadian pacemaker or clock located in the suprachiasmatic nuclei of the hypothalamus (Tolis et al., 1979). Fibers leave these nuclei to reach the ventrolateral preoptic area of the anterior hypothalamus, which is closely involved in sleep-wake control, temperature regulation and hormone secretion (Adam et al., 2000). Fibers also travel multisynaptically to the mesencephalic autonomic nuclei. From there information is distributed to every organ in the body via the sympathetic and parasympathetic systems and reaches the pineal gland to regulate the secretion of melatonin. Melatonin is a ubiquitous natural neurotransmitter-like compound produced primarily by the pineal gland. The role of endogenous melatonin in circadian rhythm disturbances and sleep disorders is well established. Thus two arms of the endogenous clock exist, one neural through the autonomic nervous system and another hormonal, via melatonin and other hormones, e.g. cortisol (Gronfier, 1998). A close, reciprocal relationship exists between the pineal and the pituitary/adrenal axis, the key element axis of stress response. Melatonin modulates the activity of this axis and the peripheral actions of corticoids.

Jet-lag refers to a variety of unpleasant symptoms that affect many individuals for several days after traveling through several time-zones. Typical symptoms are fatigue, daytime sleepiness, impaired alertness and trouble initiating and maintaining sleep. The severity of jet-lag is related to the number of time zones traveled, with the effects generally worse for eastward travel (Spitzer et al., 1999). Body clock is synchronized to recurring environmental signals conveyed by selective neural pathways and it takes several days for the external factors to shift the phase of the body clock from the time zone. All the rhythms are regulated by internal and external factors that interact (Pinter et al., 1979). Melatonin is supposed to play the major role in the pathogenesis of jetlag and attempts of therapeutic administration in order to avoid or minimize the symptoms of jetlag have given promising results, although the underlying mechanisms explaining the connection of melatonin disturbances and jetlag are unclear yet.

Melatonin appears to be somewhat beneficial, but there's much individual variation in response to this hormone. The biological rhythm disorganization caused by the rapid change of environment (and associated light/dark cues) apparently can be corrected by melatonin. The benefit is likely to be greater as more time zones are crossed and less for westward flights. However, melatonin taken before travel can actually worsen symptoms as opposed to the benefit of melatonin initiated immediately upon arrival (Herxheimer and Petrie 2003, Herxheimer and Waterhouse, 2003).

Another hormone with a possible role regulating adaptation of body to movement is thyrotropin (TSH). It is known that TSH secretion is modulated by sleep and circadian rhythmicity. It has been reported that jet-lag syndrome may be associated with a prolonged elevation of peripheral TSH levels, showing a dysrythmia of the hypothalamus-pituitary-thyroid axis (Hirschfeld *et al.*, 1996). The adaptation of this axis to any important stress man encompass, has been studied extensively and it is named sick euthyroid syndrome or low T3-T4 syndrome, reflecting the observed changes in thyroid function during a stressful situation (Tolis *et al.*, 1974).

## 3.    Violent displacement

Stress seems to be the key regulating human response to mobility. For example stress of travel preparation can contribute to jet-lag. Stress is greater when mobility is a result of violence and not of choice of the individual. For example, a war refugee is moving from one place to another, but in different situation, than the one reported above.

Pressures on the mental health of refugees are major, as the majority of parameters which affect the role of a person in society, have completely changed. These people have been exposed to a great level of stress due to cultural bereavement, problems of communication, socioeconomic factors, poverty, insecurity, racism, isolation, unemployment, loss of status and recognised role in society. As a consequence it is very difficult to manage this load of stress in the long term (Bauer *et al.*, 1994b).

Children are more than adults subjected to long-lasting anxiety. Their naivety and strong trust in stability of life is ruined and it takes great efforts and time to re-establish in them usual attitude and self-esteem (Bauer *et al.*, 1994a). Homicide actions have irreparable effects on their psychological state (Marshall, 2001). A very interesting observation was made recently by Delahanty *et al.*, who noticed that high urinary cortisol and epinephrine levels immediately after a traumatic event may be associated with increased risk for the development of subsequent acute post-traumatic stress disorder. This observation suggests a direct link between mental health and pituitary-adrenal axis dysregulation due to violence (Delahanty *et al.*, 2005). Some more intriguing data arise from other studies which documented that violence (e.g abuse) exerted during childhood has led to alterations of stress axis at adult life (Heim *et al.*, 2001). Concerning adults Bosnian, Kosovo and Kurdish have high rates of post traumatic stress disorder among their population due to "mobility" problems. As there is a lot of evidence about the relationship of emotional distress and substance abuse, the finding that many Somali refugees reported high levels of substance abuse among them is rather a disappointing than a surprising phenomenon (Sabioncello *et al.*, 2000).

A team from Sweden studied the mental and hormone profile in displaced persons versus no displaced residents. They found that refugees had much higher incidence of personality disorder, fear, depression, anxiety and hypersensitivity in comparison with the control group (Sondergaard *et al.*, 2001). These two groups differed significantly for cortisol, prolactin and b-endorphin levels, showing an exaggerated response of the stress system (Sondergaard *et al.*, 2002, Sondergaard and Theorell, 2003).

## 4.  Noise

Another very interesting point of mobility with great impact on health is noise. Noise, which is often referred to as unwanted sound, is typically characterized by the intensity, frequency, periodicity (continuous or intermittent) and duration of sound. Unwanted sound to some may be considered wanted sound by others, as in the case of loud music. The potential adverse health effects are usually classified according to the type of noise. Sudden (or impulsive) noise appears to create substantially more reaction than non-impulsive noise and intermittent noise has greater effect than louder, more continuous noise. Predictability and controllability are clearly influencing factors in individual reactions to noise. It has been found that unpredictable noise causes the most stress, but even predictable noise creates some stress and there are studies that link increased stress levels to excess noise exposure (Fong and Johnston, 2000). Individual physiological and psychological responses to noise are also influenced by susceptibility. Interestingly, more people are affected by noise exposure than any other environmental stressor.

Research has focused on noise as an auditory stressor that can produce both direct and indirect health effects. The direct health effect known to be attributable to noise is hearing loss (resulting from damage to the inner hair cells of the organ of corti) with noise exposure higher than 90 decibels. There are several non-auditory physiological effects of noise exposure including a possible increase in cardiovascular disease from elevated blood pressure and physiological reactions involving the cardiovascular endocrine system. In addition, community noise has been shown to adversely affect sleep, communication, performance and behavior, reading and memory acquisition, and mental health (Laer *et al.*, 2001).

The cost of noise in health is a very important parameter, as it makes it easier to realize the size of the problem. Despite the scarcity of data, Germany has estimated that the annual cost of noise on public health is approximately 726.4 million dollars to 2.76 billion US dollars per year for road noise, and 1.45 million US dollars per year for rail noise (European Commission, 1996).

Acute exposure to maximal sound pressure levels above 90 dB has the potential to cause inner ear hearing loss and to stimulate the sympathetic nervous system into increasing the release of adrenaline and noradrenaline (Babisch *et al.*, 2001). Noise levels above 120 dB increase cortisol in humans. One of the side-effects of chronically increased cortisol is a higher risk for cardiovascular disease and metabolic syndrome. The activation of the sympathetic and endrocrine systems drives to increments in total peripheral resistance, cardiac output, blood lipids and rheological factors (Ising *et al.*, 1999).

A cross-sectional study with about 200 females, who lived for several years in streets with low or high traffic noise, showed a significant increase of the noradrenaline excretion in subjects whose bedroom windows were facing a noisy street. Additionally, noise-disturbed persons had significantly higher noradrenaline levels than undisturbed persons (Babisch *et al.*, 1993). Chronically increased noradrenaline is known to have detrimental effects upon the heart and high cardiovascular risk.

Other reports also suggest an increased risk of hypertension and ischemic heart disease for people living in areas with road or air traffic noise at outdoor equivalent sound levels above 70 dB. In addition, some studies have found that kindergarten children had significantly higher systolic and diastolic blood pressures when exposed to noisy or very noisy environments (kindergarten and home) as compared to quiet environments. However, up to now there is no consistent evidence that chronic noise leads to hypertension and coronary disease, although data connecting these two parameters arise every day (Kawakami and Haratani, 1999, Ohlson *et al.*, 2001).

Finally, data on the action of noise to the reproductive axis are limited, but some reports mentioned a relationship between air traffic noise exposure of pregnant women in the living environment and low birth weight. However, there are virtually no data to suggest an increased risk of congenital anomalies. Unfortunately we cannot use yet, endocrine changes on occupationally or environmentally exposed people to noise as an index for future disease, due to different effects of noise on different people. However, a better follow-up of people exposed to noise is strongly recommended in order to avoid disease.

## 5.    Conclusions

In conclusion, mobility has a great impact on health and these changes seem to be mediated through hormones - neurotransmitters and mainly the stress system (hypothalamus-pituitary-adrenal axis). It seems that mobility through auditory, gustatory, olfactory, tactile and visual stimuli, modifies the function of several important areas of the brain such as mesolimbic, hypothalamus, pituitary, reticular activating system and amygdala, driving to temporary or permanent health problems, due to the abnormal secretion of many hormones and neurotransmitters (Martin *et al.*, 1979). It is clear that mobility affects in many ways and at several levels human homeostasis with adverse effects on quality and duration of life. This is an amazing area of research is wide open and a lot of work has to be done in order to understand the interaction between mobility and adaptation mechanisms. However, society has to face directly and drastically the causes of mobility problems, especially noise (because violence is a much more complicated situation…), in order to avoid impacts on health and quality of life. Society has to adopt strategies and policies which take into account the avoidance or reduction of noise exposure, as its detrimental effects on health are reported everyday.

# References

Adam, C.L., Moar, K.M., Logie, T.J., Ross, A.W., Barrett, P., Morgan, P.J., and Mercer, J.G. (2000) Photoperiod regulates growth, puberty and hypothalamic neuropeptide and receptor gene expression in female Siberian hamsters, *Endocrinology,* 141, 4349-4356.

Babisch, W., Fromme, H., Beyer, A., and Ising, H. (2001) Increased catecholamine levels in urine in subjects exposed to road traffic noise: the role of stress hormones in noise research, *Environ. Int.,* 26, 475-481

Babisch, W., Ising, H., Elwood, P.C., Sharp, D.S., and Bainton, D. (1993) Traffic noise and cardiovascular risk: the Caerphilly and Speedwell studies, second phase. Risk estimation, prevalence, and incidence of ischemic heart disease, *Arch. Environ. Healt.*, 48 ,406-413.

Bauer, M., Priebe, S., Graf, K. J., Kurten, I. and Baumgartner, A. (1994a) Psychological and endocrine abnormalities in refugees from East Germany: Part II. Serum levels of cortisol, prolactin, luteinizing hormone, follicle stimulating hormone, and testosterone, *Psychiatry Res.,* 51, 75-85.

Bauer, M., Priebe, S., Kurten, I., Graf, K.J., and Baumgartner, A. (1994b) Psychological and endocrine abnormalities in refugees from East Germany: Part I. Prolonged stress, psychopathology, and hypothalamic-pituitary-thyroid axis activity, *Psychiatry Res.,* 51, 61-73.

Breslau, N. (2001) Outcomes of posttraumatic stress disorder, *Journal of Clinical Psychiatry,* 62, 55-59.

Chrousos, G.P. (2000) The HPA axis and the stress response, *Endocr. Res.,* 26, 513-514.

Delahanty, D.L., Nugent, N.R., Christopher, N.C., and Walsh, M. (2005) Initial urinary epinephrine and cortisol levels predict acute PTSD symptoms in child trauma victims, *Psychoneuroendocrinology,* 30,121-128

Fong, S., and Johnston, M. (2000) In *Toronto Public Health,* Toronto: City of Toronto.

Gronfier, C., and Brandenberger, G. (1998) Ultradian rhythms in pituitary and adrenal hormones: their relations to sleep, *Sleep. Med. Rev.,* 2, 17-29.

Heim, D., Newport, J., Bonsall, R., Miller, A.H., and Nemeroff, C.B. (2001) Altered pituitary-adrenal axis responses to provocative challenge tests in adult survivors of childhood abuse, *Am. J. Psychiatry,* 158, 575-581

Herxheimer, A., and Petrie, K.J. (2003) Melatonin for the prevention and treatment of jet lag, in *The Cohrane Library*, Vol. 1 Update software, Oxford.

Herxheimer, A., and Waterhouse, J. (2003) The prevention and treatment of jet lag, *BMJ* 326, 296-297.

Hirschfeld, U., Moreno-Reyes, R., Akseki, E., L'Hermite-Baleriaux, M., Leproult, R., Copinschi, G., and Van Cauter, E. (1996) Progressive elevation of plasma thyrotropin during adaptation to simulated jet lag: effects of treatment with bright light or zolpidem, *J. Clin. Endocrinol. Metab.,* 81, 3270-3277.

Ising, H., Babisch, W., and Kruppa, B. (1999a) Noise-induced endocrine effects and cardiovascular risk, *Noise and Health* 1, 37-48.

Kawakami, N., and Haratani, T. (1999) Epidemiology of job stress and health in Japan: review of current evidence and future direction, *Ind. Health* 37, 174-186.

Laer, L., Kloppstech, M., Schofl, C., Sejnowski, T. J., Brabant, G. and Prank, K. (2001) Noise enhanced hormonal signal transduction through intracellular calcium oscillations, *Biophys. Chem.* 91, 157-166.

Marshall, R., Olfson, M., Hellman, J.F., Blanco, C., Guardino, M., and Struening, E.L. (2001) Comorbidity, impairment, and suicidality in subthreshold PTSD, *American Journal of Psychiatry* 158, 1467-1473.

Martin, J.B., Tolis, G., Woods, I., and Guyda, H. (1979) Failure of naloxone to influence physiological growth hormone and prolactin secretion, *Brain Res.* 168, 210-215.

McEwen, B. S. (1998) Protective and damaging effects of stress mediators, *N. Engl. J. Med.* 338, 171-179.

Ohlson, C. G., Soderfeldt, M., Soderfeldt, B., Jones, I., and Theorell, T. (2001) Stress markers in relation to job strain in human service organizations, *Psychother. Psychosom.* 70, 268-275.

Pinter, E.J., Tolis, G., Guyda, H., and Katsarkas, A. (1979) Hormonal and free fatty acid changes during strenuous flight in novices and trained personnel, *Psychoneuroendocrinology* 4, 79-82.

Ryff, C.D., and Singer, B.H. (2000) Biopsychosocial challenges of the new millennium, *Psychother. Psychosom.* 69, 170-177.

Sabioncello, A., Kocijan-Hercigonja, D., Rabatic, S., Tomasic, J., Jeren, T., Matijevic, L., Rijavec, M., and Dekaris, D. (2000) Immune, endocrine, and psychological responses in civilians displaced by war, *Psychosom. Med.,* 62, 502-508.

Sedgwick, P.M. (1998) Disorders of the sleep-wake cycle in adults, *Postgrad. Med. J.,* 74, 134-138.

Segerstrom, S.C., and Miller, G.E. (2004) Psychological stress and the human immune system: a meta-analytic study of 30 years of inquiry, *Psychol. Bull.* 130, 601-630.

Sondergaard, H.P., Ekblad, S., and Theorell, T. (2001) Self-reported life event patterns and their relation to health among recently resettled Iraqi and Kurdish refugees in Sweden, *J. Nerv. Ment. Dis.* 189, 838-845.

Sondergaard, H.P., Hansson, L.O., and Theorell, T. (2002) Elevated blood levels of dehydroepiandrosterone sulphate vary with symptom load in posttraumatic stress disorder: findings from a longitudinal study of refugees in Sweden, *Psychother. Psychosom.* 71, 298-303.

Sondergaard, H.P., and Theorell, T. (2003) A longitudinal study of hormonal reactions accompanying life events in recently resettled refugees, *Psychother. Psychosom.* 72, 49-58.

Spitzer, R.L., Terman, M., Williams, J.B., Terman, J.S., Malt, U.F., Singer, F., and Lewy, A.J. (1999) Jet lag: clinical features, validation of a new syndrome-specific scale, and lack of response to melatonin in a randomized, double-blind trial, *Am. J. Psychiatry.* 156, 1392-1396.

Stam, R., Croiset, G., Bruijnzeel, A.W., Visser, T.J., Akkermans, L.M., and Wiegant, V.M. (1999) Sex differences in long-term stress-induced colonic, behavioural and hormonal disturbances, *Life Sci.* 65, 2837-2849.

Tolis, G., Banovac, K., McKenzie, J.M., and Guyda, H. (1979) Circadian rhythms of anterior pituitary hormone secretion: effects of dexamethasone, *J. Endocrinol. Invest.* 2, 433-436.

Tolis, G., Friesen, H.G., Bowers, C.Y., and McKenzie, J.M. (1974) Glucocorticoids and thyrotropin releasing hormone (TRH) secretion, *Neuroendocrinology* 15, 245-248.

Tolis, G., Goldstein, M., and Friesen, H.G. (1973) Functional evaluation of prolactin secretion in patients with hypothalamic-pituitary disorders, *J. Clin. Invest.* 52, 783-788.

# SEDENTARISM

H. MOSHAMMER[1], E. MARTIN-DIENER[2], U. MÄDER[2] AND
B. MARTIN[2]
[1]*Institute for Environmental Health, Medical University of Vienna*
*Kinderspitalgasse 15, 1095 Vienna, AUSTRIA*
[2]*Institute of Sports Sciences,*
*Federal Office of Sports*
*Hauptstrasse 243, CH-2532 Magglingen, SWITZERLAND*

## Summary

The importance of physical activity has been well established over the last decades and a wealth of different endpoints has been identified. A dose-response-relationship could be demonstrated for most of these endpoints, most clearly for overall mortality and cardiovascular morbidity. While most of the studies have investigated the associations with overall physical activity, only very few have been able to study the independent effects of transport-related physical activity.

Physical inactivity is a world-wide public health problem. Although methodological issues still restrict the possibilities to quantify this problem in absolute terms and to carry out intercultural and international comparisons, subgroups with particularly low activity levels and changes over time can be documented. Current recommendations for physical activity underline the health enhancing effects of rather short although regular episodes of training which can easily be integrated into everyday life.

A systematic integration of data from the health and from the transport sector has not yet taken place. But there is evidence for the effectiveness of a growing number of interventions in increasing physical activity among the inactive, and transport-related physical activity has a great potential in the promotion of overall physical activity.

This overall rationale is well accepted, but the quantification of the relationships and effects remains difficult, mainly due to the need for an internationally agreed definition and measure of physical activity on the population level, the lack of data for the contribution of transport-related physical activity to overall physical activity and therefore to health. Realistic estimations of effects of transport interventions on

*P. Nicolopoulou-Stamati et al. (eds), Environmental Health Impacts of Transport and Mobility*, 135-154.
© 2005 *Springer. Printed in the Netherlands.*

transport patterns (modal shift) are essential not only for the health effects of physical activity, but also for other transport-related factors like air pollution or noise.

## 1.   Introduction

This chapter is based on papers prepared for the project "Transport Related Health Impacts, Costs and Benefits with a Particular Focus on Children" within the context of the UNECE- WHO Pan-European Programme for Transport, Health and Environment - THE PEP. The aim of the project is to provide a review on the state of the art on transport related health impacts, costs and benefits as well as to develop recommendations on political implementation strategies and also to contribute to the development of WHO-Guidelines for the economic valuation of transport related health effects.

Although the importance of regular physical activity for health is beyond any doubt on the individual as well as on the public health level, there is only a limited number of countries like Finland and Canada that have had a well-established monitoring system in place since a longer period of time. A number of countries have begun to establish national representative surveys on physical activity since the 1990s, but intercultural and international comparisons remain difficult due to methodological issues.

Nevertheless, the public health importance of regular physical activity is such that estimates of patterns of physical activity are needed, though there does not yet exist any internationally agreed definition or measure of physical activity on the population level (WHO, 2002).

## 2.   Objectives

The importance of regular physical activity for health is well established (Table 12). There is evidence for the effectiveness of a growing number of interventions in increasing physical activity among the inactive (Vuori, 1998), and transport-related physical activity has a great potential in the promotion of overall physical activity (Oja and Vuori, 2000). This overall rationale is well accepted (Figure 30), but the quantification of the relationships and effects remains difficult, mainly due to the need for an internationally agreed definition and measure of physical activity on the population level, the lack of data for the contribution of transport-related physical activity to overall physical activity and therefore to health.

The WHO (2002) cites reductions in the risk of cardiovascular disease, colon and breast cancer as well as type 2 diabetes as the most important effects, but other effects like improvements in musculoskeletal health (osteoarthritis, low back pain, osteoporosis, falls), control of body weight, reductions in symptoms of depression, anxiety and stress, and risk reduction for prostate cancer are mentioned as well.

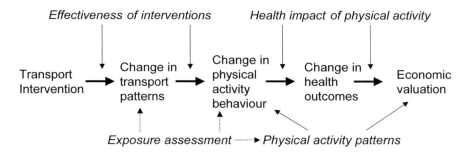

**Figure 30**. Overview of the chain from transport intervention to economic valuation of the health effects of transport-related physical activity. While the overall causal relationship (indicated by the bold arrows) is well accepted, quantification and modelling represent several challenges.

These relationships and the association with overall mortality have already been shown and commented upon in the US Surgeon General's Report on Physical Activity and Health (Department of Health and Human Services, 1996) and other encyclopaedic reviews (Marti and Hättich, 1999). More recent publications have indicated further effects on the risk of cholecystectomy (Leitzmann *et al.*, 1999) or on pancreatic cancer (Michaud *et al.*, 2001).

**Table 12**. Overview of health effects of physical activity (WHO, 2002).

| | |
|---|---|
| ⇑Life expectancy | ⇓(Pancreatic cancer) |
| ⇓Cardiovascular disease | ⇓Osteoporosis |
| ⇓Diabetes II | ⇓Symptomatic gallstone disease |
| ⇓Obesity | ⇓Depression |
| ⇓Colon cancer | ⇑Stress tolerance |
| ⇓Breast cancer | ⇑Independence in old age |
| ⇓(Prostate cancer) | |

While most of the earlier studies included only male participants, more recent research has shown that the effect size and the dose-response relationship are comparable for both genders (Oguma *et al.*, 2002). A dose-response-relationship could be demonstrated for most of the endpoints mentioned above, most clearly for overall mortality and cardiovascular morbidity (Figure 31) (Department of Health and Human Services, 1996, Marti and Hättich, 1999, Haskell, 1994).

While most of the studies have investigated the associations with overall physical activity, only very few have been able to study the independent effects of transport related physical activity (Andersen *et al.*, 2000, Hendriksen *et al.*, 2000).

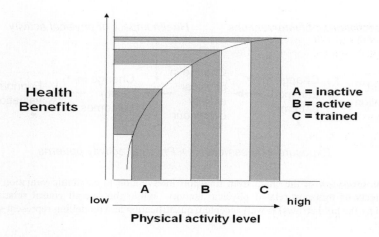

**Figure 31**. Dose-response relationship for physical activity and health (adapted from Haskell, 1994).

Hence the aim of this paper is (1) to report data and data gaps concerning activity patterns with a focus on physical activity related to transport modes, (2) to compare these data with recommendations for health-enhancing physical activity, (3) to discuss the effectiveness of interventions aimed at increasing physical activity, and (4) discuss methods to evaluate such interventions and to quantify physical activity.

## 3.    Results

### *Physical activity patterns (with a special focus on Switzerland)*

Since the Swiss Health Survey 1992, systematic information about physical activity patterns of the Swiss population has been available. The Surveys 1992 and 1997 did not yet include any items relating directly to the current activity recommendations which were only issued in 1999, but a question on sweating induced by leisure time physical activity has been included since 1992. A distinction was thus made between inactive (less than one sweating episode per week), moderately active (1 or 2 sweating episodes per week) and active (more than 3 leisure time sweating episodes) individuals (Calmonte and Kälin, 1997).

Although the interval between the first and second health survey was just five years, changes in physical activity have occurred over this short period (Lamprecht and Stamm, 1999). Whereas the proportion of those sweating several times a week as a result of physical activity remained almost constant between 1992 and 1997, striking shifts had occurred in the percentages for the moderately active and inactive groups: the proportion of inactive individuals rose by about four percentage points between 1992 and 1997 (Table 13). In other words: physical inactivity in the Swiss population has increased by over a tenth in just five years.

**Table 13**. Proportion of inactive, moderately active and active individuals in the Swiss health Surveys 1992 and 1997, based on the number of days with sweating episodes during leisure time (adapted from Lamprecht and Stamm, 1999).

| Leisure time sweating | 1992 (n=14676) | 1997 (n=12727) |
|---|---|---|
| 3 days/week and more | 26.3% | 26.9% |
| 1 to 2 days/week | 37.8% | 33.7% |
| less than 1 day/week | 35.7% | 39.4% |
| Total | 100.0% | 100.0% |

An accentuation of social differences was also observed in physical activity between 1992 and 1997. Specifically, inactivity showed a particularly sharp rise in those population groups that had already been characterised by little physical activity. The differences in activity level in respect to gender, age, linguistic region, education and household income have increased - dramatically for some indicators - between 1992 and 1997. The principal findings can be summarised as follows:

- Whereas the proportion of inactive men only increased by about two percent, inactivity rose in women - who were already less active in 1992 - by around six percent. This gender-specific difference in leisure time physical activity, which is often attributed to historical inequalities in basic conditions, has therefore not declined – as is often assumed – but rather increased. As regards the proportion of active individuals, a serious gender difference is already apparent in the 15 to 24 year age group.

- The same applies to the differences relating to age and region. Here too, inactivity rose particularly in those groups that were already fairly inactive. Although a slight increase in physical activity was noted in the 15 to 24 and 25 to 34 year age groups, inactivity has increased dramatically among those aged over 55.

- In German-speaking Switzerland, the groups of both active and inactive individuals increased slightly at the expense of the moderately active group. In the French- and Italian-speaking areas of Switzerland, on the other hand, growth was only observed in the inactive group.

- Regardless of age, sex or region, inactivity increased primarily among the least educated and those with lower household income, while minor changes only were apparent among the better educated and higher income groups.

The 1999 HEPA survey with about 1500 participants has produced the first activity prevalences that could be correlated directly to the new recommendations for health-enhancing physical activity (HEPA): 37% of the Swiss interviewees were not active at the level of the minimum recommendations and could therefore be classed as inactive. 26% reported half an hour of moderate intensity activities on most days of the week (but without any endurance type training), while a further 37% reported at

140                                H. MOSHAMMER *ET AL.*

least twenty minutes of vigorous intensity on three days of the week (Figure 32) (Martin *et al.*, 2000).

The availability of epidemiological data is an important element in the political process leading to a better recognition of the importance of HEPA on the national and international level and the current attempts for standardised measurement procedures will play an important role in this process.

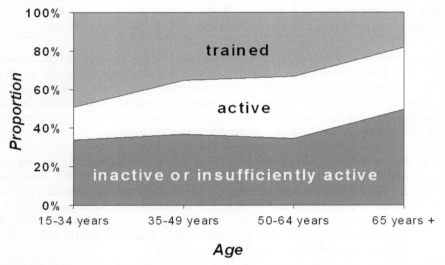

**Figure 32**. Physical activity by age group in the Swiss HEPA survey 1999. Trained individuals report at least 3 activity sessions of 20 minutes of vigorous intensity weekly; active individuals report at least half an hour of moderate intensity activity daily; inactive individuals report less or no activity at all.

These results were derived from algorithms assessing the stages of changes of the transtheoretical model (Marcus *et al.*, 1992). Target behaviour were half an hour physical activity daily with at least moderate intensity and three times twenty minutes of activities of vigorous intensity respectively. The HEPA survey 2001 has shown no difference in the prevalence of inactivity in comparison to 1999 when using the same items (Martin, 2002), but a considerably higher percentage of inactive individuals (58%) when using an alternative set of items to assess activities of both moderate and vigorous activity in a two-step-procedure (on how many days per week; with what average duration) and providing a continuous measure of physical activity.

The Swiss Health survey 1997 contained an item on daily transport by bicycle or foot. According to the data, 40.3% of the Swiss population spend 20 minutes or more per day using these forms of transport, 16.0% less than twenty minutes and 43.7% neither walk nor cycle regularly (Lamprecht and Stamm, 1999).

**Table 14.** Modal split in Switzerland according to the travel behaviour in the microcensus 2000

| Transport mode | Daily distance | Time spent | Number of trips |
|---|---|---|---|
| Walking | 4.6% | 34.3% | 40.1% |
| Cycling | 2.5% | 5.6% | 6.0% |
| Mot. individual transport | 69.5% | 43.6% | 41.6% |
| Public transport | 17.7% | 11.4% | 10.3% |
| Total (per person and day) | 48 km | 89 min | 3.6 |

According to the Swiss scientific survey of the population's travel behaviour (travel behaviour microcensus 2000: Bundesamt für Raumentwicklung and Bundesamt für Statistik, 2001), walking and cycling are important transport modes in Switzerland. 46% of trips (journey stages) per day and 40% of the time spent for travel purposes (average travelling time per day: 85 minutes) can be attributed to forms of non-motorised transport, 42% and 44% to motorised individual transport (Table 14). In 80% of all travel episodes, only one transport mode is used, in 10% of all journeys non-motorised transport is used jointly with car use, in 9% jointly with public transport. Modal split differs according to the purpose of travel (Table 15).

**Table 15.** Contribution of non-motorised forms of transport to the modal split according to the purpose of travel in the microcensus 2000

| Purpose | Walking | Cycling | Total |
|---|---|---|---|
| Commuting | 33% | 6% | 39% |
| Education | 55% | 13% | 68% |
| Shopping | 45% | 6% | 51% |
| Business transport | 23% | 3% | 26% |
| Service and accompanying transport | 24% | 2% | 26% |
| Leisure time | 42% | 6% | 48% |
| All purposes | 40% | 6% | 46% |

According to the survey results, the contribution of non-motorised transport to the modal split is more important in the age groups below 17 years and above 65 years than in the age groups between, more important in women than in men and more important in the German speaking than in the French or in the Italian speaking parts of Switzerland.

The WHO (2002) summarises statistics for physical activity, though indicating that they are derived from a number of direct and indirect data sources and a range of survey instruments and methodologies: "The global estimate for prevalence of physical inactivity among adults is 17%, ranging from 11% to 24% across subregions. Estimates for prevalence of some but insufficient activity (<2.5 hours per week of moderate activity) ranged from 31% to 51%, with a global average of 41% across the 14 subregions."

In 1999 a report was published on a study carried out in 15 member states of the European Union (EC, 1999). This survey contained items on physical activities of moderate intensity. The proportion of individuals reporting no more than 3 hours per week of such activities was 57% in the EU average, with the lowest values being 32% and 33% in Sweden and Finland. Switzerland's neighbouring countries had the following prevalences: Germany 56%, Austria 38%, France 63%, Italy 66%.

### Recommendations for health-enhancing physical activity

The first application of these results to public health were the 1995 recommendations of the US-American Centers for Disease Control CDC and the American College of Sports Medicine ACSM. They were the first recommendations to focus on activities of so called moderate intensity (Pate *et al.*, 1995) that included many activities of everyday life like brisk walking. These recommendations were also the first to acknowledge the contribution of shorter "bouts of activity" by allowing the accumulation of activity episodes of about 10 minutes or more. An attempt to include not only minimal recommendations, but also the additional effects of further or more intensive activity (WHO, 2002, Department of Health and Human Services, 1996) is the physical activity pyramid of the recommendations for health-enhancing physical activity currently in use in Switzerland (Figure 33).

## Health - Enhancing Physical Activity

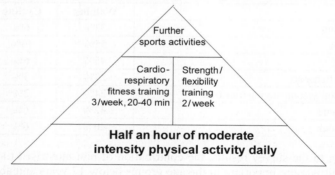

- At least half an hour a day of moderate intensity physical activity is recommended to women and men at any age.
- People who already attain this level can further increase their well-being, health, and efficiency by adding some training for cardio-respiratory fitness, strength, and flexibility.
- For people who already train regularly, further sports activities bring about additional benefits for health. Yet, the additional effect diminishes gradually.

**Figure 33**. The recommendations for health-enhancing physical activity currently in use in Switzerland.

## First estimations of health effects on the population level

Even when the differences in health outcomes between more active and less active subgroups in observational studies and the effect sizes in the more recent randomised interventional studies (Knowler *et al.*, 2002) are impressive, the exact quantification of the relationship remains a challenge due to the measurement issues discussed later. Just physiologically, physical activity is characterised by intensity, duration and frequency (Marti and Martin, 2001), and there exists no internationally agreed definition or measure of it (WHO, 2002). Therefore some arbitrary assumptions have to be made in order to apply the effect sizes obtained from the literature in effect estimations on the population level.

**Table 16**. Assumptions for effect sizes in a Swiss study on the health effects of physical activity (Martin *et al.*, 2001). Estimates for costs are in Swiss francs.

|  | Relative risk for the disease | Relative risk for mortality | Direct treatment costs | Indirect costs * |
|---|---|---|---|---|
| Cardiovascular disorders | 1.84 | 1.43 | 2239 | 2556 |
| Diabetes Type II | 1.88 | 3.00 | 3508 | 636 |
| Colon cancer | 1.90 | 1.68 | 52165 | n.d. |
| Osteoporosis | 2.00 | n.d. | 630 | n.d. |
| Breast cancer | 1.39 | 1.00 | 28490 | n.d. |
| Depression | 3.15 | n.d. | 1983 | n.d. |
| Back pain | 1.36 | n.d. | 739 | 1126 |
| Hypertension | 1.47 | 1.00 | 693 | n.d. |

n.d.: no data
* only productivity loss

An example of the assumptions to be made for effect sizes is shown in Table 16. By assuming a proportion of physically inactive individuals of 37%, this study has estimated that in a number of 1.4 million cases of disease, almost 2.000 deaths and direct treatment costs of 1.6 billion Swiss francs (1.1 billion Euro) are caused by physical inactivity in Switzerland each year (Martin *et al.*, 2001). Using exactly the same assumptions except for a proportion of inactive individuals of 58% identical with the WHO global estimate (2002) and according to most recent findings also more likely for Switzerland (Martin, 2002), the respective estimates would be 1.9 million cases of disease, 2.700 deaths and direct treatment costs of 2.2 billion Swiss francs (1.5 billion Euro).

The global estimations of WHO (2002) indicate that physical inactivity causes about 10–16% of cases each of breast cancer, colon and rectal cancers and diabetes mellitus, and about 22% of ischaemic heart disease, resulting in 1.9 million deaths and 19 million DALYs (disability-adjusted life years).

## The effectiveness of interventions to increase physical activity

In general, most experience regarding design and feasibility of intervention studies and the effects of these programs is available for interventions on the individual and group level. There is good evidence that interventions on these levels can increase physical activity among the inactive. There is also good evidence that interventions in the worksite setting are effective to increase physical activity.

There seem to be virtually no studies assessing the impact of interventions targeting transport policies and environmental changes on physical activity – neither on their effect to increase individual physical activity behaviour in general, and particularly not regarding their potential to reach the inactive segment of a population. An observed increase in bicycle use for example is far less relevant for public health if it occurs in individuals already physically active in other ways than if the same increase could be observed in a previously inactive group.

The outstanding importance of physical activity as a health resource and for the prevention of various chronic diseases has extensively been documented (U.S. Department of Health and Human Services, 1996). In the last decade, the recommendations for health enhancing physical activity (HEPA) have focussed on moderate intensity physical activity as part of an active lifestyle. In Switzerland, national authorities have adapted the international recommendations for health enhancing physical activity, which are defined as the accumulation of 30 minutes or more of moderate intensity activities on most, preferably all days of the week, or alternatively at least 20 minutes of vigorous exercise on three or more days of the week (Bundesamt für Sport (BASPO) *et al.*, 2002). Nevertheless, one third of the Swiss adult population does not meet any one of these two recommendations (Martin *et al.*, 1999, Martin 2002). In Switzerland, the need for interventions to promote physical activity has been recognised and has influenced the political agenda-setting (BASPO, 2000). From the public health point of view, it is most important to reach individuals and sub-groups with the lowest baseline activity. Four levels of interventions (Table 17) to increase physical activity have been described (King, 1994).

**Table 17.** Levels of intervention

| Level | Framework | Examples |
|-------|-----------|----------|
| 1 | Legislative / policy | Bike paths, compulsory school PE, countryside access |
| 2 | Organisational / environmental | Workplace, local council initiatives, mass media approaches, environmental prompts |
| 3 | Inter-personal | Teaching approaches, provision of classes, peer lead groups |
| 4 | Individual | Consultations, fitness assessment, written materials |

Interventions such as transport policies and environmental changes which promote or facilitate lifestyle activities like walking and cycling are of particular interest. They are located mainly on level 1, to some degree on level 2. There is consensus among physical activity promotion experts (e.g. King, 1998, Sallis and Owen, 1999, Sallis *et al.*, 1998), that there is a great potential in this type of interventions to have an impact on public health, because entire populations of inactive people can be influenced.

### *Effects of physical activity interventions: Swiss and international experiences*

The importance of regular physical activity for health is well known in Switzerland and physical activity promotion activities are generally appreciated by the general population. Nevertheless, the prevalence of physical inactivity is still rising in the country. A growing number of large scale physical activity promotion projects have begun to target physically inactive individuals.

First intervention studies indicate both the possibilities and the limitations of interventions for behavioural change. The *"Office in Motion"* study used the ecological or settings approach in a white collar worksite intervention and has shown encouraging changes in previously inactive individuals. Energy expenditure through moderate intensity activities increased particularly in those intervention worksites with the lowest activity levels at baseline (Titze *et al.*, 2001). The one year follow up results suggest, that mainly changes in active commuting were maintained (unpublished results). A randomised controlled study was conducted in the primary care setting *("Active upon Advice")* offering a counselling session to increase lifestyle activities to the intervention group and an advise-only treatment to the controls. Almost half of the initially inactive patients were regularly active after one year – though both in the intervention and control group (Jimmy and Martin).

These experiences from Switzerland demonstrate that behavioural change in previously inactive individuals is possible. Expectations concerning the size of the intervention effect have to be realistic. Promotion of health enhancing physical activity remains a challenge. However, to our knowledge, no evaluation of changes in individual physical activity behaviour through environmental or policy interventions to promote cycling and walking has been conducted in Switzerland so far.

A recent systematic review summarises the effectiveness of the different types of interventions to increase physical activity (Kahn *et al.*, 2002). There is good evidence, that individually adapted interventions or community wide campaigns are effective. Environmental and policy level interventions were defined as interventions "creating or enhancing access to places for physical activity combined with informational outreach". Ten studies were included in the review, all from the US. Eight studies were worksite interventions. The intervention programs involved changes such as provision of walking trails, building exercise facilities or provision of access to nearby facilities, all supported by informational and motivational measures on the individual level. Outcome measures were changes in aerobic capacity, energy expenditure, self-reported physical activity, frequency of physical activity or exercise scores. All studies were effective to increase physical activity. The authors conclude, that there is strong evidence that interventions "creating or enhancing access to places for physical activity combined with informational outreach" are effective in increasing physical activity.

However, it seems that only a minority of these interventions included changes of the physical environment or even transport policies to encourage walking and cycling. On a website summarising the results of this review, it is announced, that the accumulation of experience regarding "Transportation policy and infrastructure changes to promote non-motorised transit" and "Urban planning approaches - zoning and land use" is in progress (www.thecommunityguide.com).

Earlier reviews specifically on environmental and policy interventions (Sallis *et al.*, 1998) confirm that there are only few studies available on this level of interventions, even though studies in a broader sense were also included.

Two European studies on worksite interventions were not included in the recent review on effectiveness (Kahn *et al.*, 2002). A promotion project in an industrial plant in Finland demonstrated, that walking or cycling to work can be successfully promoted, also among those not active on a regular basis (Oja *et al.*, 1998). A limitation of this study is the uncontrolled design. In a randomised controlled trial in Scotland self-help material for active commuting to work was delivered to the intervention group (Mutrie *et al.*, 2002). In the intervention group, the increase of walking to work was significantly higher than in the control group. No difference between the groups could be observed for cycling to work. The authors conclude, that before cycling can be promoted on the individual level, the necessary improvements on the environmental and policy level must be realised. These two studies promoted cycling and walking to work – however, also these interventions were located on the individual and organisational level and did not touch transportation policies or environmental changes.

Obviously, also on the international level, knowledge is scarce on the effectiveness of interventions targeting policy and environmental changes to increase physical activity. A current research focus is the assessment of environmental determinants for cycling and walking in cross-sectional studies (e.g. Pikora *et al.*, 2003).

Without any doubt, there are many policies and interventions developed and realised by urban planning and transport agencies, which have the potential to increase physical activity levels also among those not active on a regular basis. Just one example is the "Cycle City Odense" in Denmark where with more than 60 initiatives cycling increased at the end of the 20th and the beginning of the 21st century, while in the rest of the country cycling decreased (www.cycleby.dk/english.asp). But it seems that such interventions have never been evaluated for a physical activity outcome on the level of individuals.

A little more data exist on the practicability of mobility plans introduced by single enterprises (Employer Transport Plans - ETP). Rye (1999) argues that "selling the concept of the ETP to employers is a challenging task which will succeed only with the application of considerable resources over a number of years and that, even then, it will succeed only in a limited number of cases". Nevertheless several case studies (e.g. Cleary and McClintock, 2000) underline the great potential of such measures introduced on the local level and in close cooperation with single enterprises.

## Methods and sources of physical activity assessment

The measurement of physical activity presents several challenges, in particular with respect to transport and health. First of all, there exists no internationally agreed definition or measure of physical activity on the population level (WHO, 2002). Then, while associations between overall physical activity and health are very clear, the contribution of transport-related physical activity to overall physical activity is difficult to assess, not least because of the different methods and data sources for those two aspects.

The ideal instrument for the assessment of physical activity should give a global assessment of physical activity, be easily applicable, comparable between population groups, result in good distinction between more and less active individuals or groups and should provide information that is relevant for interventions. Obviously, such an instrument does not exist. The different possibilities for the assessment of physical activity have already been outlined in the US Surgeon General's Report on Physical Activity and Health (Department of Health and Human Services, 1996). Since then, only GPS (global positioning system) has been added to the list (Table 18), though this device is not yet in routine use.

**Table 18**. Physical activity assessment methods

| Surveying | Task specific diary |
|---|---|
| | Recall questionnaire |
| | Quantitative history |
| | Global self-report |
| Monitoring | Behavioural observation |
| | Job classification |
| | Heart rate monitor |
| | Motion sensors |
| | Calorimetry |
| | GPS |
| | Doubly labelled water |

Except for job classifications which only give a very rough estimate of physical activity, most possibilities in the "monitoring" group are methodically rather demanding. Laboratory methods such as direct and indirect calorimetry or doubly labelled water technique are generally considered accurate in measuring energy expenditure of physical activity. Nevertheless, both methods have limitations. While the doubly labelled water technique cannot be used to examine short-term physical activity (Starling *et al.*, 1999), indirect calorimetry is not suitable to measure energy expenditure in free-living humans during longer periods. In addition, these methods are expensive and not suitable for epidemiological studies. The only method in this group that has already been used in representative studies are accelerometers as a particular kind of motion sensors (Sjöström, 2002).

Self-completed or interviewer-administered questionnaires are often used to assess physical activity. These instruments record information on duration, intensity and type of activities performed over a certain period of time. Detailed physical activity recall questionnaires show the best results in validity studies and also allow the identification of the contribution of different forms of activity to overall activity usually expressed as energy expenditure (Bernstein *et al.*, 1998). However, their utilisation in large representative surveys is limited by their sheer size. Therefore, short physical activity questionnaires are most widely used, though their validity is clearly poorer than that of their larger counterparts (Sequeira *et al.*, 1995, Mäder *et al.*, 2002). Even by respecting the necessary procedures for trans-cultural adaptation, questionnaires in general are prone to differential use not only between countries, but also between different cultural groups within the same country (Martin and Mäder, 2002).

A standardised and internationally comparable physical activity questionnaire (IPAQ) has been developed (www.ipaq.ki.se) and only recently was evaluated in representative surveys (De Bourdeaudhuij *et al.*, 2003, Craig *et al.*, 2003, Hallal and Victoria, 2004).

The Swiss Health Survey is a large (nearly 20.000 participants in 2002) representative population survey conducted by the Swiss Federal Statistical Office every five years and is also the most reliable and the most important data source for health-related physical activity in Switzerland on a national scale. It has used an item on sweat episodes in physical activities during leisure time since 1992, a secondary analysis of the respective data was published by Calmonte and Kälin (1997). The Swiss Health Survey 1997 included several more items on physical activity and has given researchers the opportunity to study the associations between physical activity behaviour and several health variables in more detail (Martin *et al.*, 2000). Among others, it contained an item on daily transport by bicycle or on foot. The Swiss Health Survey 2002 encompassed additional questions on activities of moderate intensity.

In a supplementary smaller study, the Swiss HEPA survey 1999, activities of moderate intensity and intention for behavioural change were assessed for the first time (Martin *et al.*, 2000). The Swiss HEPA survey 2001, again carried out in a sample of 1500 participants, has allowed researchers to compare the prevalence data with those of two years before and has also used a new set of less suggestive items (Martin, 2002).

The Swiss Household Panel (www.swisspanel.ch) carried out annually since 1999 in more than 5000 households also contains some items on physical activity and allows the observation of spontaneous change in activity patterns in the same individuals as well as the cross-sectional and longitudinal associations with potentials predictors on the individual and on the household level.

## *Walking and cycling assessment*

For the measurement of travel behaviour and also transport-related physical activity, methods used are diaries, questionnaires, manned or automated counting stations and as a new approach in some studies, but not yet in routine use for monitoring also GPS. While diaries, questionnaires and devices like GPS provide data on individuals, counting stations can only provide aggregated information.

The most important data source on the national level is the scientific survey of the population's travel behaviour conducted by Bundesamt für Raumentwicklung and Bundesamt für Statistik (2001, travel behaviour microcensus) carried out every five years since 1974.

The 2000 microcensus included nearly 30.000 participants selected in a random-procedure who were asked about every travel distance of more than 25 meters during the preceding day with respect to distance and time travelled, transport mode and purpose.

Similar surveys are carried out by the Federal Statistical Office for transport and tourism and by the University of St. Gallen for the travel market in Switzerland, but an integration of the respective results has not yet taken place. None of the surveys mentioned includes information on overall physical activity.

Systematic counting station networks on a national level exist only for motorised transport. Some counting on cycling is carried out by the foundation "Cycling in Switzerland" (www.cycling-in-switzerland.ch) and used for modelling, but an actual national cycling counting network is only in the planning stage. Some of the bigger cities in Switzerland like for example Bern or Zurich use systematic counting for walking and cycling, but a national counting station network for walking does not yet exist.

An integration of data from surveys and counting stations has not yet taken place on a larger scale.

## 4.    Conclusions

### *Research needs*
- In general, most experience regarding design and feasibility of intervention studies and the effects of these programs is available for interventions on the individual and group level. There is good evidence that interventions on these levels can increase physical activity among the inactive.

- There is good evidence that interventions on the worksite setting are effective to increase physical activity.

- There seem to be virtually no studies assessing the impact of interventions targeting transport policies and environmental changes on physical activity – either on their effect to increase individual physical activity behaviour in general, and particularly not regarding their potential to reach the inactive segment of a population. An observed increase in bicycle use for example is far less relevant for public health if it occurs in individuals already physically active in other ways than if the same increase could be observed in a previously inactive group.

- In general it is well established that urban populations are on average less active than rural populations. This difference might even be more pronounced in developing countries (Torun *et al.*, 2002), But is also influenced by gender, jobs, and socio-economic status. An easier access to motorised traffic might contribute to increased rate of sedentarism in urban areas but the absolute amount of this contribution is difficult to assess.

- For school children it has been shown that active commuting to school contributes relevantly to their overall physical activity so that transport policies that successfully encourage active commuting to school would have an impact on overall physical activity and rate of sedentarism in this age group (Tudor-Locke *et al.*, 2002)

- It is difficult to quantify and compare intervention effects because of different measures for physical activity and the respective outcomes of the studies.

Therefore:

- Examples of good practice have to be documented.

- Baseline evaluations before the start of policy and environmental interventions are essential to assess any changes in individual physical activity behaviour.

Elements of a longitudinal monitoring system based on surveys have been established in Switzerland for both overall physical activity and walking and cycling as transport-related physical activity. An integration of the data from both sources has not yet taken place.

The fact that these monitoring systems rely solely on questionnaire data imposes limitations on inter-cultural comparisons both within the country and with other countries. The use of objective measurements like accelerometry might open up new possibilities in this respect.

## *Policy recommendations*

Physical inactivity is a world-wide public health problem. While quantifying the impact of transport patterns and the built environment on activity behaviour is still a matter of research, the general fact that sustainable transport modes attribute to a healthier lifestyle is evident.

In general, there are not enough financial resources in the health and physical activity sector to conduct environmental interventions. Therefore collaborations with experts from the traffic and environmental sector are necessary. The interventions of these partners should be evaluated also regarding individual physical activity behaviour.

## References

Andersen, L.B., Schnohr, P., Schroll, M., and Hein, H.O. (2000) All-cause mortality associated with physical activity during leisure time, work, sports, and cycling to work, *Arch. Intern. Med.* 160, 1621-1628.

Bernstein, M., Sloutskis, D., Kumanyika, S., Sparti, A., Schutz, Y., and Morabia, A. (1998) Databased approach for developing a physical activity frequency questionnaire, *Am. J. Epidemiol.* 147, 147-154.

Bundesamt für Raumentwicklung, and Bundesamt für Statistik (2001) Mobilität in der Schweiz, Ergebnisse des Mikrozensus 2000 zum Verkehrsverhalten.: Bundesamt für Raumentwicklung, Bundesamt für Statistik, Bern and Neuenburg.

Bundesamt für Sport BASPO (2000) The Swiss Federal Governments Concept for a National Sports Policy, November 30th 2000, BASPO, Magglingen.

Bundesamt für Sport BASPO, Bundesamt für Gesundheit BAG, Gesundheitsförderung Schweiz, and Netzwerk Gesundheit und Bewegung Schweiz (2002) Gesundheitswirksame Bewegung. Ein Grundsatzdokument, BASPO, Magglingen.

Bundesamt für Sport BASPO, Bundesamt für Gesundheit BAG, Schweizerische Beratungsstelle für Unfallverhütung bfu, Schweizerische Unfallversicherungsanstalt Suva, Abteilung für Medizinische Ökonomie des Instituts für Sozial- und Präventivmedizin und des Universitätsspitals Zürich, and Netzwerk Gesundheit und Bewegung Schweiz (2001) Volkswirtschaftlicher Nutzen der Gesundheitseffekte der körperlichen Aktivität: erste Schätzungen für die Schweiz, *Schweizerische Zeitschrift für Sportmedizin und Sporttraumatologie* 49, 84-86.

Calmonte, R., and Kälin, W. (1997) Körperliche Aktivität und Gesundheit in der Schweizer Bevölkerung. Eine Sekundäranalyse der Daten aus der Schweizerischen Gesundheitsbefragung 1992, Institute for Social and Preventive Medicine, Bern.

Cleary, J., and McClintock, H. (2000) Evaluation of the Cycle Challenge project: a case study of the Nottingham Cycle-Friendly Employers' project, *Transport Policy* 7, 117-125.

Craig, C.L., Marshall, A.L., Sjostrom, M., Bauman, A.E., Booth, M.L., Ainsworth, B.E., Pratt, M., Ekelund, U., Yngve, A., Sallis, J.F., and Oja, P. (2003) International physical activity questionnaire: 12-country reliability and validity, *Med. Sci. Sports. Exerc.* 35, 1381-1395.

De Bourdeaudhuij, I., Sallis, J.F., and Saelens, B.E. (2003) Environmental correlates of physical activity in a sample of Belgian adults, *Am. J. Health Promot.* 18, 83-92.

Department of Health and Human Services (1996) Physical activity and health: A report of the Surgeon General, U.S. Department of Health and Human Services, Centers for Disease Control and Prevention, National Center for Chronic Disease Prevention and Health Promotion, Atlanta, GA.

European Commission, Directorate-General for Employment, Industrial Relations and Social Affairs (1999) A pan-EU survey on consumer attitudes to physical activity, body-weight and health, Office for Official Publications of the European Communities, Luxembourg.

Hallal, P.C., and Victoria, C.G. (2004) Reliability and validity of the International Physical Activity Questionnaire (IPAQ), *Med. Sci. Sports Exerc.* 36, 556.

Haskell, W.L. (1994) Health consequences of physical activity: understanding and challenges regarding dose-response, *Med. Sci. Sports Exerc.* 26, 649-660.

Hendriksen, I.J.M., Zuiderveld, B., Kemper, H.C.G., and Bezemer, P.D. (2000) Effect of commuter cycling on physical performance of male and female employees, *Med. Sci. Sports Exerc.* 32, 504-510.

Jimmy, G., and Martin, B.W. (in preparation) Implementation and effectiveness of a primary care based physical activity counselling scheme.

Kahn, E.B., Ramsey, L.T., Brownson R.C., Heath, G.W., Howze, E.H., Powell, K.E., Stone, E.J., Rajab, M.W., Corso, P., and the Task Force on Community Preventive Services (2002) The effectiveness of interventions to increase physical activity: a systematic review, *American Journal of Preventive Medicine* 22, 73-107.

King, A.C. (1994) Clinical and Community Interventions to Promote and Support Physical Activity Participation, in R. Dishman (ed.), *Advances in exercise adherence*, Human Kinetics Pub., p185.

King, A.C. (1998) How to promote physical activity in a community: research experiences from US highlighting different community approaches, *Patient Education and Counselling* 33, S3-S12.

Knowler, W.C., Barrett-Connor, E., Fowler, S.E., Hamman, R.F., Lachin, J.M., Walker, E.A., and Nathan, D.M. (2002) Diabetes Prevention Program Research Group. Reduction in the Incidence of Type 2 diabetes with lifestyle intervention or metformin, *N. Engl. J. Med.* 346, 393-403.

Lamprecht, M., and Stamm, H.P. (1999) Bewegung, Sport und Gesundheit in der Schweizer Bevölkerung. Eine Sekundäranalyse der Daten aus der Schweizerischen Gesundheitsbefragung 1997 im Auftrag des Bundesamtes für Sport. Forschungsbericht, L&S Sozialforschung und Beratung AG, Zurich, (Short version available in German, French, Italian and English).

Leitzmann, M.F., Rimm, E., Willet, W.C., Spiegelman, D., Grodstein, F., Stampfer, M.J., Colditz, G.A., and Giovannucci, E. (1999) Recreational physical activity and the risk of cholecystectomy in women, *N. Engl. J. Med.* 341, 777-784.

Mäder, U., Martin, B., Schutz, Y., Bernstein, M., and Marti, B. (2002) Physiological validation study of five widely-used epidemiological physical activity short questionnaires, based on heart rate monitoring, accelerometry and indirect calorimetry. Research report, Federal Office of Sports, Institute of Sport Sciences, Magglingen.

Marcus, B.H., Selby, V.C., Niaura, R.S., and Rossi, J.S. (1992) Self-efficacy and the stages of exercise behaviour change, *Research Quarterly for Exercise and Sport* 63, 60-66.

Marti, B., and Hättich, A. (1999) Bewegung – Sport – Gesundheit: epidemiologisches Kompendium, Haupt, Bern, Stuttgart, Wien.

Marti, B., and Martin, B.W. (2001) Sportliches Training oder Bewegung im Alltag zur Optimierung von Gesundheit und Lebensqualität? *Therapeutische Umschau* 58, 189-195.

Martin, B.W. (2002) Physical activity related attitudes, knowledge and behaviour in the Swiss population: comparison of the HEPA Surveys 2001 and 1999, *Schweiz. Z. Sportmed. Sporttraumatol.* 50, 164-168.

Martin, B.W., and Mäder, U. (2002) Körperliches Aktivitätsverhalten in der Schweiz. In: G. Samitz, and G. Mensink (eds.), Körperliche Aktivität in Prävention und Therapie. Evidenzbasierter Leitfaden für Klinik und Praxis, Marseille Verlag GmbH, München.

Martin, B.W., Beeler, I., Szucs, T., Smala, A.M., Brügger, O., Casparis, C., Allenbach, R., Raeber, P.A., and Marti, B. (2001) Economic benefits of the health-enhancing effects of physical activity: first estimates for Switzerland. Scientific position statement of the Swiss Federal Office of Sports, Swiss Federal Office of Public Heal, Swiss Council for Accident Prevention, Swiss National Accident Insurance Organisation (SUVA), Department of Medical Economics of the Institute of Social and Preventive Medicine and the University

Hospital of Zurich and the Network HEPA Switzerland, *Schweiz. Z. Sportmed. Sporttraumatol.* 49, 131-133.

Martin, B.W., Jimmy, G., and Marti, B. (2001) Bewegungsförderung bei Inaktiven: Eine Herausforderung auch in der Schweiz, *Therapeutische Umschau* 58, 196-201.

Martin, B.W., Lamprecht, M., Calmonte, R., Raeber, P.A., and Marti, B. (2000) Körperliche Aktivität in der Schweizer Bevölkerung: Niveau und Zusammenhänge mit der Gesundheit. Gemeinsame wissenschaftliche Stellungnahme von Bundesamt für Sport (BASPO), Bundesamt für Gesundheit (BAG), Bundesamt für Statistik (BFS) und Netzwerk Gesundheit und Bewegung Schweiz, *Schweiz. Z. Sportmed. Sporttraumatol.* 48, 87-88.

Martin, B.W., Mäder, U., and Calmonte, R. (1999) Einstellung, Wissen und Verhalten der Schweizer Bevölkerung bezüglich körperlicher Aktivität: Resultate aus dem Bewegungssurvey 1999, *Schweiz. Z. Sportmed. Sporttraumatol.* 47, 165-169.

Michaud, D.S., Giovannucci, E., Willett, W.C., Colditz, G.A., Stampfer, M.J., and Fuchs, C.S. (2001) Physical activity, obesity, height, and the risk of pancreatic cancer, *JAMA* 286, 921-929.

Mutrie, N., Carney, C., Blamey, A., Crawford, F., Aitchison, T., and Withelaw, A. (2002) "Walk in to Work Out": a randomised controlled trial of a selfhelp intervention to promote active commuting, *J. Epidemiol. Community Health* 56, 407-412.

Oguma, Y., Sesso, H.D., Paffenbarger, R.S., and Lee, I.M. (2002) Physical activity and all cause mortality in women: a review of the evidence, *Br. J. Sports. Med.* 36, 162-72.

Oja, P., and Vuori, I. (2000) *Promoting of transport walking and cycling in Europe: strategy directions*, UKK Institute, Tampere, Finland, ISSBN 951-9101-34-9.

Oja, P., Vuori, I., and Paronen, O. (1998) Daily walking and cycling to work: their utility as health-enhancing physical activity, *Patient Education and Counselling* 33, S87-S98.

Pate, R.R., Pratt, M., Blair, S.N., Haskell, W.L., Macera, C.A., Bouchard, C., Buchner, D., Ettinger, W., Heath, G.W., and King, A.C. (1995) Physical activity and public health: a recommendation from the Centers for Disease Control and Prevention and the American College of Sports Medicine, *JAMA* 273, 402–407.

Pikora, T., Giles-Corti, B., Bull, F., Jamrozik, K., and Donovan, R. (2003) Developing a framework for assessment of the environmental determinants of walking and cycling, *Social Science & Medicine* 56, 1693-1703.

Rye, T. (1999) Employer attitudes to employer transport plans: a comparison of UK and Dutch experience, *Transport Policy* 6, 183-196.

Sallis, J.F., and Owen, N. (1999) *Physical activity and behavioral medicine*, Sage Publications, London.

Sallis, J.F., Bauman, A., and Pratt, M. (1998) Environmental and policy interventions to promote physical activity, *American Journal of Preventive Medicine* 15, 379-397.

Sequeira, M.M., Rickenbach, M., Wietlisbach, V., Tullen, B., and Schutz, Y. (1995) Physical activity assessment using a pedometer and its comparison with a questionnaire in a large population survey, *Am. J. Epidemiol.* 142, 989-99.

Sjöström, M. (2002) Level and pattern of physical activity in the population, Abstract in S. Miilunpalo, and Tulimäki (eds.) International Symposium on Health-Enhancing Physical Activity (HEPA). Evidence-based Promotion of Physical Activity. Helsinki, Finland, September 1-2, 2002. Book of Abstracts, UKK Institute Tampere, p 67.

Starling, R.D., Matthews, D.E., Ades, P.A., and Poehlman, E.T. (1999) Assessment of physical activity in older individuals: a doubly labeled water study, *J. Appl. Physiol.* 86, 2090-2096.

Titze, S., Martin, B.W., Seiler, R., Stronegger, W., and Marti, B. (2001) Effects of a lifestyle physical activity intervention on stages of change and energy expenditure in sedentary employees, *Psychology of Sport and Exercise* 2, 103-116.

Torun, B., Stein, A.D., Schroeder, D., Grajeda, R., Conlisk, A., Rodriguez, M., Mendez, H., and Martorell R. (2002) Rural-to-urban migration and cardiovascular disease risk factors in young Guatemalan adults, *Int. J. Epidemiol.* 31, 218-226.

Tudor-Locke, C., Neff, L.J., Ainsworth, B.E., Addy, C.L., and Popkin, B.M. (2002) Omission of active commuting to school and the prevalence of children's health-related physical activity levels: the Russian longitudinal monitoring study, *Child-Care-Health-Dev.* 28, 507-512.

Vuori, I. (1998) Does physical activity enhance health? *Patient Education and Commuting* 33, S95-S103.

WHO (2002) *The World Health Report 2002: Reducing risks, promoting healthy life*, World Health Organisation WHO, Geneva.

# MOBILITY AND TRANSPORT IN THE JOHANNESBURG PLAN OF IMPLEMENTATION AND NATIONAL ENVIRONMENTAL ACTION PLANS

L. HENS
*Department of Human Ecology,*
*Faculty of Medicine and Pharmacy,*
*Vrije Universiteit Brussel, Laarbeeklaan 103,*
*B-1090 Brussels, BELGIUM*

## Summary

Mobility and transport result in both direct and indirect environmental health effects. The direct effects entail the health consequences of air pollution, noise, traffic accidents and sedentarism. The indirect health effects result from global changes and the use of resources. This latter links mobility and transport up with general pollution of water, soil and air and the production of waste.

The paper analyses the responses of the international policy community to deal with the environmental health problems resulting from increasing traffic and mobility. Two groups of documents are analysed: the outcome of the World Summit on Sustainable Development (Johannesburg, South Africa, 2002) and the by 2003 published National Environmental Health Action Plans (NEHAPs).

The Johannesburg Plan of Implementation (JPoI) provides different anchor points for a mobility and transport policy from which environment and health can benefit. These include:
- (-) in the chapter on consumption and production: improved technical performance and safer, affordable and energy-efficient transport systems;
- (-) in the chapter on natural resources: support of the Kyoto and Montreal Protocols;
- (-) in the chapter on health: preventive, promotive and curative programmes on traffic-pollution related diseases and the phasing out of lead in gasoline.

Also partnerships launched in Johannesburg target specific aspects of health and mobility. The paper discusses the example of lead-poisoning action.

*P. Nicolopoulou-Stamati et al. (eds), Environmental Health Impacts of Transport and Mobility,* 155-170.

The objectives on transport in 9 NEHAPs are inventoried. The analysis shows that European countries merely aim at reducing the emissions and noise from traffic and are determined to bring down the mortality and injury risks resulting from accidents. Only few plans pay attention to the problems resulting from sedentarism. Traffic, environment and health are policy fields that cover the problems in this paper. In spite of international efforts to integrate these elements, in daily business they operate mainly within divided structures in most European Countries. Their integration is a major defeat for the years to come.

## 1.  Introduction

(1)  *Air pollutants:* traffic emissions contribute to CO, $SO_2$, $NO_x$, volatile organic compounds, ozone and (fine) particulates in the air. On a short term basis these pollutants are known to increase the number of deaths, hospital admissions and emergency visits, in particular for respiratory and cardiovascular problems. Of particular concern are fine particulates ($PM_{2.5 \text{ and smaller}}$). In Europe and the Asian part of Russia an estimated 100,000 people (out of approximately 10 million deaths) a year die prematurely as a result of exposure to air pollutants (Dora and Phillips, 2000). In Germany particulate pollution causes an estimated number of 25,000 deaths a year (UPI, 1999). In the long term, the carcinogenic potential of a number of these pollutants is significant.

(2)  *Noise:* road traffic has become the most important source of exposure to noise in the urban environment. Airplane noise threatens the quality of life of people living around airports. Environmental noise causes hearing impairment, hypertension, ischaemic heart disease, annoyance, decreased learning performance in children at school, sleep disturbance, changes in EEG, heart rate, hormone levels and the mood next day. Heart rate changes, annoyance and sleep disturbance have been reported at exposure levels as low as 40-45 dB(A).

(3)  *Road traffic accidents* (RTAs): in the WHO European Region approximately 120,000 people die every year and more than 2.5 million are injured as a result of road traffic collisions (Dora and Phillips, 2000). In Germany, the annual number of deaths from road traffic accidents is 8,758, or approximately one third of the number of premature deaths as a result of exposure to (fine) particulates (UPI, 1999). Pedestrians are the most vulnerable. Road traffic accidents also are not distributed equally between different social groups. In the UK, a child in the lowest socio-economic group is six times more likely to be killed or seriously injured than a child in the highest group (TRL, 2000). For Germany it was shown that, based on the total number of hours of life lost through RTAs and the size of the German vehicle fleet, the authors estimate that in the ten-year average life of a car, each car is responsible for 820 hours of lost life and 2,800 hours of handicapped life. This compares to the average amount of use of a car in its ten-year life, which is 2,400 hours (UPI, 1999).

(4)  *Sedentarism:* the attributable fraction of mortality from physical inactivity is estimated to range between 5 and 10 percent of the total mortality in the WHO European Region (WHO, 2002). This is equivalent to a few hundred thousands of deaths a year. In Belgium, over 30 percent of the car trips are done to bridge distances that are shorter than 1 km and could easily be done on foot or by bicycle. Using cars for these short-distance trips not only increases traffic unnecessarily, but contributes in a major way to the sedentary lifestyle of Europeans.

In particular the effects of atmospheric pollution and noise have been extensively reviewed in this book. Traffic also has a number of indirect environmental health effects. The main ones are the following:

(1)  *Climate changes:* world-wide, emissions of $CO_2$ from all transport sectors are currently approximately 22 per cent of global carbon emissions from fossil fuel use (IPCC, 2000). Carbon dioxide emissions in the European Union (EU) increased by 47 per cent between 1985 and 2001. Other sectors increased by 4.2 per cent. More than 30 per cent of final energy in the EU is now consumed by transport. Road transport is the main cause of this increase and contributed 84 per cent of the $CO_2$ emissions from transport in 1998. In the same year, EU greenhouse gas emissions from international transport (aviation and shipping) amounted to 5 per cent of the total emissions in the Union. Aviation emissions are expected to rise dramatically in future years and to account for 20 per cent of the greenhouse gas emissions by the year 2020 (Whitelegg, 2003). The forecast is that the greenhouse gas emissions from transport in the EU will be 39 per cent above the 1990 levels by 2010 (IEA, 2001). In this way Europe will continue to contribute to greenhouse gas effects also on health (premature deaths, increased tropospheric ozone problems, geographically altered spreading of biological vector diseases such as malaria) (McMichael *et al.*, 1996).

(2)  *Resource use:* the energy needed to produce a car is comparable to the amount of energy the car consumes during its life-span. An average car weighs about 1.14 ton, most of it steel and plastic (UPI, 1999). All this material has to be extracted from the ground and/or processed from other materials. At every stage of its life cycle of sourcing, manufacturing and assembly, there are transport and energy costs and waste produced. In this way each car is responsible for 25 ton of waste (UPI, 1999). The other half of the energy a car consumes is oil. These days the world consumption of oil totalises some 74 million barrels a day (IEA, 2001). Over 50 per cent of this are transportation fuels and their share is rising steeper than that of any other sector. A full materials analysis should also include the effects of extracting and transporting crude oil. Oil spills in oceans are an aspect of this (including accidental spillage and routine washing out of tanks whilst at sea). This amounts to 13 litres of crude oil deposited in the oceans for every car. Land contamination is also a problem and a source of environmental and political crisis in the land of the Ogoni in Nigeria (Babalola, 2002). Also land use is to be addressed. In Germany the total land used for parking and road is 3,800 $km^2$ or 200 $m^2$ of

land that is allocated to an average car. 3,800 km$^2$ is 60 per cent higher than the total land required for all housing space for every German citizen (UPI, 1999). In this way traffic and mobility are linked up with the general pollution of air, water and soil and the other effects resulting from resource use. The health effects resulting from these impacts are real, but currently not quantified.

While during the last decade a lot of progress has been made in making the environmental health consequences more objective and transparent, one might question the policy impacts of these data. Current traffic policy in the EU and the member states is largely independent of environmental health data and largely driven by more strict technical emission regulations, that are expected to reduce the total amount of traffic-related emissions even when the car park space increases. Is this what we really want for the future? Most probably, the important questions are not about engineering, but about ways of living (Peñalosa, 2003). This will necessitate a new paradigm of mobility and transport. What will this look like?

As the problem has important international and world-wide aspects, international environmental policy should be able to draw the main contours of this new paradigm. The question is of particular interest also because international environmental policy was accessible for and sensitive to questions of environmental health. From the Montreal Protocol (1987) on the reduction of ozone layer depleting substances to the Stockholm Convention (2001) on the phasing of the 20 most important persistent organic pollutants, there is an impressive record of agreements that have largely been reached by environmental health concerns (see Table 19 for a selected list of these agreements).

**Table 19**. Major international agreements on health and environment.

| | |
|---|---|
| 1. | Montreal Protocol on substances that deplete the ozone layer (September 1987) |
| 2. | Basel Convention on the control of transboundary movements of hazardous wastes and their disposal (March 1989) |
| 3. | UN Conference on Environment and Development, Rio de Janeiro, Brazil (June 1992) |
| 4. | Millennium Development Goals (September 2000) |
| 5. | Bahia Declaration on Chemical Safety (October 2000) |
| 6. | Stockholm Convention on Persistent Organic Pollutants (May 2001) |

In this paper the aspects of environmental health and mobility are discussed as they emerge from two groups of recent international policy documents:

(1) the 'political declaration' (JPD) and the 'Plan of Implementation' (JPoI) that were agreed upon during the World Summit on Sustainable Development (WSSD) or commonly called the Johannesburg Summit, as it was organised in Johannesburg, South Africa from August 26$^{th}$ until September 4$^{th}$, 2002.
(2) the National Environmental Health Action Plans (NEHAPs). These are the result of a decision of the second interministerial Conference on Health and

Environment (Helsinki, Finland, 1994). On that occasion the ministers decided that each country in the area of the World Health Organisation, Regional Office for Europe should prepare a national plan to act on environmental health issues. The first plans were published in 2003.

## 2. World Summit on Sustainable Development (WSSD)

The main driver behind the WSSD was the documented finding that although progress was made with the implementation of sustainable development (SD) during the past decade, this progress was insufficient and too slow to curb the main development of unsustainability. Therefore the conference focussed on how to (more effectively and more efficiently) implement Rio's Agenda 21, a world-wide action plan for SD that resulted from the UNCED Conference in Rio de Janeiro, Brazil, in June 1992.

Part of this discussion has to do with setting priorities. In this context UN Secretary-General Kofi Annan launched *WEHAB* - a world-wide focus on issues (inter)related to *Water, Energy, Health, Agriculture* and *Biodiversity*. This means that health in a setting that combines environmental health issues with other world-wide health problems as HIV-AIDS is at the core of SD. Figure 34 relates health as understood in an SD context with the four other WEHAB areas. It is clear from this analysis that, although traffic and mobility are not mentioned explicitly, both the direct (e.g. urban air pollution) and indirect (e.g. climate change, water pollution) effects mentioned in the introduction of this paper can be associated with this analysis. They cluster in the 'health and energy' package.

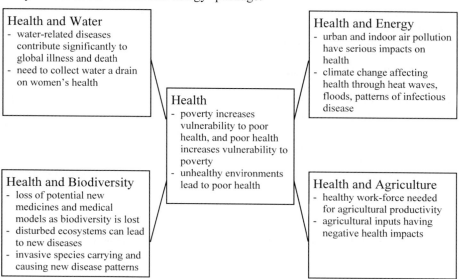

**Figure 34**. Health in a context of sustainable development as defined by the WSSD-WEHAB (Water, Energy, Health, Agriculture, Biodiversity) priorities.

The WSSD resulted in three main outputs:

(1) The political declaration is at the same time an overarching and concluding document pointing to the main issues addressed by the conference and echoing the attachment of the international community to SD.

(2) The 'Plan of Implementation of the World Summit on Sustainable Development' (JPoI) that builds further on the achievements made during the last decade and focuses on the political alleviation of the main sustainability problems as they exist in 2002.

(3) Partnerships: that are an open invitation from the UN and its member countries to the civil society to put SD into practice.

A more extensive introduction to WSSD and its immediate results can be found in Hens and Nath (2003). In the following sections the health aspects of traffic and mobility as they appear in these main WSSD outputs will be analysed.

### 2.1.    The Political Declaration

This document includes 69 articles grouped in 6 sections. In section 4, which is on "The Johannesburg Commitment on Sustainable development", health and energy are mentioned as general concerns in the following way: "We welcome the focus of the Johannesburg Commitment on the basic requirements of human dignity, access to clean water and sanitation, energy, health care, food security and biodiversity..."

The political declaration sets the Johannesburg Summit scene. However, it is currently unclear what the political declaration adds to the international policy on traffic and environmental health.

### 2.2.    The Plan of Implementation (JPoI)

Apart from an introduction, the JPoI includes 9 chapters that are listed in Table 20. The table also indicates to what extent core terms in the health and mobility debate are mentioned. Transport is mentioned 5 times in the plan. Among the direct health effects only air pollution is mentioned in the JPoI. Noise has no entry. Injuries and sedentarism are mentioned in the health chapter. The main link to the issue is through 'energy' that is mentioned 45 times in total. Although many energy references in the JPoI have no or most indirect links with the mobility and health issue.

An analysis by chapter shows that three chapters are of core importance:

*Chapter III: Changing unsustainable patterns of consumption and production*

The way societies produce and consume was already in Agenda 21 identified as main fundamental causes of unsustainability and environmental degradation, in particular in industrialised countries. Moreover, in preparing the WSSD, it was clear that particularly this area made very little progress.

The JPoI addresses a selected number of most important items in the production and consumption debate: 10-year work programmes to shift towards sustainable production and consumption; enhance corporate environmental and social responsibility and accountability; energy; transport; waste; and chemicals.

**Table 20**. Mobility and health as mentioned in the different chapters of the Johannesburg Plan of Implementation (JPoI).

| | Chapter number in the JPoI | 1 | 2 | 3 | 4 | 5 | 6 | 7 | 8 | 9 | 10 | 11 | Total number of mentions |
|---|---|---|---|---|---|---|---|---|---|---|---|---|---|
| | transport | 0 | 0 | 4 | 1 | 0 | 0 | 0 | 1 | 0 | 0 | 0 | 6 |
| | transport and health | 0 | 0 | 0 | 0 | 0 | 0 | 0 | 0 | 0 | 0 | 0 | 0 |
| | energy | 0 | 13 | 59 | 5 | 0 | 1 | 5 | 7 | 2 | 0 | 1 | 93 |
| | energy and health | 0 | 0 | 0 | 0 | 0 | 0 | 0 | 0 | 0 | 0 | 0 | 0 |
| | air pollution | 0 | 0 | 0 | 3 | 0 | 0 | 0 | 0 | 0 | 0 | 0 | 3 |
| | air pollution and health | 0 | 0 | 0 | 0 | 0 | 0 | 0 | 0 | 0 | 0 | 0 | 0 |
| Number of mentions | Chapter number in the JPoI | 1 | 2 | 3 | 4 | 5 | 6 | 7 | 8 | 9 | 10 | 11 | Total number of mentions |
| | noise | 0 | 0 | 0 | 0 | 0 | 0 | 0 | 0 | 0 | 0 | 0 | 0 |
| | noise and health | 0 | 0 | 0 | 0 | 0 | 0 | 0 | 0 | 0 | 0 | 0 | 0 |
| | atmospheric pollution | 0 | 0 | 0 | 0 | 0 | 0 | 0 | 0 | 0 | 0 | 0 | 0 |
| | accidents | 0 | 0 | 0 | 0 | | | | | | | | |
| | climate changes | 0 | 0 | 0 | 0 | 0 | 0 | 0 | 0 | 0 | 0 | 0 | 0 |
| | resource use | 0 | 0 | 0 | 0 | 0 | 0 | 0 | 0 | 0 | 0 | 0 | 0 |

Both the energy and the transport articles are relevant for the discussion of this paper. Article 20 is important as, without mentioning transport explicitly, it refers to more energy efficiency, advanced energy technologies, alternative technologies, increased use of renewable energy, advanced and cleaner fossil fuel technologies, reduction of flaring and venting of gas associated with crude oil production and policies to reduce energy market distortions. In general, although also policies and instruments to realise these aims are mentioned, the article has a strong scientific-technical focus. In relation to this discussion, it was noticeable that one of the companies that organised an outspoken presence at the WSSD was BMW presenting at large its hydrogen fuelled car.

Article 21 is about transport systems and services. Sustainable transport is safe, affordable, efficient, more energy-efficient; it reduces pollution, congestion and adverse health effects; it limits urban sprawls. In this article there is a clear focus on both direct and indirect health effects of the current traffic system. Sustainable transport strategies should improve urban air quality and reduce greenhouse gas emissions. This needs better vehicle technologies that are more environmentally sound, affordable and socially acceptable. Sustainable transport also means investment and partnerships in multi-modal transportation systems. This includes public mass transportation and better transport in rural areas.

*Chapter IV: Protecting and managing the natural resource base of economic and social development*

This is the longest and most comprehensive chapter of the JPoI. It deals with how to limit the increasing impact that human societies have on the integrity of ecosystems that provide the essential resources and services for them.

The chapter includes sections on: drinking water; oceans, seas, islands and coastal areas; fishery; risk assessment and disaster management; greenhouse gas emissions and the Kyoto Protocol; agriculture; desertification; mountain ecosystems; sustainable tourism; biodiversity; forests and trees; and mining of minerals and metals.

Among these, article 38 on efforts to ensure the putting into force of the Kyoto Protocol to the United Nations Framework Convention on Climate Change, is of particular importance for the mobility and health discussion.

The Kyoto Protocol is a central element of discussion at the WSSD. After the announcement that the United states and later also Australia would not join the protocol, at the beginning  of the summit, it was uncertain whether sufficient countries with an emission load that was significant enough would join the protocol to put it into action. The positive attitude of Russia towards the protocol was a positive element. Moreover, during the conference, leading developing countries including Brazil, India and South Africa advocated a regime with stronger greenhouse gas reduction targets.

Article 38 refers to a set of general, national actions to support and intensify the Kyoto Protocol. These include: technical and financial assistance and capacity building to developing countries and countries with economies in transition; building of scientific and technological capabilities and networks; systematic observations and monitoring of the earth's atmosphere; and special attention to the Arctic environment.

The JPoI does not mention traffic specifically in relation to Kyoto, neither does it provide a sectoral analysis of the problems of implementation. Nevertheless, it is obvious that in many industrialised countries the current evolution of the emissions mainly by car and air traffic are perpendicular to the Kyoto targets. In the EU the greenhouse gas emissions from transport are estimated to increase by 39 per cent between 1990 and 2010 (IEA, 2001). This is almost impossible to match with an overall (sectors) decrease of 5.5 per cent. Moreover, one should take into account that traffic evolves as a sector that sociologically is more difficult to change than other sectors contributing to air pollution.

Article 39 deals with air pollution in a wider context than just the greenhouse gasses. It calls for a reduction of (transboundary) air pollution and its main effects, including acid deposition and stratospheric ozone depletion. This should be done by strengthening the capacities of developing countries and countries with economies in transition to measure, among others, health impacts. Moreover, the regime of the Montreal Protocol on Substances that Deplete the Ozone Layer should be further supported and elaborated.

It is obvious that the contribution of car traffic to urban environments in particular and of air traffic in transboundary air pollution are implicated by this article.

*Chapter VI: Health and sustainable development*

Johannesburg favours a wide interpretation of health in an SD context. The JPoI links health up with article 1 of the Rio Declaration on Environment and Development stating that 'human beings ... are entitled to a healthy and productive life, in harmony with nature'. To achieve the goals of sustainable development, health is linked up with, among others, poverty eradication, particular attention to vulnerable groups (unborns, children, elderly, diseased), traditional medicine knowledge and practice, occupational health, environmental health and the HIV-AIDS pandemic.

The chapter addresses implicitly traffic and mobility where it advocates '... preventive, promotive and curative programmes to address non-communicable diseases and conditions, such as cardiovascular diseases, cancer, ..., respiratory disease, injuries, ... and associated risk factors, including ... lack of physical activity' (article 54[o]). The same applies to paragraph 54(k) that advocates the launch of international capacity building initiatives that assess health and environment linkages and use this knowledge to create more effective policy responses to environmental threats to human health.

Traffic and mobility are addressed explicitly in article 56 which deals with reducing respiratory diseases in children by *inter alia* phasing out lead in gasoline and the use of cleaner fuels.

Overall, this chapter mentions explicitly, apart from some specific noise impacts, all direct impacts of traffic on health. It calls, moreover, for specific action on tetraethyl- and tetramethyl lead and cleaner fuels. It advocates more collaborative research and monitoring on environmental health effects.

### 2.3.   *Partnerships*

Partnerships for SD, conceptualised during the third PrepCom, are the third main outcome of the WSSD. The concept was mooted in response to this question: 'how best could civil society make project-wise contribution to the implementation of SD?'. However, there is at present a good deal of confusion over the concept of such partnerships and their *modus operandi*.

Type II partnerships cover the main issues of the JPoI, including WEHAB; but also networking on SD in science, education and decision-making; finance, trade and technology transfer; sustainable production and consumption; urbanisation and other areas.

At the WSSD, 218 partnerships on which the United Nations had agreed, were announced. However, the initiative is still open for contributions. By June 2003, 266 partnerships were announced at the website of the Summit (UN, 2003).

In particular the health and energy sections include initiatives that are relevant for the mobility and environmental health discussion. Table 21 lists the most relevant type II partnerships in this context. Note that the ones on cleaner fuels, LPG and lead directly fit into the objectives of article 56 of the JPoI. The pan-European programme on transport, health and environment originates from earlier work on the issue by the regional office of WHO-Europe.

**Table 21**. Type II partnerships and leading partners on mobility and health by May 2003.

| Type II Partnership on mobility and Health | Leading Partners |
| --- | --- |
| Global Partnership towards Cleaner Fuels | International Fuel Quality Center, Brussels, Belgium |
| The LPG Challenge | UNDP, World LP Gas Association, New York |
| Global Lead Initiative | Alliance to End Childhood Lead Poisoning, Washington |
| Transport, Health and Environment Pan-European Program (The PEP) | UN/ECE and WHO Secretariats, Geneva |

The list in Table 21 offers a modal illustration of the main actors in the type II partnerships: industry (mainly through its sectoral, thematic or other international networks), NGOs and United Nations-related organisations. The 'Alliance to end childhood lead poisoning' is an example that illustrates how civil society can actually contribute to implementing the JPoI targets. This Washington-based organisation is the main driver behind the 'Global Lead Network'. Box 5 summarises their actions. Their initiatives provide an illustration on how civil society can help to realise targets of sustainable development.

**Box 5**. Basic facts on the 'Alliance to end childhood lead poisoning'.

---

Problem statement
Lead poisoning is a global problem that can be completely prevented. Preventing lead poisoning is a vital component of sustainability and fully in line with the targets of the JPoI.

Prevention actions should be based on six principles:
1. Prevention
2. Interdisciplinary approach
3. Co-ordination at international, regional, national and local levels
4. Public awareness and community involvement
5. Priority-based approach
6. Life cycle management

Action
1. World-wide phasing-out of leaded gasoline
2. Promoting a transition to clean fuels

Global lead network
Provides resources and facilities information exchange and collaboration among those working on prevention of lead exposure around the world. A website exists that is designed to assist agencies, NGOs, researchers, concerned citizens, the private sector, ... all stakeholders interested in developing and implementing solutions to lead poisoning and pollution, as well as to other environmental, health and sustainability problems

Website
http://www.globalleadnet.org/

---

## 3.    National Environmental Health Action Plans (NEHAPs)

### 3.1.    *Origin and context*

The JPoI is only one policy document relating traffic and mobility to environmental health. Other international plans that aim at reaching an environmental quality that does not harm environmental health are listed in Table 22. Among them are the National Environmental Health Action Plans (NEHAPs) and the Child Environmental Health Action Plans (CEHAPs). These are WHO-Europe initiatives, established in the framework of the International Interministerial Conferences on Environmental Health. The first one was organised in Frankfurt, Germany, in 1989. It aimed at situating health in a sustainable development context and resulted in a 'European Protocol on Environment and Health'.

The Second Interministerial Conference was organised in Helsinki, Finland in 1994. On that occasion the activities were linked up with chapter 8 on health and environment of Agenda 21. This was facilitated most by the report 'Our Planet, our health' (WHO, 1992) that addressed environment and health in a comprehensive way and in a sustainable development context. The novelty in this discussion was the explicit linking of the formerly separated areas of environmental protection and health promotion. This resulted in a vision on health stating that political, economic, social, cultural and environmental factors all are influential for health and wellbeing.

**Table 22**. International plans targeted to environmental health.

| Policy Plan | Target | Organisation(s) |
|---|---|---|
| Pan European Plan (PEP) on Health and Environment | Reducing the health effects resulting from environmental exposure to traffic impacts. | WHO, LINECE |
| National Environmental Health Action Plans and Child Environmental Health Plans | Implementation of health targets formulated in Agenda 21. | Inter-ministerial conferences on Environmental Health - WHO Europe |
| Strategy for Sustainable Development | One of the four basic aims: improvement of human health, quality of life, positive employment effects of environmental policy, public participation and environmental education. | OECD |
| Sixth Environmental Action Programme | Contributes to the high level of quality of life and social well-being for the citizens, by establishing an environment in which pollution does not reach a level which results in adverse effects on public health and the environment. | EU |
| European Strategy for Environmental Health "SCALE"-initiative | Integration of information on the state of the environment and human health. The strategy aims at filling the knowledge gaps on environmental health. | EU |

The Second Interministerial Conference also agreed on the establishment of National Environment and Health Action Plans (NEHAPs). These are instruments to integrate environmental protection and health promotion into political programmes. In June 1999, environment and health ministers committed themselves to endorsing and strongly supporting the implementation of NEHAPs in the London Declaration, at the Third Ministrial Conference on Environment and Health. By 2002, 43 NEHAPs have been developed of which 12 are available at the website of WHO-Europe (http://www.who.dk/envhealthpolicy/Plans/20020809_1). These plans will be evaluated at the Fourth Ministerial Conference on Environment and Health, to be held in Budapest (Hungary) in June 2004.

## 3.2. *Mobility and traffic*

Table 23 overviews the mobility and traffic objectives in 9 NEHAPs that are accessible at the WHO-Europe website by May 2003. In all these plans traffic is a cross-sectoral issue. It is referred to in sections such as air pollution or health risks. However, the issue is explicitly mentioned in 6 out of these 9 plans.

**Table 23**. Mobility and traffic related objectives in 9 National Environment and Health Action Plans (NEHAPs).

| Country | Latest Update | Chapter on transport or mobility | Objectives |
|---|---|---|---|
| Bulgaria | May, 2002 | 5.3 Transportation | (-) To implement integrated transportation policy.<br>(-) To reduce gaseous and particulate emissions.<br>(-) To reduce traffic noise and pollution.<br>(-) To safeguard the right of pedestrians.<br>(-) To promote the development of railways. |
| Czech Republic | ? | 6.5 Transport | (-) To monitor environmental parameters and population health indicators.<br>(-) To assess relationships between parameters.<br>(-) To assess health risks.<br>(-) To establish priorities and activities. |
| Finland | 1997 | - | Traffic and transport are mentioned but there are no specific objectives. |
| Germany | June, 1999 | - | Traffic and transport are mentioned but there are no specific objectives. |
| Malta | 1997 | 5.3 Transport | (-) To reduce road traffic injuries, disabilities and deaths by 25% by 2000 compared to 1990.<br>(-) To reduce gaseous and particulate emissions.<br>(-) To abate noise from traffic.<br>(-) To set and enforce speed limits.<br>(-) To check the road worthiness of all vehicles.<br>(-) To safeguard the rights of pedestrians.<br>(-) To regulate traffic to reduce accidents, pollution and noise.<br>(-) To thoroughly investigate car accidents. |
| Poland | ? | 2.5.3 Transportation | (-) Reduction of noise and emission of gaseous pollutants and dust.<br>(-) Reduction of the risk of road accidents, especially fatality cases.<br>(-) Decrease of exposure of urban populations to transportation noise. |
| Slovak Republic | ? | - | Traffic and mobility are mentioned but there are no specific objectives. |

**Table 23**, continued.

| Country | Latest Update | Chapter on transport or mobility | Objectives |
|---------|---------------|----------------------------------|------------|
| Switzerland | August, 2003 | Mobility and well being | (-) By he year 2002, 80% of the population will know about the relationships between motorised traffic, emissions and the effects on human health. <br>(-) Reduction of emissions to such on extent that the impact threshold levels of the Ordinance on Air Pollution Control can be respected. <br>(-) By 2007, the proportion of journeys by bicycle will have doubled for commuting, shopping and leisure as compared to 1995. |
| UK | July, 2003 | Personal transport | Encourage production, marketing, purchase and use of vehicles that are more fuel efficient. |

The majority of these address:
(-)    reduction of emissions,
(-)    reduction of exposure to noise,
(-)    minimisation of the physical risks resulting from traffic accidents.

Note that the Swiss plan, that more than other plans focuses on 'wellbeing' rather than on indicators for illness, is the only one to mention objectives on bicycling and physical activity.

The issue also needs to be cross-linked with other plans that exist in the different countries. An interesting example of that is the UK, where NEHAP is an integrated part of the 'Strategy for sustainable development for the United Kingdom' (UK Government, 1999). In other countries the NEHAPs should be read together with National Plans for Sustainable Development and/or mobility plans and/or environmental plans.

## 4.    Discussion

The review of the health aspects in the JPoI and a selected number of NEHAPs shows that mobility and transport is a recognised focal area. European countries target merely reduction of air pollutants and noise emissions in combination with a reduction of accident risk. This is to a lesser extent the case in the WSSD documents. Only in a limited number of NEHAPs are the targets quantified and timed, which is a necessary condition for a plan.

The health effects resulting from the exposure to emissions from traffic are at the interface of three policy areas: health, environment and traffic. An integrated, comprehensive approach is necessary to handle problems in this field. With their attention for NEHAPs the Ministrial Conferences on Environment and Health have addressed environment and health integrated in one political programme. The next logical step is the integration with traffic policies. In most countries this integration still needs to be addressed.

A similar problem applies to the integration of NEHAPs with the Agenda 21 and JPoI processes. In daily practice the areas of environment and health, including their social and economic dimensions, remain separate fields of policy decision-making.

Finally, more commitment and control on the implementation of these plans might significantly enhance the credibility of these most valuable instruments of policy-making.

## References

Babalola, K.W. (2002) *Combating the Impacts of Desertification in Nigeria: a Case Study of Northern States*, Masters Thesis, Masters Programme in Human Ecology, Vrije Universiteit Brussel, Belgium.

Dora, C., and Phillips, M. (eds) (2000) *Transport, Environment and Health*, WHO regional publications, European Series No. 89, WHO Regional Office for Europe, Copenhagen, Denmark, http://www.euro.who.int/document/e72015.pdf (Dec. 2003).

Global Lead Network (2003) http://www.globalleadnet.org (Dec. 2003).

Hens, L., and Nath, B. (2003) The Johannesburg Conference, *Environment, Development and Sustainability* 5 (1-2), 7-39.

IEA - International Energy Agency (2001) *World Energy Outlook 2001*, International Energy Agency, Paris, France, http://www.iea.org/ (dec. 2003).

IPCC - International Panel on Climate Changes (2000) *Methodological and Technological Issues in Technology Transfer*, Cambridge University Press, Cambridge, UK.

McMichael, A.J., Haines, A., Slooff, R., and Kovats, S. (1996) *Climate Change and Human Health*, WHO, Geneva, Switzerland.

Peñalosa, E. (2003) Foreword, in J. Whitelegg, and G. Haq (eds), *World Transport: Policy and Practice*, Earthscan Publications Ltd., London, UK.

TRL - Transport Research Laboratory (2000) *Estimating Global Road Fatalities*, Crowthorne, UK, http://www.trl.co.uk/ (dec. 2003).

UK Government (1999) *A Better Quality of Life: a Strategy for Sustainable Development for the UK*, The Stationery office, London, UK.

UN - United Nations (2002) Plan of Implementation of the World Summit on Sustainable Development, in *UN Report of the World Summit on Sustainable Development*, Johannesburg, South Africa, 26 Aug. - 4 Sept. 2002, A/CONF.199/20, New-York, pp. 6-72, http://www.johannesburgsummit.org/html/documents/documents.html (Dec. 2003).

UN - United Nations (2003) *Partnerships for Sustainable Development - Summary Analysis*, http://www.un.org/esa/sustdev/partnerships/summary_analysis.html (Dec. 2003).

UPI - Umwelt und Prognose Institut (1999) *Oeko-Bilanzen von Fahrzeugen*, UPI Bericht Nr. 25, 6 Auflage, May, Heidelberg, Germany.

Whitelegg, J. (2003) Transport in the European Union: time to decide, in N.P. Lowe, and B.J. Gleeson (eds), *Making Urban Transport Sustainable*, Palgrave-Macmillan, Basingstoke, UK.

WHO - World Health Organisation (1992) *Our Planet, Our Health*, WHO Commission on Health and Environment, Geneva, Switzerland.

WHO - World Health Organisation (2002) *Reducing Risks, Promoting Healthy Life*, World Health Report, Geneva, Switzerland, http://www.who.int/whr/2002/en/ (Dec. 2003).

# INTEGRATING HEALTH CONCERNS INTO TRANSPORT POLICIES: *FROM THE CHARTER ON TRANSPORT, ENVIRONMENT AND HEALTH TO THE TRANSPORT, HEALTH AND ENVIRONMENT PAN-EUROPEAN PROGRAMME*

F. RACIOPPI[1] AND C. DORA[2]

[1]*European Centre for Environment and Health, World Health Organization Regional Office for Europe, Via Francesco Crispi 10, 00187 Rome – ITALY*
[2]*World Health Organization, 20 Avenue Appia, 1211 Geneva 27, SWITZERLAND*

## Summary

This paper describes international policy developments in the WHO European Region over the last five years, to promote the integration of health concerns into transportation and land use policies. It presents an overview of the evidence about the main effects that transport has on health, and summarizes the main objectives and policy directions set out in the Charter on Transport, Environment and Health (adopted in 1999 at the Third Ministerial Conference on Environment and Health). It also describes the political process that led to the development, jointly with the United Nations Economic Commission for Europe - UNECE - of the Transport, Health and Environment Pan European Programme - THE PEP - (adopted in 2002 at the Second High-level Meeting on Transport, Environment and Health), and presents the policy directions to achieve transport patterns that are sustainable for health and the environment. Finally, the paper presents examples of actions that have been triggered by the Charter and THE PEP.

## 1. Introduction

Present transportation patterns raise concerns about the environmental sustainability of on-going trends, e.g. with respect to objectives related to the reduction of emissions of green-house gases from the transport sector, or the achievement of compliance with air quality standards (EEA, 2002). Increasingly, transport trends are posing questions also about the toll in terms of health effects – and their related costs – that societies are called to pay for the benefit of enjoying the convenience of

*P. Nicolopoulou-Stamati et al. (eds), Environmental Health Impacts of Transport and Mobility, 171-177.*
© 2005 *Springer. Printed in the Netherlands.*

transportation systems, which are largely based on road transport, both for passengers and freight transportation.

When at the end of the 1990s the World Health Organization (WHO) started its work in the area of transport, health and the environment, the picture of the health impacts of transport had started to reveal its complexities. Next to 'historical' and in a way 'familiar' effects, such as those caused by road traffic accidents and air pollution, new ones started to emerge, such as those related to noise, physical inactivity, and psychosocial effects and social inequalities.

The development of a more comprehensive understanding of the health effects of transport paints a picture, where the impacts of transport activities for the WHO European Region can be summarized as follows (WHO, 2000).

- Approximately 120,000 people die every year and more than 2.5 million are injured as a result of road traffic collisions.
- The number of people who die prematurely as a result of their exposure to air pollutants (using particulate matter $PM_{10}/PM_{2.5}$ as an indicator) is in the order of 100,000 per year (WHO, 2001).
- Road traffic has become the most important source of exposure to noise in the urban environment.
- The attributable fraction of mortality from physical inactivity is estimated to range between 5 and 10 per cent of the total mortality in the European Region (WHO, 2002a). This is equivalent to a few hundred thousand deaths per year, considering that the number of deaths in the Region is approximately 10 million. The increasing substitution of trips that could be done on foot or by bicycle and are now done by motorized means is a major contributor to the increasing prevalence of sedentary lifestyle among the European population.

The above effects are unequally distributed across the European Region, with the eastern part bearing a comparatively higher disease burden than the western one, owing to a rapidly increasing motorization that is not accompanied by technological improvements and adequate policy settings. For example, a recent assessment on air pollution in the Newly Independent States (NIS) indicated that annual mean values of total suspended particles (TSP) ranging from 100 to 400 micrograms per cubic meter are quite common in bigger cities of the NIS, where transport is held responsible for up to 75 per cent of the total emission of selected pollutants, and is solely responsible for air toxins like benzo(a)pyrene and soot (WHO, 2002b). In addition, average mortality rates from road traffic accidents are double in the NIS compared to Nordic countries: i.e. Denmark, Finland, Iceland, Norway and Sweden (WHO, Health for All Data base, 2002).

When economic valuations are applied to some of these impacts, the costs of the so-called 'external' effects of transport (i.e. those effects that fall on society rather than on those who originate them), are estimated in the order of 10 per cent of the GDP of western European countries (INFRAS and IWW, 2000), i.e. the same order of magnitude of the estimated contribution of the transport sector to the growth of the economy.

The analysis of the effects of transport on health pointed also at important limitations in the way these issues had been addressed by policy makers, often separately from each other. Only in recent years have there been signs of change, such as those obtained when consideration of health impacts was given in cost-benefit analyses of transport infrastructures carried out in Norway. In this study, the costs and benefits of further developing infrastructures for cyclists and pedestrians in three different cities were calculated, including some health effects resulting from expected increases in physical activity, in addition to more 'traditional' variables. Results indicated large net cost-benefit ratios for the three cities considered, providing a solid argument to the economic soundness of investing in infrastructures for cyclists and pedestrians (Saelensminde, 2002).

## 2. The Charter on Transport, Environment and Health

The notion that transport-related activities are responsible for a significant share of the burden of ill health in the Region has contributed to raising awareness of the importance of the issue and catalysed political support to mitigate and prevent these negative environmental and health effects. The negotiation of the 'Charter on Transport, Environment and Health' (WHO, 1999), which was adopted at the Third Ministerial Conference on Environment and Health (London, 16-18 June 1999), brought together representatives of ministries of transport, environment and health, along with inter-governmental and non-governmental organizations.

The Charter aims at placing health considerations (along with environmental ones) firmly on the agenda of transport policy makers. It contains health targets for reducing injuries, air and noise pollution and increasing opportunities for physical exercise through walking, cycling and use of public transport; principles for 'transport sustainable for health and the environment'; and a plan of action for its implementation. The Plan of action contains commitments for Member States and requests for supportive actions from WHO and other international organizations on the following aspects:

a. Integration of environment and health requirements and targets in transport and land use policies and plans
b. Promotion of modes of transport and land use planning which have the best public health impacts
c. Health and environmental impact assessments
d. Economic aspects of transport, environment and health
e. Special care of groups at higher risk
f. Risks to public health not yet clearly quantified
g. Indicators and monitoring
h. Pilot actions and research
i. Public participation, public awareness, information
j. Countries in transition and countries with severe problems concerning transport-related health effects

The Charter also established a steering group of Member States, inter-governmental organizations and non-government organizations to facilitate its implementation.

### 3. From the Charter on Transport, Health and the Environment to the Transport, Health and Environment Pan-European Programme (THE PEP)

An important aspect of the implementation of the Charter was related to the promotion of enhanced co-operation between WHO and other intergovernmental bodies active in the fields of transport and environment, in particular the United Nations Economic Commission for Europe (UNECE). More specifically, the Charter called on WHO and UNECE to:

> *Provide an overview of relevant existing agreements and legal instruments, with a view to improving and harmonizing their implementation and further developing them as needed.*

In line with this mandate, WHO and UNECE directed their efforts towards the development of an overview of instruments relevant to transport, environment and health and recommendations for further steps, whose conclusions were discussed at the First High-Level Meeting on Transport, Environment and Health (Geneva, 4 May 2001). The overview recommended, *inter alia,* starting negotiations for a framework convention on transport, environment and health, focusing on the integration of the transport, environment and health sectors, and on transport-related environment and health problems in urban areas (UNECE and WHO, 2001).

The First High-Level Meeting decided that further preparatory work was necessary to be able to decide whether to start negotiations for a framework convention, and that such a decision would be taken at the Second High-Level Meeting (5 July 2002).

Following the decisions taken at the first High-level Meeting on Transport, Environment and Health, a Joint WHO/UNECE Ad Hoc Expert Group on Transport, Environment and Health was established to carry out the additional background work.

The Second High-level Meeting on Transport, Environment and Health decided to streamline and consolidate the activities undertaken at the national and international levels under the auspices of UNECE and WHO under a single new non-legally binding policy framework: the Transport, Health and Environment Pan-European Programme (THE PEP) (UNECE and WHO 2002a). The Meeting decided also that negotiations towards a framework convention on transport, environment and health, to guide the direction of international policy actions to integrate policies in the three sectors, seemed to be premature and could be re-examined at the next High-level Meeting, taking place not later than 2007, in the light of progress of THE PEP.

THE PEP aims at making progress towards the achievement of transport patterns that are sustainable for health and the environment by focusing work at the pan-European level on those priorities where further work of the international community is most needed and could make the biggest impact, namely: integration of environmental and health aspects into transport policies and decisions; shift of the demand for transport towards more sustainable mobility; urban transport issues. Special attention will be paid to the needs of the NIS and south-eastern European countries as well as to areas that are particularly sensitive from an environmental point of view. THE PEP has also been launched at the World Summit on Sustainable Development as one of the type II partnerships for health and sustainable development (UNECE and WHO, 2002b).

The implementation of THE PEP is overseen by a steering committee of representatives of Member States, Inter-governmental organizations and non-governmental organizations. WHO and UNECE provide secretariat functions to the steering committee.

## 4.    Examples of actions triggered by the Charter and THE PEP

The momentum created by the Charter, and more recently by THE PEP, prompted action on several fronts, encompassing international policy developments, effects on national approaches and organizational settings, and progress on methods and research for health impact assessment, including the development (presently underway) of guidelines for carrying out health impact assessments of transportation policies and of their health effects through changed levels of walking and cycling.

In addition, the analysis of transport-related health and environment issues produced by WHO and UNECE inspired the development of a European Parliament resolution on transport and health, calling for 'a stronger integration of health considerations into transport policies, including carrying out health impact assessment of major transport projects' as well as for actions promoting walking and cycling (European Parliament, 2002).

Importantly, the Charter and THE PEP have been highly effective in mobilizing the interest of some European countries towards supporting further developments in the methods and tools for carrying out economic valuations of transport-related health effects, and improving the estimates of the external costs of transport, by taking better into account the impacts of air pollution, noise, lack of physical activity, as well as psycho-social effects. This work represents the extension of a landmark study performed in Switzerland, Austria and France as background to the Charter, which shed new light on the cost of transport-related health effects of air pollution (Künzli et al., 2000). The study, which involved also the Netherlands, Sweden and Malta, reviewed the state of the art with respect to methods for assessing the exposures, effects and costs of transport-related health impacts; and identified research gaps and recommended next steps in research and policy action, to inform also the development of the Children's Environment and Health Action Plan for Europe, adopted at the Fourth Ministerial Conference on Environment and Health, held in Budapest in 2004.

Other actions that are being proposed for implementation during the period 2003-2005 and whose endorsement was sought from the first meeting of THE PEP steering committee (Geneva, 10-11 April 2003) include the establishment of a web-based clearing house on transport, environment and health; the elaboration and implementation of urban plans for transport sustainable for health and the environment, and the establishment of a set of indicators to monitor the integration of environmental and health aspects into transport policies and the impact of these policies on health and the environment.


## 5.    Conclusions

Both the Charter and THE PEP are non-legally binding policy instruments, and as such are limited in their implementation by the lack of adequately strong enforcement tools and allocation of resources for implementation at the international and national level, and by competition with other (legally-binding) international processes, such as the integration of the European Union *acquis* into national policies by new EU member States. In spite of these important limitations, they are contributing to making progress at the pan-European level on several aspects of the integration of health and environmental concerns into transport policies.

In particular, THE PEP is now providing a discussion platform that is co-owned by the three involved sectors, places, transport, health and environment on an equal footing, and involves relevant inter-governmental organizations and non-governmental organizations. In addition, the set of actions proposed for THE PEP implementation has the potential of promoting cross-sectoral integration of policies, and of creating opportunities for the exchange of experiences, the dissemination of state-of-the-art knowledge, capacity building, networking and the forging of new partnerships across the European Region. The web-based clearing house on transport, environment and health is particularly helpful for the achievement of these objectives. Also the development of a set of indicators to monitor the integration of environmental and health aspects into transport policies and the impact of these policies on health and the environment would help Member States to better assess their performance towards the achievement of transport sustainable for health and the environment and to direct their efforts to areas where more progress needs to be made.


## References

EEA - European Environment Agency (2002) Paving the way for EU enlargement – Indicators of transport and environment integration – TERM 2002, *Environmental Issue report No 32 Luxembourg: Office for Official Publications of the European Communities.*

European Parliament (2002) European Parliament resolution on the impact of transport on health (2001/2067(INI)). Final A5-0014/2002, 22 January 2002, adopted on 28 February 2002.

INFRAS and IWW (2000) External costs of transport (accidents, environmental and congestion costs) in western Europe, University of Karlsruhe, Zurich.

Künzli, N., Kaiser, R., Medina, S., Studnicka, M., Chanel, O, Filliger, P., Herry, M., Horak, F. Jr., Puybonnieux-Texier, V., Quénel, P., Schneider, J., Seethaler, R., Vergnaud, J-C., and Sommer, H. (2000) Public-health impact of outdoor and traffic-related air pollution: a European assessment, *Lancet* 356, 795–801.

Saelensminde, K. (2002) *Walking and cycling track networks in Norwegian cities – Cost-benefit analyses including health effects and external costs of road traffic*. Institute of Transport Economics, Norway TOI report 567/2002 – (Norwegian language)

UNECE and WHO (2001) *Overview of instruments relevant to transport, environment and health and recommendations for further steps* - synthesis report document ECE/AC.21/2001/1 - EUR/00/5026094/1 http://www.euro.who.int/document/trt/advreport1.pdf (accessed on 7 April 2003).

UNECE and WHO (2002a) *Transport, Health and Environment Pan-European Programme* Document ECE/AC.21/2002/9 EUR/02/5040828/9 http://www.the-pep.org (accessed on 24 March 2003).

UNECE and WHO (2002b) *Transport, Health and Environment Pan-European Programme* (THE PEP) http://www.johannesburgsummit.org/html/sustainable_dev/p2_health/2508_ transport_health_env_pep.pdf (accessed on 19 December 2002)

WHO Regional Office for Europe (1999) *Charter on Transport, Environment and Health,* Third Ministerial Conference on Environment and Health, London, 16–18 June 1999. Copenhagen, WHO Regional Office for Europe. http://www.euro.who.int/document/peh-ehp/charter_transporte.pdf (accessed on 13 June 2002)

WHO Regional Office for Europe (2000). Transport, Environment and Health, edited by C. Dora and M. Phillips WHO Regional Publications, *European series, no.* 89, Copenhagen, WHO Regional Office for Europe. http://www.euro.who.int/document/e72015.pdf (accessed on 11 May 2005).

WHO European Centre for Environment and Health (2001) Health impact assessment of air pollution in the WHO European Region, Copenhagen, WHO Regional Office for Europe.

WHO (2002a) Reducing risks, promoting healthy life, the World Health Report 2002, Geneva, World Health Organization. http://www.who.int/whr/2002/en/ accessed on 19 December 2002.

WHO Regional Office for Europe (2002b) NIS Environment Strategy - Background paper: Pollution Prevention and Control Section - Reducing urban air pollution, Copenhagen, WHO Regional Office for Europe.

# MOBILITY AND HEALTH IN BULGARIA

S. STOYANOV AND E. TERLEMESIAN
*University of Chemical Technology and Metallurgy,*
*Ecology Centre, 8, blvd. "Kl. Ohridski",*
*1756 Sofia, BULGARIA*

## Summary

The country report aims at outlining briefly the state and the specific environmental and health impacts of transport in Bulgaria. Environmental impacts of the main branches of transport in a group of 98 people are ranked and the greatest weight of impact is attributed to road transport. In the last decade the trend of increase in the total number of vehicles is established together with a reduction of total emissions and emissions produced by transport. Specific economic development linked with the period of transition along with the limitations connected with the international obligations undertaken by the country for the reduction of emissions are seen as a reason for explanation. The health status of the Bulgarian population is depicted. Morbidity rate expressed by the visits to medical centers shows a prevailing contribution of the respiratory diseases, followed by cardiovascular and nervous disorders. Cancer rate in Bulgaria shows an alarming trend of continuous increase during the last decade. Cardiovascular diseases are the first cause of death with a commitment of 66.6 per cent, followed by cancer (13.8 per cent) and respiratory diseases.

No evidence is found concerning the size of effects to be quantified in relation to the air pollutants of transport. Two case studies are described: Impact of mobility on the environmental health in Sofia and the correlation between noise and respiratory diseases in Sofia.

## 1. Introduction

The invasion of up-to-date road, air, rail and water transport during the past century caused unpredictable environmental and health impacts. The short-sighted approach to the environmental impact of transport has created a lot of natural, cultural and human health damage. Road transport is one of the main generators of air and noise pollution and accidents. Air toxins from motor vehicles along with industrial air

*P. Nicolopoulou-Stamati et al. (eds), Environmental Health Impacts of Transport and Mobility, 179-198.*

pollutants affect millions of people. The problem is more severe for the developing countries and for countries under transition such as Bulgaria. Careful study and extensive systematic research of the state of travelling in each country is necessary in order to avoid damage in the future.

The main aim of this report is to outline briefly the state and the specific environmental and health impacts of transport in Bulgaria.

Some main outlines for Bulgaria, which will give a clearer basis for the problem under discussion, are the following (BSRB, 2001, 2002). The territory of the country is 110,912 $km^2$ and the population 7,974,050. The density of population is 71.9 $people/km^2$. The average life expectancy is 68.1 years for males and 75.3 for females, which is 6 years less than the average in the EU. The number of people over 60 years is 22.4 per cent.

## 2.    The transport sector in Bulgaria

### 2.1.    Water transport

Water transport in Bulgaria is under continued development. The two main seaports are in Varna and Bourgas. At present the country possesses 110 sea cargo vessels with a capacity of over 1.8 million register tons. Predominantly sea transport is for bulk cargoes like ore, coal, chemicals and agricultural products, wooden materials and petrol. The only inland water transport is in Danube River along the boundary with Romania. Several small ports are situated along the Bulgarian bank. At present the country has 37 cargo ships and 260 platforms.

### 2.2.    Air transport

Air transport is relatively negligible for the country. The main airport is in Sofia and two small civil airports are situated near Varna and Bourgas. There are also several military airports. Sofia airport is situated in the city itself.

### 2.3.    Railway transport

Nowadays the total length of railway track in Bulgaria is 6,607 $km$, mostly electrified. A quarter of the lines are double. Though railway transport has second place in Bulgarian transport the level of mechanization, speed and safety are still not satisfactory.

### 2.4.    Road transport

Road transportation is the leading mode of transport for people and goods in Bulgaria. The total length of Bulgarian roads is 36,535 $km$ including 250 $km$ of highways. Road transport is giving rise to the main environmental health problems.

## 3. Environmental impact of transport in Bulgaria

### 3.1. Environmental impact of transport as perceived by the people

The environmental impact of transport is continuously increasing and during the last decade became a significant problem. In order to select the main environmental impacts of different types of transport in Bulgaria a questionnaire investigation was performed by us among different categories (age, education, professions, sex, etc.) of 98 people. The people gave their opinion by ranking the environmental impacts of different types of transport and different indicators using a scale of ten grades. Using the methodology developed by Stoyanov (1993) the evaluated average grades are given in Table 24. The estimated concordance coefficient $w_c = 0.68$ of the group opinion is proved by the rank correlation methods (Kendall, 1957) with a level of significance $\alpha = 0.05$. It is noticeable that according the public opinion the highest environmental impact grade is given to road transport (level nine). The highest grades among the pollution indicators of automotive transport are given to air pollution (grade nine), noise (grade eight) and car accidents (grade eight). This is the reason why we focus our attention in the present report mainly on the environmental impact of road transport.

**Table 24**. Grade of environmental impact of different kinds of transport (10 grade scale).

| | Grade of impact | | | | | | | |
| | Total | Noise | Air pollution | Water pollution | Accidents | Solid waste | Landscape destruction | Terrain occupation |
|---|---|---|---|---|---|---|---|---|
| Road transport | 9 | 8 | 9 | 2 | 8 | 1 | 5 | 2 |
| Air transport | 4 | 9 | 4 | 2 | 2 | 3 | 2 | 3 |
| Railway transport | 4 | 4 | 2 | 2 | 1 | 2 | 5 | 3 |
| Water transport | 3 | 2 | 1 | 5 | 1 | 4 | 1 | 1 |

### 3.2. Environmental impact of air transport

Bulgarian civil airports are situated in Sofia, Varna, Bourgas and Plovdiv. The average number of take offs and landings per twenty four hours in Bulgarian civil airports is given in Table 25.

**Table 25**. Average number of take offs and landings per twenty four hours in Bulgarian civil airports.

| Airport | Winter time From 30.10 to 30.03 | Summer time From 30.03 to 30.10 |
|---|---|---|
| Sofia | 100 - 105 | 115 - 120 |
| Varna | 2 - 3 | 40 - 45 |
| Bourgas | 2 - 3 | 40 - 45 |
| Plovdiv | 1 - 2 | 10 - 12 |

Source: Ministry of Transport and Communications, 2003, unpublished data.

The principal Bulgarian airport is Sofia airport which is situated in the town. The main environmental problem is concerned with noise pollution (Petkov, 2001). About 61,564 citizens of Sofia are exposed to average noise pollution above the norms (60 - 80 *dBA*, by day, summer time). The total number of Sofia citizens is about 1.18 million. The plan is to move the runway to an eastern direction and to change the aircraft park (EPMS, 2000a). After the reconstruction of Sofia airport the forecast about the number of Sofia citizens exposed to noise above the daily norm will be less than 1,000 people in 2008 and 4,400 people during the day and 2,600 at night in 2018. Expectations are that the noise impact will be reduced by at least ten times.

### 3.3.    *Environmental impact of railway transport*

The main environmental impacts of railway transport in Bulgaria are given in Table 26. Having in mind that for one *km* of railway between 3000 to 5000 $m^3$ broken stones are necessary, the estimated destruction of natural resources is about 26 *mil.* $m^3$ stones for the total length of 6600 *km* of Bulgarian railway tracks. If the average width of the railway is 10 *m,* the agricultural land losses are about 66 $km^2$.

**Table 26**. Environmental impact of railway transport.

| No | Environmental impact | Kind of pollution | Impact |
|----|----------------------|-------------------|--------|
| 1. | Landscape and cultural heritage changes | - Landscape changes | * Biodiversity impact<br>* Social impact |
| 2. | Agricultural land destruction | - Decreasing the agricultural land | * Biodiversity impact<br>* Economical impact<br>* Social impact |
| 3. | Soil pollution | - Ferrous dust<br>- Lubricating oils<br>- Pesticides/herbicides<br>- Dissipation of goods (coal, ore, oil products, etc.) | * Health impact |
| 4. | Water pollution | - Washing the wagons and engines<br>- WC<br>- Chemicals removing grass | * Health impact |
| 5. | Noise | - Noise | * Health impact |
| 6. | Fires | - Dust<br>- Air pollutants<br>- Landscape damages | * Air pollution<br>* Health impact<br>* Biodiversity impact<br>* Economic impact |
| 7. | Risk creation | - Tunnels, bridges, supporting walls, crossings, rails | * Health impact<br>* Increased risk |
| 8. | Destruction of natural resources | - Natural equilibrium | * Biodiversity impact |

The negative environmental impact of railway transport can be significantly reduced if the following actions are performed (Tashev and Petkov, 2001):

1)   change the cast iron brakes with synthetic or composite materials;
2)   change the diesel fuel with castor fuel;
3)   change lubricating oil with chlorine-free motor oils;
4)   build anti-noise walls (plantations, banks);
5)   use noise-free gears;
6)   use pesticides/herbicides only selectively.

## 3.4.   Environmental impact of water transport

The environmental impact of water transport concerns mainly water pollution (grade five). The pollution of the Black Sea and the Danube River is a significant international problem. Some international programs and regulations exist in order to reduce the fast growth of pollution and damage including these caused by water transport.

## 3.5.   Environmental impact of road transport

### 3.5.1.   National fleet of motor vehicles

The national car park by 01 January 2002 and the new registered cars in 2001 are given in Table 27.

**Table 27**. National fleet of motor vehicles

|  | Total | Motorcycles | Personal cars | Lorries | Buses | Others |
|---|---|---|---|---|---|---|
| Registered by 01.01.2002 | 3.373.857 100% | 526.046 15.59% | 2.085.730 61.82% | 312.050 9.25% | 42.870 1.27% | 407.162 12.07% |
| New registered in 2001 | 138.741 100% | 3.725 2.68% | 117.333 84.57% | 13.488 9.72% | 1.307 0.94% | 2.888 2.09% |

Source: Annual Report for Control of Automobile Transport, Ministry of Internal Affairs (2002)

Distribution of motor vehicles in Bulgaria (except motorcycles and others) in respect of the car age is shown in Figure 35. The majority of car age is over twenty years, which is the worst indicator for the environmental impact of road transport. The variety of different makes of cars in Bulgaria is also very big.

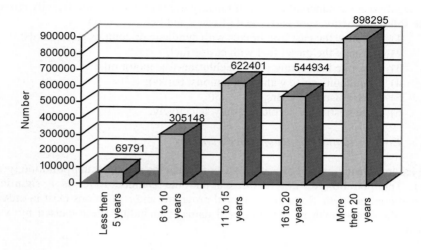

**Figure 35**. Car age in Bulgaria (2001).

Source: Annual Report for Control of Automobile Transport, Ministry of Internal Affairs (2002).

### 3.5.2.  Car accidents

As is shown in Table 24 car accidents are ranked with grade eight. The mortality rate in car accidents based on data from the Ministry of Internal Affairs (MIA) is given in Figure 36. The tendency is for reduction, but still the rate is very high. The number of people killed in car accidents per month expressed as an average for thirteen years (1990 to 2002) is shown in Figure 37. It is obvious that the summer months are the most dangerous ones.

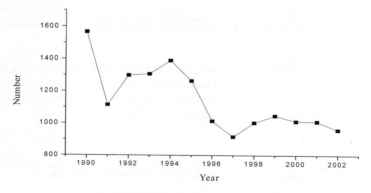

**Figure 36**. Mortality rate by car accidents.

Source: Annual Report for Control of Automobile Transport, Ministry of Internal Affairs (2002).

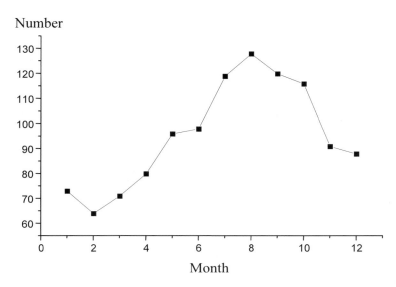

**Figure 37**. Average number of people killed in car accidents for thirteen years (1990 to 2002) per month.

Source: Annual Report for Control of Automobile Transport, Ministry of Internal Affairs (2002).

*3.5.3.    Road transport emissions*

Pollutant emissions into the atmosphere of Bulgaria are given in Table 28.

**Table 28**. Total emissions into the atmosphere of Bulgaria.

| Year | $SO_2$ kt/year | $NO_2$ kt/year | NMVOC kt/year | CO kt/year | Cd t/year | Pb t/year | PAH t/year | Diox.+ Fur. g/year |
|------|------|------|------|------|------|------|------|------|
| 1994 | 1.482.4 | 325.2 | 432.6 | 1031.6 | - | - | - | |
| 1995 | 1.477.0 | 265.0 | 351.0 | 832.0 | - | - | 521.43 | 456.00 |
| 1996 | 1.420.4 | 259.0 | 309.1 | 623.0 | - | - | 487.51 | 340.91 |
| 1997 | 1.359.8 | 224.7 | 281.6 | 517.8 | 14.23 | 231.34 | 364.35 | 309.48 |
| 1998 | 1.251.0 | 223.6 | 295.7 | 678.3 | 14.87 | 250.78 | 384.02 | 288.43 |
| 1999 | 961.6 | 228.6 | 282.9 | 640.7 | 13.57 | 223.51 | 286.00 | 245.28 |
| 2000 | 982.0 | 185.3 | 287.9 | 706.3 | 10.99 | 213.36 | 118.08 | 232.53 |

Source: State of Environment in Republic of Bulgaria, Annual Bulletins for 1998, 1999 and 2000, MoEW.

Transport emissions in Bulgaria are estimated on the basis of fuel consumption. Correct estimate is made up to 1999. The large number of newly-created companies trading in fuels after 1999 still creates difficulties in the correct statistical information concerning the amount of fuels sold.

**Table 29**. Number of registered cars and transport vehicle emissions in Bulgaria.

Emissions, *kt*

| Year | Total number of cars | Total Emissions | $SO_x$ | VOC | NOx Road transport | NOx Other transport | CO Road transport | CO Other transport | Pb* |
|---|---|---|---|---|---|---|---|---|---|
| 1990 | 2.322.152 | 658.2 | 26.6 | 79.2 | 136.6 | 29.3 | 371.9 | 14.6 | 199.2 |
| 1991 | 2.404.799 | 414.9 | 30.9 | 49.1 | 68.8 | 32.7 | 219.6 | 13.8 | - |
| 1992 | 2.504.991 | 466.0 | 26.5 | 58.0 | 68.5 | 26.9 | 256.6 | 11.5 | - |
| 1993 | 2.645.964 | 526.0 | 25.5 | 66.7 | 86.9 | 24.6 | 311.7 | 10.6 | - |
| 1994 | 2.762.062 | 488.8 | 22.1 | 66.2 | 66.3 | 22.8 | 306.2 | 9.9 | - |
| 1995 | 2.847.608 | 576.3 | 24.1 | 76.0 | 86.6 | 23.4 | 356.1 | 10.1 | 153.3 |
| 1996 | 2.925.735 | 510.1 | 25.4 | 64.7 | 83.2 | 25.2 | 299.9 | 11.7 | 135.9 |
| 1997 | 2.961.221 | 318.7 | 23.3 | 43.0 | 59.2 | 24.0 | 159.6 | 9.6 | 86.3 |
| 1998 | 3.045.626 | 441.1 | 31.4 | 55.5 | 59.7 | 34.5 | 246.2 | 13.8 | 109.3 |
| 1999 | 3.162.560 | 424.0 | 34.2 | 44.6 | 52.6 | 38.9 | 238.1 | 15.6 | 98.5 |

\* Pb emissions are expressed in tons.
Source: BSRB, 2001, 2002; Green book, MoEW, 1999.

Trends in road transport emissions of the main pollutants for the period 1990 to 1999 are given in Table 29. Reduction of pollution is caused by decreasing of total consumption of fuels by the transport sector and increase of unleaded gasoline consumption. For example the fuel burned by automobile transport in 1998 is 798.9 *kt*. It is reduced down to 783.7 *kt* in 1999. The portion of unleaded fuel for 1999 is 148.9 *kt* compared to 138.0 *kt* for 1998.

The national annual transport emissions in *kt* for the main pollutants CO and NOx are shown in Table 29 for the period from 1990 to 1999. They are divided into two parts - emissions from road transport and from other types of transport (air, water, real and supporting). The increased emissions from the other types of transport are due to the increased consumption of diesel fuel - from 689.5 *kt* for 1998 to 777.5 *kt* for 1999.

Another most commonly generated urban pollutant is ozone, which is produced in the atmosphere as a result of photochemical processes. At national level ozone is controlled at fourteen monitoring sites with automatic stations. Ten of them are situated in cities. Exceeding of norms is established mainly in the cities with a highly developed industry of nitrogen fertilizers. In the town of Sofia for the period 1994 - 2000 exceeding of the maximum allowable concentrations (MAC) is reported only twice - in August 1994 and in October 1999. In spite of that Regulation No 8/1999 on the limit values for ozone in the ambient air revised MAC and increased 8h average from 100 $\mu g/m^3$ to 110 $\mu g/m^3$ and 1h average - from 160 $\mu g/m^3$ to 180 $\mu g/m^3$.

The transport sector is a significant source of generation of greenhouse gases and especially $CO_2$. The trends in the total $CO_2$ emissions for Bulgaria and particularly from transport, are given in Figure 38. A promising tendency of decrease is noticed.

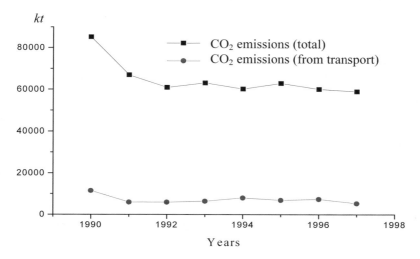

**Figure 38**. Total $CO_2$ emissions and $CO_2$ emissions from transport.

Source: Project for National Action Plan for Climate Change in Bulgaria, 1999.

The main reasons for the bad environmental characteristics of Bulgarian transport are: physical aging of the motor fleet, flaying stock, rail motor stock, ships etc.; usage of non suitable fuels; bad road conditions; bad driving culture; non observation of legislative measures.

## 4. Health effects from the environmental impact of transport

Health status concerning selected diseases, is given in Table 30 for people over eighteen and in Table 31 for children under eighteen.

In both age groups in the period studied there is no evidence of an increase in the total number of visits to medical centers. Children visited doctors twice more often than adults. Comparison shows that respiratory diseases prevail, followed by cardiovascular and nervous system diseases. In the group of children airway and lung function incidents are three times more frequent - 67.2 per cent of the total visits in 2000 - are due to respiratory problems, compared to those of adults with 22.0 per cent of visits.

**Table 30**. Registered diseases by selected types per 1000 people over eighteen in Bulgaria according to visits to medical centres.

| Year | Total | Cancer | Immune system | Blood and blood forming organs | Nervous system | Cardio - vascular | Respiratory system |
|------|-------|--------|---------------|--------------------------------|----------------|-------------------|--------------------|
| 1994 | 1051.3 | 9.7  | 20.6 | 2.2 | 166.6 | 113.0 | 268.0 |
| 1995 | 1052.9 | 9.8  | 21.4 | 2.1 | 164.8 | 114.5 | 272.8 |
| 1996 | 1172.5 | 11.3 | 25.3 | 2.4 | 176.9 | 134.6 | 313.8 |
| 1997 | 1075.7 | 11.2 | 26.0 | 2.8 | 160.6 | 137.1 | 280.1 |
| 1998 | 1035.6 | 11.4 | 26.6 | 3.6 | 158.6 | 157.4 | 228.4 |
| 1999 | 1081.7 | 11.4 | 27.7 | 3.6 | 156.8 | 161.4 | 245.6 |
| 2000 | 1048.6 | 11.4 | 27.7 | 3.6 | 158.7 | 160.2 | 231.2 |

Source: Health Handbook, NSI, 1995 – 2001.

**Table 31**. Registered diseases by selected types per 1000 children under 18 in Bulgaria according to the visits in medical centers.

| Year | Total | Cancer | Immune system | Blood and blood forming organs | Nervous system | Cardio - vascular | Respiratory |
|------|-------|--------|---------------|--------------------------------|----------------|-------------------|-------------|
| 1994 | 2.349.1 | 1.9 | 5.7 | 2.6 | 174.2 | 9.9  | 1.626.4 |
| 1995 | 2.396.0 | 1.9 | 5.6 | 2.5 | 168.3 | 9.3  | 1.698.6 |
| 1996 | 2.068.0 | 1.6 | 6.4 | 2.3 | 132.0 | 7.5  | 1.458.4 |
| 1997 | 2.319.7 | 2.6 | 7.6 | 2.7 | 150.0 | 13.0 | 1.621.6 |
| 1998 | 2.382.6 | 3.2 | 8.6 | 3.5 | 157.7 | 19.6 | 1.605.3 |
| 1999 | 2.516.4 | 3.1 | 8.8 | 3.4 | 158.5 | 18.7 | 1.599.0 |
| 2000 | 2.498.0 | 3.4 | 9.2 | 3.8 | 166.7 | 18.5 | 1.679.4 |

Source: Health Handbook, NSI, 1995 – 2001.

Mortality rate in Bulgaria attributed to selected diseases is shown in Table 32.

**Table 32**. Mortality rate in Bulgaria per 100.000 people from all age groups.

| Year | Mortality rate | | | |
|------|-------|--------|--------------------------|----------------------|
|      | Total | Cancer | Cardiovascular diseases | Respiratory diseases |
| 1990 | 1 245.7 | 173.6 | 766.6 | 74.0 |
| 1995 | 1 364.1 | 191.9 | 867.6 | 62.8 |
| 2000 | 1 408.6 | 187.8 | 933.8 | 55.1 |

Source: Health Handbook, NSI, 1995 – 2001

Cardiovascular diseases are the first cause of death with a commitment of 66.6 per cent, followed by cancer (13.8 per cent) and respiratory diseases. Total mortality and cardiovascular mortality rates in the period have a tendency to increase, while mortality due to respiratory diseases has dropped. Mortality rate of children in Bulgaria is 14/1000 - for towns and 18/100 - for villages which is twice higher than in the EU.

During the last decade the cancer rates of total registered cases per 100,000 people show alarming trends of continuous increase (see Figure 39).

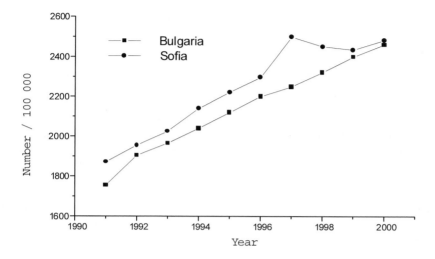

**Figure 39**. Cancer rate for 100.000 people in Bulgaria and in Sofia.

Source: Health Handbook, NSI, 1995 – 2001.

The size of effects could not be quantified in relation to the air pollutants of transport, but having in mind the main releases from the vehicles and their known effects, correlation could be suspected. Various other factors such as genetic inheritance, social and economic factors, including style of life, environmental pollutants from industry etc. can also be attributed to the disease etiology. According to the Ministry of Health (Annual Report, 2002) the main causes of illnesses in Bulgaria are: bad social and economic conditions - 40 per cent; environmental impact - 30 per cent, genetic inheritance - 20 per cent and lack of medical assistance - 10 per cent.

## 5.    Mobility and environmental health in Sofia (Case Study)

### 5.1.    General information

Sofia is the capital of Bulgaria. Its population is approximately 1.2 million people. It is an industrialized city with heavy traffic. According to the National Environmental Strategy for 2001 – 2006, special attention is necessary to be paid to traffic emissions. For Sofia road traffic is one of the leading problems.

Motor vehicles registered in Sofia by 31 December 2000 (Sofia in Numbers, 2001) are presented in Table 33.

**Table 33**. Motor vehicles registered in Sofia by 31 December 2000.

| Type of the vehicles | With gasoline (no) | With diesel fuel (no) |
|---|---|---|
| Passenger cars | 419 974 | 22 995 |
| Trucks | 26 161 | 20 278 |
| Buses | 1 692 | 3 396 |
| Total: | 447 827 | 46 669 |

Source: Sofia in Numbers, 2000.

Trend of increase in the number of passenger cars is established. In the beginning of 1999 it numbered 427,060 (Baynova, 2002). The number of cars per 1000 people in 2000 was 406.4, which means one car per 2.6 citizens of the city, or on average one car per family. This is the same as the average in EU cities. The expected increase in the number of cars by 2005 is foreseen not to exceed 460,000 – 498,000.

Total number of bus lines maintained by the Municipality of Sofia has enhanced from 68 in 1998 to 76 in 2000 (Georgiev and Manolov, 1999). The total route length in 2000 was 836 *km*. There is evidence of an increase in the number of buses by 2005 of 1.1 - 2.0 per cent.

## 5.2.    Air pollution

There are three main sources of air pollution in cities: industry, motor vehicles and burning of fuels for power generation and heating. Air pollution caused by urban transport is inherently linked to the increased use of motor vehicles. Car exhaust contains NOx, CO, $CO_2$, volatile organic compounds (VOC), particles and lead.

Total emissions formed in Sofia are shown in Table 34. It is evident that a slight increase in emissions is recorded in 1996 with a decrease afterwards. The same trend is maintained during the years 2000 and 2001. This is the result of the international obligations for the reduction of emissions taken by the governments in Götheborg (1989) and Kyoto, (1997).

**Table 34**. Total emissions in Sofia.

| Year | Emissions, *kt* | | | | |
|---|---|---|---|---|---|
| | NOx | CO | Pb* | $SO_2$ | Particles |
| 1994 | 16.58 | 112.31 | 32.46 | 16.43 | 71.85 |
| 1995 | 17.58 | 123.75 | 35.19 | 16.73 | 75.75 |
| 1996 | 19.09 | 138.57 | 43.92 | 20.78 | 76.21 |
| 1997 | 19.04 | 126.97 | 31.64 | 16.95 | 73.36 |
| 1998 | 18.98 | 117.65 | 33.20 | 17.14 | 63.84 |

* Pb emissions are expressed in tons.
Source: EPMS, 2000a.

Sofia is one of the biggest territorial emitters of hazardous gases into the atmosphere in comparison to most of the other cities. This concerns mainly nitrogen oxides - about 8.8 per cent from the total emissions in the country come from Sofia, non methane volatile compounds (NMVOC) - about 11.1 per cent from the total, carbon oxide - about 29.6 per cent and particles - 39.5 per cent (data for 1998). In spite of the fact that industry has been in crises, industrial emissions predominate, compared to those of transport. As an exception, emissions from the transport of nitrogen oxides (from 48 to 52 per cent), carbon oxide (from 39 per cent to 45 per cent) and lead droplets (from 46 to 55 per cent) are comparable to the industrial emissions.

According to the information supplied by the Municipality of Sofia, road transport emissions rose continuously in the period 1994 - 1998 (see Table 35).

**Table 35**. Road transport emissions in Sofia.

| Year | Emissions, *kt* | | | | |
|------|------|------|------|------|------|
| | NOx | CO | Pb* | SO$_2$ | Particles |
| 1994 | 8.06 | 44.1 | 14.9 | 0.71 | 0.70 |
| 1995 | 9.11 | 49.9 | 16.8 | 0.80 | 0.79 |
| 1996 | 9.37 | 51.3 | 17.3 | 0.82 | 0.81 |
| 1997 | 9.48 | 51.9 | 17.5 | 0.83 | 0.82 |
| 1998 | 9.79 | 53.6 | 18.0 | 0.86 | 0.85 |

* Pb emissions are expressed in tons.
Source: EPMS, 2000a.

In addition traffic emissions of polycyclic aromatic hydrocarbons (PAH) in Sofia in 1998 composed 12 per cent of the total PAH emissions for the country, dioxins – 9.6 per cent and polychlorinated biphenyls (PCBs) – 98.6 per cent or a total of 9108 *g*.

Emissions are determined as a result of model investigations, cited in the Environmental Program of the Municipality of Sofia, (EPMS, 2000a). According to these data transport is the main source of pollution of the ground layer of the streets in the central part of the city, which are burdened by heavy traffic. In their vicinities a lot of dwellings, children's gardens, schools and health centers are situated.

### 5.3. Review on the environmental health investigations for Sofia

Health status investigation of the population of Sofia, published in the Environmental Program of the Municipality of Sofia, (EPMS, 2000b) has found a correlation between six pollutants (particulate, sulfur dioxide, nitrogen oxide, hydrogen sulfide, phenol and lead droplets) and respiratory diseases, including acute infections of the upper airways, acute bronchitis, chronic pharyngitis, pneumonia, chronic bronchitis and asthma. Pollutants are expressed for 1998 with their mean annual concentrations in the ambient air in twenty one geographic areas of the town. Respiratory disease rates per 100,000 persons are included for the same sites and the same year. A total correlation matrix between the single parameters is calculated and the significant correlation factors are determined. Significant correlation ($\alpha = 0.05$) is established between:

- Respiratory diseases and sulfur dioxide;
- Chronic pharyngitis and phenol plus sulfur dioxide;
- Chronic bronchitis and particulates;
- Acute bronchitis and phenol.

Cross-sectional study on the influence of all pollutants on the respiratory diseases and asthma, carried out by using regression analysis, showed that sulfur dioxide, hydrogen sulfide and lead are the most significant factors (EPMS, 2000b). Consequently, the correlation studies based on data from Sofia confirmed that all the pollutants investigated could be considered very important for the respiratory system and five of them come from transport.

The link between the ambient air pollution, including the pollutant mix routinely measured by the National System for Monitoring and the morbidity rate of the population of Sofia, is studied for the period 1983 -1985 (Nikiforov $et$ $al.$, 1999). Morbidity is reported by the cases of visits to medical centers of the population in three age groups (less then one, one to fifteen and over fifteen). Six districts of the town are compared. The authors established higher degree and frequency of the respiratory problems in the industrial and the central districts with a higher severity of traffic pollution.

A study was carried out in 2001 by two scientific panels from the National Center of Hygiene and the Sofia Institute of Hygiene and Epidemiology for the establishment the compliance with the requirements of Ordinance No 9/1999 for the limit values of $SO_2$, $NO_2$, $PM_{10}$ and lead in the ambient air and their concentrations in two sites in the central part of the city with heavy traffic (Nikiforov, 2001). The authors found that due to heavy traffic the mean annual pollutant concentrations at both sites exceed from three to six times the norms. They established as well a trend for the reduction of $NO_2$ concentrations in 2000 and 2001 as a result of the reorganization of public transport and the improvement of pavements.

A study on the influence of air pollutants on the health of the population in the big cities (Sofia, Varna) showed that they enhance the morbidity rate of respiratory diseases, bronchial asthma, skin diseases, organs of sense, blood forming organs, cardiovascular morbidity and reproductivity (Nikiforov, 2000). The most sensitive members of the population are children. Two types of districts are defined in Sofia where children's morbidity rate is higher:

* Five districts in the center of Sofia with severe traffic: Vazrajdane, Oborishte, Sredec, Triaditca, Garata;
* Eight industrial regions - Iana, Kremikovci, Serdika, Nadejda, Poduene, Iskar, Novi Iskar and Vrabnitca.

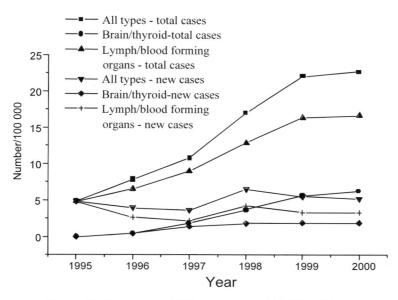

**Figure 40**. Cancer rate of different types per 100.000 children.
*Source: Health Handbook, NSI, 1995 – 2001.*

Cancer rates of different types (per 100,000 children under eighteen) are demonstrated in Figure 40. Trends of increase in the period are depicted. Slight picks are formed for cancer rates of lymph and blood-forming organs. The relative risk of cancer (RR) for the population of Sofia for the period 1995-2000, related to the relative risk of cancer (accepted as 1.00) of the population of Bulgaria in the same period is shown in Table 36.

**Table 36**. RR for the population of Sofia.

| People of all ages | | Children | | | | | |
|---|---|---|---|---|---|---|---|
| All types of cancer | | All types of cancer | | Brain / thyroid cancer | | Lymph / blood forming organs | |
| All registered cases | New registered cases | All registered cases | New registered cases | All registered cases | New registered cases | All registered cases | New registered cases |
| 1.05 | 0.99 | 0.83 | 0.31 | 0.21 | 0.63 | 0.48 | 0.78 |

Data from the table show that for the period studied the RR of cancer for the people of all ages in Sofia exceeds a little the RR of the country (RR = 1.05). In spite of the increasing trends of cancer for children shown in Figure 40, in general the relative risk of cancer for children in Sofia is much smaller, even for the most risky types: brain and thyroid cancer and cancer of lymph and blood-forming organs. This exception could be explained by the fact that there are some industrial towns with specific very polluting industries, which impose higher relative risk of cancer for the population living around and especially for children, exceeding the prevalent risk in the rest of the country. As an example, the towns of Bourgas with the biggest oil refinery in Bulgaria and production of petrochemicals, Plovdiv – with metallurgical plant for production of nonferrous metals, as well Varna and Russe – with complexes of chemical industries could be shown.

According to the National Environmental Strategy for 2001 - 2006 and Environmental Program of the Municipality of Sofia, mitigation measures for the reduction of transport emissions are under development. They are listed below:

* Legislation measures are accepted for limitation emissions of VOC during storage, loading or unloading and transport of petroleum and diesel fuel (Regulation No 16, 1999) and for the norms for content of sulfur, lead and other compounds, hazardous for the environment (Regulation No 17, 1999);
* Production and use of ethylated petroleum, containing lead should be stopped by the end of 2003;
* Process of replacement of diesel engines type Euro-1 by more environmental friendly engines Euro-2 is carried out;
* Replacement of the old buses by new is in process;
* By the year of 2000 the first part of a subway is in operation and the second part is under construction;
* To prevent import of old cars (more than five years old) from Europe from the beginning of 2003 import taxes for old cars are increased;
* Harmonization with EU standards for control of emissions from car engines;
* To strengthen requirements about registration, annual technical inspection, licensing the technical and emission control bodies (Regulation No 32, 1999 and Regulation No 33, 1999 of the Ministry of Transport);
* Enlargement of the pedestrian zone in the center of the town;
* Reduction of congestion by the construction of underground parking.


## 6.    Correlation between noise and respiratory diseases in Sofia (Case Study)

The noise generated by cars and lorries is estimated to be one of the most annoying impacts of the road transport (grade eight, Table 24). In the present case study we are trying to compare the urban noise level in different regions in Sofia and respiratory diseases.

Sofia is one of the noisiest towns in Europe. According to the investigations of the European Environmental Agency 47 per cent of the people in Sofia are subjected to 70 *dBA* noise. Forty-nine monitoring sites are situated in different regions in the city. In six of them over 60 *dBA* of noise are measured constantly. The results observed during the last seven years are relatively stable (Chuchkova and Simeonova, 2001). Two main streets are with the highest level of noise (blvd. "Tzarigradsko shose" -75.6 *dBA* and "Slivnitca" blvd. – 73.8 *dBA*).

To reduce the impact of noise caused by transport, the Municipality should take abatement measures including: building barriers and "green walls", organizing "one way" streets and optimizing traffic lights, changing the window and door types, construction of building design, etc.

The main difficulty to assess the impact of transport on human health is how to select and to prove the claim that some diseases are caused by the impact of transport.

On the basis of average level of noise during three years in the non-industrial districts of Sofia and the respiratory system morbidity rates (see Table 37) the coefficients of linear correlation $R$ are estimated. The correlation graph is given in Figure 41.

All correlation coefficients are significant with a level of significance $\alpha = 0.05$, except for noise and morbidity in 1997, which is significant for $\alpha = 0.10$.

The respiratory system diseases could be caused by the integrated impact of transport emissions and other reasons. The noise in the non-industrial districts shown in Table 37 is mostly caused by transport. It is clear that level of noise and traffic intensity are directly connected. If the level of noise is high, the level of pollutant emissions will also be high, since the pollutant emissions are proportional to traffic intensity. Hence, the correlation between noise level and air pollution and further – the respiratory diseases is to be expected. It is confirmed by the positive and significant linear correlation coefficients.

Additionally, noise also affects human health directly.

**Table 37**. Average annual noise level and respiratory system morbidity rate per 100,000 people for 1994, 1997 and 1998 in different districts of Sofia.

| No | District of Sofia | Average noise per year in dBA | | | Respiratory system morbidity rate | | |
|----|------------------|------|------|------|------|------|------|
| | Year | 1994 | 1997 | 1998 | 1994 | 1997 | 1998 |
| 1. | Sredec | 70.1 | 70.9 | 71.6 | 62.33 | 69.68 | 47.48 |
| 2. | Krasno selo | 67.7 | 65.0 | 66.9 | 31.34 | 36.57 | 31.65 |
| 3. | Vazrajdane | 66.2 | 67.5 | 70.5 | 53.55 | 58.16 | 53.82 |
| 5. | Serdika | 71.4 | 73.8 | 72.3 | 65.82 | 48.27 | 47.61 |
| 6. | Poduene | 65.8 | 66.4 | 68.0 | 79.98 | 123.46 | 91.61 |
| 7. | Slatina | 60.8 | 59.2 | 59.9 | 48.33 | 47.28 | 39.99 |
| 8. | Izgrev | 63.5 | 68.7 | 70.7 | 55.78 | 67.23 | 56.38 |
| 9. | Lozenec | 64.5 | 52.4 | 64.8 | 71.58 | 45.17 | 40.17 |
| 10. | Triaditca | 69.8 | 68.4 | 69.8 | 50.53 | 70.29 | 57.70 |
| 13. | Nadejda | 60.4 | 63.0 | 62.6 | 56.84 | 94.31 | 70.26 |
| 15. | Mladost | 59.4 | 60.1 | 65.7 | 52.15 | 59.06 | 55.16 |
| 18. | Ovcha kupel | 54.3 | 51.6 | 53.8 | 183.48 | 126.80 | 114.84 |
| 19. | Lulin | 62.1 | 65.8 | 66.2 | 71.51 | 80.74 | 69.28 |
| * | Total | | | | 48.60 | 70.82 | 61.08 |

Source: EPMS, 2000.

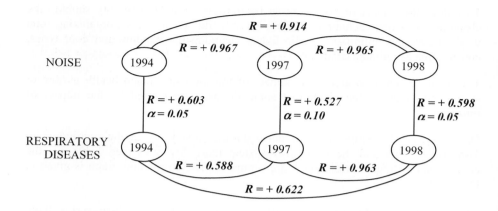

**Figure 41**. Linear correlation between noise and number of respiratory diseases per 100,000 people in Sofia for the years 1994, 1997 and 1998.


## Conclusion

In the period investigated, Bulgaria changed its political and economic system. The economic stress accompanying transition, along with many other things, reflected on the motor vehicle fleet of the country. The number of cars increased because of the import of cheap second-hand cars from Europe, which were used together with the ten – fifteen year-old socialist cars. In spite of the increase in the number of cars, their physical aging as well as use of low grade fuels, bad road conditions, bad driving culture and lack of respect for legislative measures, in the period general trends of reduction of the total emissions into the air were observed, including emissions from transport. Explanation of this fact is connected with the reduction in industrial activity, bankruptcy of many factories and abstention from driving.

Some positive trends are established. The Bulgarian government accepted the international obligations from Kyoto for reduction of greenhouse emissions. The process of harmonization of the Bulgarian legislation with EU law occurs, entailing increased consumption of unleaded petroleum, increased control on emissions from car engines, better technical inspection etc.

No evidence was established in the period for the increase of the total visits to medical centers, both for adults and children. It was observed that respiratory diseases prevail, followed by cardiovascular and nervous system diseases. Cancer rates in Bulgaria show alarming trends of continuous increase during the last decade. Cardiovascular diseases are the first cause of death with about 66 per cent commitment in the last few years. The size of effects could not be quantified in relation to the air pollutants of transport, but correlation could be suspected.

Two case studies on the environmental health of the people of Sofia are involved. Review on the environmental health status investigations showed correlation between emissions and noise and health. In spite of the increasing trends of cancer for children in the capital, it is established that for them the RR of cancer is lower, even for the most risky types. The fact is explained by the stronger impact on the health of specific industrial pollutants in selected areas of the country.

The linear correlation coefficients between urban noise and respiratory diseases in Sofia are significant with a level of significance $\alpha = 0.05$. The results show that the respiratory system morbidity is mainly caused by the impact of the emissions of street transport, the intensity of which is indirectly expressed by noise. Additionally noise also affects human health directly.

## References

Annual Report for Control of Automobile Transport (2002) Ministry of Internal Affairs, Sofia.

Annual Report of the Ministry of Health (2002) Sofia.

Annual Report of the State of the Environment in 1994, Green Book (1995) Council of the Ministers of the Republic of Bulgaria, Sofia.

Baynova, A. (2002) Impact of leaded gasoline on the human body, in Handbook "Can I drive with an unleaded gasoline?", "Time - Ecoprojects" Foundation, Sofia.

Bulgarian Statistical Reference Books (BSRB), (2001, 2002), National Statistical Institute, Sofia.

Chuchkova, M., and Simeonova, S. (2001) Methods for the assessment of the mean exceeding over permissible noise levels in the functional urban areas, *Envir. Emg. and Envir. Protection* 2, 12-16.

Environmental Program of the Municipality of Sofia (EPMS) (2000a) Polyconsult ECO and DEPSC-SM, vol. 1, Municipality of Sofia, Sofia.

Environmental Program of the Municipality of Sofia (EPMS) (2000b) Polyconsult ECO and DEPSC-SM, vol. 2, Municipality of Sofia, Sofia.

Georgiev I., and Manolov M. (1999) *Ecology and Sustainable Development*, University Publisher "Economy", University of National and World Economy, Sofia.

Green Book (1999) Ministry of Environment and Waters (MoEW), Sofia.

Health Handbook, (1995 - 2001) National Center for Health Information at the Ministry of Health, National Statistical Institute (NSI), Sofia.

Kendall, M.G. (1957) Rank Correlation Methods, Griffin, London, pp.109-111.

Nikiforov, B., Breugelmans, O., Brunekreef, B., Fletcher, T., Houthuijs, D., Jazwiec-Kanyion, B., Lebret, E., Leonardi, G.S., Pinter, A., Roemer, W., Rudnai, P., Slotova, K., Spichalova, A., and Surdu, S. (1999) Respiratory health, *Epidemiology* 10 Suppl. 1.4.

Nikiforov, B. (2000) Environmental Health Priorities and the Role of Prevention Health Services in Transition, *Balkan Public Health Care Series, Public Health in Transition*, 51-62.

Nikiforov, B. (2001) Project for compliance with the requirements of Ordinance No 9/1999, Unpublished paper, Sofia.

Petkov, T. (2001) Evaluation of aspects of aviation noise pollution due to Sofia airport activities, *Environ. Eng. and Environ. Protection* 1, 24-29.

Project for National Action Plan for Climate Change in Bulgaria (1999) MoEW, Sofia.

Regulation 8 on limit values for ozone in the ambient air, *SG, No 46/1999.*

Regulation 9 on limit values for sulfur dioxide, nitrogen dioxide, particulate matters and lead in the ambient air, *SG, No 46/1999.*

Regulation 16 on the reduction of VOS emissions from storage, loadingor unloading and transport of petrol, *SG, No 75/1999.*

Regulation 17 on limit values for content of lead, sulfur and other harmful substances in liquid fuels, *SG, No 97/1999.*

Regulation 32 on the periodic inspections for technical conditions of vehicles, *SG, No 74/1999.*

Regulation 33 on the public transportation of passengers and goods in Republic of Bulgaria, *SG, No 101/1999.*

Report on the State of the Environment in 1996, Green Book (1999) Council of Ministers of the Republic of Bulgaria, Sofia,.

State of the Environment in the Republic of Bulgaria. Annual Bulletin for 1998 (1999) MoEW, Executive Environmental Agency, Sofia.

State of the Environment in the Republic of Bulgaria. Annual Bulletin for 1999 (2000) MoEW, Executive Environmental Agency, Sofia.

State of the Environment in the Republic of Bulgaria. Annual Bulletin for 2000 (2001) MoEW, Executive Environmental Agency, Sofia.

Sofia in Numbers (2001) Report of the Municipality of Sofia, Sofia.

Stoyanov S. (1993) Optimization Technique, Technika, Sofia, 270-275.

Tashev, A., and Petkov, J. (2001) Rail transport and environmental protection. Achievements of European countries, *Environ. Eng. and Environ. Protection* 2, 4-11.

# HEALTH AND MOBILITY IN FLANDERS (BELGIUM)

P. LAMMAR AND L. HENS
*Department of Human Ecology,*
*Faculty of Medicine and Pharmacy,*
*Vrije Universiteit Brussel, Laarbeeklaan 103,*
*B-1090 Brussels, BELGIUM*

## Summary

This report focuses on the main health and environment aspects related to mobility. The study area, Flanders, with almost 6 million inhabitants and on average 440 persons per square kilometer, is one of the most densely populated regions in Europe. Traffic and transport affect the environment by largely contributing to the global air pollution. Traffic emissions significantly contribute to environmental problems such as acidification, photochemical air pollution and greenhouse effect. Traffic and transport also cause adverse health effects influencing the quality of life of the population. The health effects of traffic originate from three important categories: accidents, air pollution and noise pollution. Each year traffic accidents in Flanders cause about 900 deaths, 7,000 seriously injured and 39,000 slightly injured people. This report mainly focuses on the categories of air and noise pollution. The health impact of the air emissions in terms of lung impairments and diseases (heart diseases, cancer and mental disorders) is very important. The current levels of benzene and benzo(a)pyrene in ambient air are responsible for an extra 43 cancer patients per year. The current ozone concentrations are responsible for 900 extra deaths, about 800 hospital admissions, 1 million asthma attacks and 24 million symptom days a year. At least 1,095 extra deaths could be attributed to the current $PM_{10}$ concentrations. Lifetime exposure to current $PM_{10}$ concentrations would result in the loss of one healthy life year per person. Apart from air pollution, motor vehicles also cause significant noise pollution. This not only affects the quality of life, but also health by causing sleep disturbance, irritation and concentration problems among others. In the long run, noise may result in hearing impairment and cardiovascular diseases. The potential number of noise-induced awakenings from road and railway traffic in Flanders amounted to 2,129,000 a night in 1998. Air traffic badly affected the sleep and health quality of at least 26,400 people in 2001. In 1998, 951,000 Flemish inhabitants or 16.1 per cent were potentially seriously affected by the high noise levels from road traffic. In addition, about 80,000 people or 1.2 per cent experienced serious nuisance from railway traffic. About 1,800,000

*P. Nicolopoulou-Stamati et al. (eds), Environmental Health Impacts of Transport and Mobility, 199-222.*

Flemish inhabitants or 30 per cent are exposed to noise levels exceeding 65 dB in 2001, which may result in health effects such as increased blood pressure and cardiovascular diseases.

## 1.   Introduction

In Belgium there are three Regions: the Flemish Region, the Brussels-Capital Region and the Walloon Region. This report focuses on the Flemish region, also called Flanders, as the study area. As Flanders is the most densely populated region of Belgium the traffic effects are most visible. Flanders, with a population of 5,952,552 inhabitants and a surface area of 13,522 km$^2$, is located in the north of Belgium. The traffic network is very dense with a road network of 67,884 km (849 km motorways, 5,400 km regional roads, 635 km provincial roads and 61,000 km local roads), a railway network of 1,718 km and a navigable waterway network of 1076 km. These figures indicate that Flanders has the most dense motorway and railway network of Europe: 62 m motorway/km$^2$ (European average: 15 m/km$^2$) and 127 m railway/km$^2$ (European average: 47 m/km$^2$). The total road density is 5020 m/km$^2$ and the waterway density 80 m/km$^2$. The car fleet contains 3,328,657 motor vehicles (cars, buses, lorries and motorcycles), with a very large share of diesel cars (43.8 per cent) (VMM, 2002a).

The traffic and transport sector in Flanders causes an important pressure on health and environment. The main traffic pollutants are:
- Inorganic gases : sulphur dioxide ($SO_2$), nitrogen oxides ($NO_x$), carbon oxides ($CO_x$), nitrous oxide ($N_2O$) and ammonia ($NH_3$).
- Volatile organic compounds (VOC) : benzene, toluene, xylenes etc.
- Polycyclic aromatic hydrocarbons (PAH) such as benzo(a)pyrene
- Heavy metals : lead, copper, etc.
- Particulate matter (PM).

The emission of the traffic pollutants results in environmental problems such as acidification ($NO_x$, $SO_2$, $NH_3$) photochemical air pollution ($NO_x$, VOC) and greenhouse effects ($CO_2$, $CH_4$, $N_2O$).

Health effects are caused directly through accidents resulting in injuries and possibly impairments and disabilities. In the year 2000 about 47,000 people were injured in traffic accidents in Flanders. Of these 47,000 traffic victims 900 died, 7,000 got seriously injured and 39,000 slightly injured. With 13.6 deaths/100,000 inhabitants in 1999 Flanders shows a poor record compared to the EU-average of 11.1 deaths/100,000 inhabitants. For 2000 the record is even worse: 14.7 deaths/100,000 inhabitants (BIVV, 2000). Indirect health effects are caused by air and noise pollution.

This country report mainly focuses on the environmental and health effects resulting from air and noise pollution.

This pollution is a threat to, in particular, vulnerable population groups such as pregnant women, elderly people, people with respiratory or heart diseases, children and people exposed to high concentrations in their professional environment.

The health effects of the different traffic pollutants, the concentration levels and health impact in Flanders and the various guidelines are summarised in the appendix.

## 2.    Mobility and environmental issues

### 2.1.    $CO_2$ emissions (contribution to greenhouse effect)

The main greenhouse gases are $CO_2$ (83 per cent), $CH_4$ (8 per cent) and $N_2O$ (8 per cent). The emission of greenhouse gases in Flanders increased from more than 83 Mtonnes $CO_2$-equivalents in 1990 to almost 93 Mtonnes in 2001. This is an increase of 11.4 per cent. This increase is almost completely due to the increased $CO_2$ emissions (+13 per cent during the period 1990-2001). The traffic and transport sector (+3.6 Mtonnes $CO_2$-eq) is largely responsible for the higher emissions of greenhouse gases in the period 1990-2001. The contribution of traffic and transport to the total $CO_2$ emissions rose from 17.3 per cent in 1990 to 19.3 per cent in 2001. The contribution of traffic and transport to the total $N_2O$ emissions also increased from 2.3 per cent to 6.1 per cent as a result of the increase in the number of catalytic converters. Anticipating the Kyoto-protocol, Flanders decided to stabilize the $CO_2$ emissions at the 1990 level by 2005 (short-term objective). However, the emission of greenhouse gases by traffic continues to increase and the evolution is parallel with the evolution of the number of personkilometers (VMM, 2002a).  The greenhouse gases have no direct impact on health.

### 2.2.    $SO_2$ and $NO_x$ emissions and contribution to acidification

Emissions of sulphur and nitrogen compounds cause acidification. Sulphur oxides are mainly emitted by diesel traffic, incinerators and industry. Nitrogen oxides originate from traffic and other combustion processes. Ammonia is released by agriculture. Deposition of these compounds damages ecosystems, corrodes materials and erodes buildings (VMM, 2002a).

The total amount of acidifying emissions in Flanders decreased by 36 per cent to 10,945 million acid equivalents in 2001 from 17,071 million acid equivalents in 1990. This is mainly due to the decrease of $SO_2$ emissions. The $NO_2$ emissions are just below the 1990 level. Since 1998, $NO_2$ has become the most important component of the acidifying emissions, replacing $SO_2$.

The contribution of the traffic and transport sector to the total of acidifying emissions slightly decreased by 1 per cent between 1990 and 2001, mainly as a result of measures to decrease the $SO_2$ emissions. The emissions of acidifying pollutants by passenger cars in 2000 are for the first time lower than in 1990. In

2001 there is also a large decrease for freight transport but the emissions are still higher than in 1990 (VMM, 2002a).

The $NO_x$ emissions from traffic and transport amount to 90 ktonnes in 2001 (VMM, 2002a). Road traffic is the main source of $NO_x$ in Flanders, emitting about 52 per cent of the total $NO_x$ emissions in 2000 (VMM, 2002b). The EU-limit value for $NO_2$ (200 $\mu g/m^3$) wasn't exceeded in any of the 38 stations of the permanent monitoring network in 2001 (VMM, 2002b). The highest annual average $NO_2$ concentration, measured in Antwerp and Borgerhout, was 48 $\mu g/m^3$ in 2001. The annual averages were slightly higher than in 2000 (VMM, 2002b).

The contribution of traffic and transport to the emission of sulphur dioxides is 2 per cent in 2000 (VMM, 2002b). The $SO_2$ emissions from traffic and transport amount to 3,272 tonnes in 2001, same as in 2000. By limiting the sulphur content in petrol and diesel, as a consequence of Phase 1 of the European directive 98/70/EEC the $SO_2$ emissions substantially decreased in 2000. Phase 2 will result in a further decrease of the sulphur content by 2005 (VMM, 2002a). Because of the decrease of $SO_2$ emissions in Belgium as well as in the neighbouring countries during the last decade, there were no winter smog episodes with high air pollution of $SO_2$ during the period 2000-2001. The EU-limit values for $SO_2$ (80 $\mu g/m^3$ (50 percentile) and 250 $\mu g/m^3$ (98 percentile) as median of the daily mean values taken throughout the year) weren't exceeded in any of the measuring stations in Flanders during the period 2001-2002. The highest P50 and P98-values were respectively 23 $\mu g/m^3$ and 95 $\mu g/m^3$ and were measured in the Antwerp region. The guideline value for the annual average $SO_2$ concentrations (40 $\mu g/m^3$) wasn't exceeded either. The highest annual average was 29 $\mu g/m^3$ (VMM, 2002b).

The total acidifying deposition decreased by 22.3 per cent in Flanders from on average 5,928 acid equivalents/ha.y in 1990 to 4,605 acid equivalents/ha.y in 2001. The values, however, greatly vary within Flanders. The highest values are reported in the proximity of (large) cities, major traffic routes and in agricultural areas with intensive stock breeding. 45.4 per cent of the total acidifying deposition in 2001 has its origin in emission sources from outside Flanders. The contribution within the region comes from agriculture (28.9 per cent), traffic and transport (9.9 per cent), industry (6.8 per cent) and the energy sector (4.4 per cent). The emissions by high chimneys decreased much more than the emissions by low sources (e.g. traffic and transport). Emissions by low sources are transported over shorter distances and lead more easily to depositions in the country of origin (VMM, 2002a).

## 2.3.    *Contribution to ground-level ozone*

At ground level, ozone is a secondary pollutant because it is produced in a complicated chemical reaction initiated by sunlight and not emitted directly into the atmosphere. The contribution of traffic to ground-level ozone is most important since the main precursors, nitrogen oxides and volatile organic compounds, are emitted by motor vehicles in large amounts. In Flanders, the number of days with ozone excesses continues to increase, as well as the background values of ozone. The evolution of the number of days with a maximum 8-hourly value higher than 120 $\mu g/m^3$ has kept increasing slightly since 1993 (VMM, 2002a).

In 2001 there were 38 days on which the EU-threshold value for the protection of human health (110 $\mu g/m^3$) was exceeded. In 2000 this figure was 23 days. The number of ozone days (days with ozone concentrations > 180 $\mu g/m^3$) almost doubled from 6 days in 2000 to 11 days in 2001. The highest ozone concentration in 2001 was 250 $\mu g/m^3$. Trends are difficult to interpret because ozone concentrations are largely influenced by weather conditions and in particular by the amount of sunshine in spring and summer (VMM, 2001; VMM, 2002b).

The emissions of ozone precursors by passenger cars were for the first time in 1999 lower than in 1990. In 2001 there was also a large decrease for freight transport as compared to the year before but the emissions are still higher than in 1990 (VMM, 2002a).

## 2.4    *Contribution to ambient carcinogens*

Volatile organic compounds (VOC) and polycyclic aromatic hydrocarbons (PAH) entail some important carcinogenic traffic pollutants. VOC are compounds with benzene as a basic element and polycyclic aromatic hydrocarbons are organic compounds composed of 2 or more benzene rings.

Road traffic and petrol stations are responsible for 32 per cent of the total NMVOC (non-methane VOC) emissions in 2001 (VMM, 2002b). The most important compound is benzene, emitted up to 87.6 per cent by the traffic and transport sector in 2001. Petrol stations are responsible for an additional 8.6 per cent. The NMVOC emissions of the traffic sector were 14 per cent lower in 2001 than in 1990 and 8 per cent lower in 2001 than in 2000. In 2001 the NMVOC-emissions from traffic and transport amounted to 51 ktonnes (VMM 2002a).

The contribution of the traffic and transport sector to the total of PAH emissions is 21.5 per cent in 2001. The contribution of road construction is almost negligible since 1995 because tar asphalt is not used anymore (VMM, 2002a).

VOC emissions show a gradual decrease in Flanders. Between 1990 and 2001 the average benzene concentration in Flanders decreased from 3.9 $\mu g/m^3$ to 1.0 $\mu g/m^3$. In 2001 the concentrations were similar to those of 2000 (VMM, 2002a; VMM, 2002b). The decrease of NMVOC and benzene emissions by traffic and transport are the result of the larger share of diesel cars within the car fleet and the implementation of the AUTO-OIL program. This programme enforces traffic emission limits on a European level, determines the volatility and the maximum content of aromatics and benzene in petrol and enforces the use of catalytic converters. The VOC emissions from petrol stations will be further reduced by the introduction of the vapour recovery system promoted by the European regulations and BAT (VMM, 2001; VMM, 2002a).

The highest benzene concentrations are observed in urban areas with high traffic intensity. The highest annual average benzene concentration in one of the measuring stations in 2001 was 2.0 $\mu g/m^3$ (VMM, 2002b). The benzene concentrations in ambient air are already considerably lower than the medium-term objective of 5 $\mu g/m^3$ (VMM, 2002a).

High personal exposure is mainly due to exposure during traffic. For instance, concentrations of more than 3,000 $\mu g/m^3$ have been measured during filling up. Busy crossings (100 $\mu g/m^3$ and more) and car-parks (68.2 $\mu g/m^3$) are also characterized by higher benzene concentrations (Neumeier, 1993). Benzene is also shown to accumulate indoors posing specific problems to indoor air pollution (VITO, 1999).

Benzo(a)pyrene (BaP) is best known for its toxicity and dispersion and acts as PAH-indicator. In cities or in the proximity of industrial activities the BaP concentrations can rise to values between 10 $ng/m^3$ and 20 $\mu g/m^3$ (VMM, 1998). The short-term objective (2002) of 1 ng $BaP/m^3$ was already achieved in 1997. In 2001 the measurements vary between 0.23 and 0.89 $ng/m^3$ (measuring of small volumes) and between 0.38 and 0.54 $ng/m^3$ (measuring of large volumes) (VMM, 2002a). The decreasing trend in PAH-emissions stopped in 2001 (VMM, 2002b).

Dioxins are also known carcinogens. The share of the traffic and transport sector in the total emissions is rather limited (0.6 per cent) (VMM, 2001).

### 2.5.  *Noise*

Noise pollution is problematic in areas that combine a high population density with a dense road network, which is the case in Flanders. Traffic is the main source of noise pollution (56 per cent) in Flanders, as well as indoor, outdoor and in the workplace. Road traffic is responsible for 3/4 of the nuisance and air traffic for 1/4. The contribution by rail traffic is minimal (1.2 per cent). The affected area is still growing as a consequence of the widespread ribbon building in Flanders. It is expected that, if no particular measures are taken, the noise pollution will further increase because, at first, motor vehicles will not be spectacularly more silent in the near future and, secondly, the car fleet still increases. Noise is known to have a negative impact on the quality of life in particular in the cities (VMM, 1998; VMM, 2002a).

The number of people potentially seriously affected by sound is an adequate indicator of noise impact (VMM, 2002a). In 1996 road traffic was responsible for exposing more than 50 per cent of the population to noise levels above 55 dB(A) and approximately 27 per cent to noise levels above 65 dB(A), a level at which serious disturbance occurs (Duchamps *et al.*, 1997). In 2001, road traffic exposed 30 per cent of the general population to noise levels above 65 dB(A) during daytime. This percentage is much higher than the EU average (13-20 per cent). The increase in traffic intensity resulted in an increase of the noise level by 1 dB between 1996 and 2001, although this increase in noise level was partly compensated by an average decrease in noise emissions by individual vehicles. The proportionally stronger growth of traffic intensity, but especially the growing share of lorries on busy roads, explains the increase in the number of people exposed to noise levels above 65 dB(A) (VMM, 2002a). To protect the majority of the population from serious nuisance during daytime the noise level should not exceed 55 dB(A) according to the WHO. For moderate nuisance the figure is 50 dB(A). At night 45 dB(A) should not be exceeded, with a maximum noise level of 30 dB(A) in the bedroom, to avoid sleep disturbance (WHO, 1999). The ambition of the Flemish policy for the period 2003-2007 is 0 per cent exposure to road traffic noise above 65 dB(A) by 2020 (VMM, 2002a).

The number of people exposed to noise levels above 60 dB(A) from air traffic decreased significantly from 1999 to 2000 because of the introduction of less noisy airplanes. This evolution was accelerated through the limitation of the accepted quota per night flight. In 2001, the number of affected people continues to decrease because of the further introduction of less noisy airplanes and the decrease in the number of flights due to the general malaise in the air traffic sector, including the bankruptcy of the national air company SABENA and 11 September 2001 (VMM, 2002a). However, the concentration of night flights over the North of Brussels has caused a high level of disturbance to the people living in this area. This forced politicians to decide on a more dispersed distribution of night flights above Brussels. This decision will negatively affect the decreasing trend in the number of disturbed people.

### 2.6.   Other problems

Other important mobility-related environmental problems are waste production and energy use. Motor vehicles contribute to the waste problem. 266,000 passenger cars were put out of use during the year 2000. It is estimated that about 183,000 cars are exported as second-hand cars, especially to Eastern Europe. The other 83,000, or about 70,500 tonnes, were removed as wrecks to a shredder to recover metals (ferro) and other materials (non-ferro). The share of the ferro-fraction in a car wreck is 65-70 per cent; the share of the non-ferro-fraction is 5-10 per cent. This means that 70 to 80 per cent of a car's weight is recycled. The remaining 20 to 30 per cent is dumped or incinerated. 75 per cent of the car wrecks was reused or recycled in 2000, not yet reaching the objective of 80 per cent set for 2005 (VMM, 2001). In addition, tyres produce each year 27,000 tonnes of waste. So-called Life Cycle Impacts (the impact taking into account the whole life cycle of the vehicle) for certain vehicle categories are almost as important as the impact during use (Int Panis et al., 2001).

The total energy use by traffic and transport was 209.9 PJ in 2001, an increase by 0.7 per cent as compared to the year 2000. The average increase for the 1995-2000 period was 2.4 per cent. The majority is used by road traffic (more than 96 per cent).

About half of the energy a car uses is needed during the construction phase. There is a close relationship between the energy use and the $CO_2$ emissions by the combustion of fossil fuels (VMM, 2001).

## 3.    Health effects of the environmental problems

### 3.1.    Lung impairment and diseases (including asthma and fibrosis)

The traffic pollutants $NO_x$, $SO_2$, $PM_{10}$ and $O_3$ all affect the respiratory system.

Inhalation of nitrogen dioxide ($NO_2$) may result in irritation of the bronchial tubes. Long-term exposure causes coughing, increased sensitivity to infections, decreased lung function, increased occurrence of acute respiratory illness and symptoms, damage to the lung tissue (at high exposure levels) and paralysis of the smooth muscles of the cilia in the nose and bronchial tubes. This may harm the self-cleaning capacity of the respiratory tract and give carcinogenic and mutagenic pollutants and harmful micro-organisms the possibility to penetrate without resistance (Janssens and Hens, 1997; Harrison, 2001; VMM, 2002b). The effects vary with the duration of the exposure (short-term or long-term). Effects in healthy people resulting from short-term exposure will only be noticed under very high $NO_2$ concentrations (>1880 $\mu g/m^3$). People suffering from asthma or chronic lung diseases will experience adverse effects on the respiratory function at lower exposure levels (1 to 2 hour exposure to concentrations of 375 tot 565 $\mu g/m^3$ $NO_2$).

Sulphur dioxide ($SO_2$) emissions generally affect the upper bronchial tubes, causing irritation and constriction of the bronchi. High concentrations contribute to respiratory symptoms, reduced lung function, increased mucous secretion and rises in hospital admissions, especially among people with asthma or chronic lung diseases (Harrison, 2001; VMM, 2002b; Pilotto et al., 1997). The health effects are caused by absorption of $SO_2$ in the mucous membranes of the nose and the upper bronchial tubes and by deposition of sulphate aerosols in the bronchial tubes. Epidemiological studies have shown that small reversible decreases in lung function may occur among children from 250-450 $\mu g/m^3$ and an increased mortality from 500-1000 $\mu g/m^3$ (VMM, 2002b). Asthma patients have a higher risk on an asthma attack when exposed to $SO_2$ during pollution episodes (Koren, 1996; EEA, 1997; VMM, 2002a; Harrison, 2001). Long term exposure can lead to bronchitis and lung emphysema (Koren, 1996; EEA, 1997).

Particulate matter is one of the most important air pollutants causing adverse health effects. Toxic effects of particulate matter are mainly attributed to particles with an aerodynamic diameter of 10 $\mu m$ ($PM_{10}$) or smaller. These small particles can reach the alveoli and block the lungs. Effects are associated both with short-term exposure (hours, days) to high concentrations as with chronic exposure (years) to low concentrations. There is no safe threshold value (Dora, 1999; Schwela, 2000; Janssens and Hens, 1997). The health effects of $PM_{10}$ range from an increased frequency and severity of respiratory problems to an increased risk of premature death. The smaller particles penetrate deep in the lungs (VMM, 2002a). They are able to disturb the mucus drainage in the bronchial tubes by mechanical or toxic action, provoke breathing problems and increase the sensitivity to respiratory infections (VMM, 2000; VMM, 2002a). A short increase in $PM_{10}$-concentrations is estimated to be responsible for 7 to 10 per cent, and even to 20 per cent in the most polluted cities, of all respiratory complaints among children (EEA, 1996). High $PM_{10}$-concentrations are connected to higher respiratory mortality among elderly people and to asthma attacks and decreased lung function among children (VMM, 2000). There is consistent evidence from the United States and Europe that hospital admissions and emergency room attendances for respiratory complaints and asthma are related to ambient levels of particulate matter. Studies in the United Kingdom indicate that a 10 $\mu g/m^3$ rise in $PM_{10}$ is associated with between 1.5 and 5 per cent increase in hospital admissions or attendances (Harrison, 2001). In Flanders, $PM_{10}$-concentrations show an increasing trend since 1999, especially in urban areas (VMM, 2000; VMM, 2002a). This can mainly be attributed to traffic by an increase in the number of motor vehicles and the greater share of diesel vehicles (VMM, 2000).

There is consistent experimental evidence that ozone has an effect on health, with established dose-response effects on a number of lung function parameters at concentrations close to those in the ambient air. Ozone is a powerful oxidizing agent, causing direct cellular damage by damaging the anti-oxidant mechanisms in cells lining the airway walls. Prolonged exposure to high doses may result in persistent inflammation of the bronchioles and scarring (fibrosis) of the lung tissue (Harrison, 2001). At lower ozone concentrations, the following health effects are reported: decreased lung capacity, inflammations, hypersensitivity of the bronchial tubes, irritation of the mucous membranes, irritation of the eyes, nose and throat, coughing, headache, dizziness, shortness of breath, nausea, general malaise and chest discomfort. The effects worsen with increasing concentrations, sensitivity of exposed people and with physical activity during exposure (Harrison, 2001; VMM, 1998; Janssens and Hens, 1997; VMM, 2002b). The current ozone concentrations are responsible for 900 extra deaths, about 800 hospital admissions, 1 million asthma attacks and 24 million symptom days a year (Ministerie van de Vlaamse Gemeenschap, 2001).

## 3.2. *Health failure*

Sulphur dioxide, particulate matter and ozone show significant associations with deaths and hospital admissions, especially among patients with heart and lung diseases (Harrison, 2001; VMM, 2000; Dockerey and Pope, 1994; Pope *et al.*, 1995). The association between nitrogen dioxide and mortality or hospital admissions is weaker (Harrison, 2001).

US and European studies have demonstrated consistent relationships between $PM_{10}$ and daily mortality, with - depending on the study - a 0.5 to 1 per cent rise in mortality for every 10 $\mu g/m^3$ increase in $PM_{10}$ (Harrison, 2001) (see also Table 38).

An extrapolation of international risk assessment results indicates that 3 additional people a day may die as a result of exposure of the population to the current ambient concentrations of particulate matter in Flanders (VMM, 1999).

According to the Flemish report on the environment and nature, one healthy life year per person is lost at lifetime exposure to the current $PM_{10}$-concentrations (VMM, 2002b).

**Table 38.** Dose-response coefficients from short-term ecological time series studies for $PM_{10}$, sulphur dioxide and ozone (From Committee on the Medical Effects of Air Pollution 'Quantification of the Effects of Air Pollution on Health in the United Kingdom', London: Department of Health (1998)) (Harrison, 2001).

| Pollutant | Health outcome | Change per 10 µg/m³ increase in pollutant |
|-----------|----------------|-------------------------------------------|
| $PM_{10}$ | Deaths brought forward (all causes) | + 0.75 % (24 h mean) |
|           | Respiratory hospital admissions | + 0.8 % (24 h mean) |
| Sulphur dioxide | Deaths brought forward (all causes) | + 0.6 % (24 h mean) |
|           | Respiratory hospital admissions | + 0.5 % (24 h mean) |
| Ozone | Deaths brought forward (all causes) | + 0.6 % (8 h mean) |
|           | Respiratory hospital admissions | + 0.7 % (8 h mean) |

Although the estimated health effect is most impacting for particulate matter (see Table 38), sulphur dioxide and ozone also show significant associations with deaths and hospital admissions (Harrison, 2001).

The WHO (2000) estimates the number of deaths in European cities due to long-term exposure to air pollution generated by traffic to be between 36,000 and 129,000 each year. This assumes that around 35 per cent of the deaths attributed to particulate matter pollution are due to traffic air pollution (WHO, 2000a). The WHO (2000) estimates the number of deaths in the European WHO-region due to total air pollution between 102,000 and 368,000 each year (WHO, 2000a). The same analysis also estimates that particulate matter accounts for 6,000 to 10,000 additional hospital admissions for respiratory diseases in European cities every year (WHO, 2000a). A study in Austria, France and Switzerland assessed the health impact due to air pollutants from traffic, to result in an estimated 40,000 deaths each year or 6 per cent of the total mortality. About half of all mortality caused by air pollution was attributed to motorised traffic, accounting also for: more than 25,000 new cases of chronic bronchitis (adults); more than 290,000 episodes of bronchitis (children); more than 0.5 million asthma attacks; and more than 16 million person-days of restricted activities. The study also attributes about twice as many deaths to air pollution than to accidents (Künzli et al., 2000). According to a report of the APHEIS project, nearly 3,000 premature deaths in assessed cities (or 9 premature deaths per 100,000 inhabitants) could be prevented annually if long-term exposure of particulate air pollution was reduced to the target level, set by the European Commission for 2005, in all 26 (European, except Tel Aviv) cities participating in the project (Medina et al., 2002).

### 3.3. Cancer

Volatile organic compounds (VOC) and polycyclic aromatic hydrocarbons (PAH) are known carcinogens. The attachment of other (carcinogenic) pollutants such as heavy metals and volatile organic compounds to particulate matter offers an additional problem due to the ability to penetrate deep into the respiratory system (VMM, 2000).

Exposure to benzene, a group 1 carcinogen, is associated with different types of blood cancer of the lymphatic system (Van Larebeke, 2002). There is a proven causal association with acute non-lymphocytic leukaemia in humans (Harrison, 2001). The chronic toxicity of benzene is attributed to its metabolites. Most sensitive for benzene toxicity is the bone marrow of which the biology is suppressed with reductions in red cell, white cell and blood platelet production as a result (aplastic anaemia) (WHO, 1987; WHO, 1995; Harrison, 2001). For carcinogens there is no safe threshold value. With lifelong exposure, the risk of developing leukaemia at a benzene concentration of 1 $\mu g/m^3$ is 6 per million according to the WHO (VMM, 2001). This means that for 2001, according to the WHO risk factor, traffic causes almost 36 extra leukaemia patients per year in Flanders (VMM, 2001).

Many of the polycyclic aromatic hydrocarbons, together with their metabolites and nitro-derivatives, are carcinogens. Several studies indicate that PAH cause cancer in almost all exposed tissues. Most data concern lung, skin and digestive system cancers. Most intensively studied among the PAH is benzo(a)pyrene (BaP). An excess of lung cancer (for example a relative risk around 1.5 in truck drivers) in workers exposed to high concentrations of vehicle exhausts has been demonstrated (Hayes *et al.*, 1989). Most studies show an increased relative risk between 1.5 and 2.0 (Harrison, 2001). It has been estimated that lifetime exposure to one $ng/m^3$ BaP results in 0.3 to 1.4 deaths each year per 10,000 people exposed (Harrison, 2001). For the BaP concentrations of 1998 in the ambient air and food, 7 respectively 0.065 extra cancer deaths were calculated per year in Flanders (VMM, 1998). In 2001, the BaP concentrations were at the same level (or slightly higher, depending on the measurement station) as those of 1998 (VMM, 2002a). Therefore we might assume that the risk calculation for 1998 also applies to 2001. With regard to PAH, increased attention is paid to a number of secondary pollutants (pollutants formed by further reaction of primary pollutants e.g. nitro-PAH). These products are dangerous as they often show higher mutagenic and carcinogenic properties than the primary pollutants. An example is the presence of 3-nitrobenzanthrone, a highly carcinogenic compound, in the exhaust fumes of cars and ambient air (VMM, 2002b). Its effects on humans are currently not quantified.

### 3.4. Haematological effects

Carbon monoxide, nitrogen oxides, sulphur dioxide, lead, particulate matter, volatile organic compounds as benzene and noise are associated with haematological and/or cardiovascular effects.

When carbon monoxide (CO) enters the lungs it is quickly absorbed into the blood. Carbon monoxide exerts its toxic effect by binding avidly to haemoglobin, thereby reducing the oxygen-carrying capacity of the blood. At lower concentrations, as they result from traffic, CO affects the cerebral function, heart function and lung capacity, all of which are sensitive to lower blood oxygen concentrations (Harrison, 2001). Other effects include dizziness, headache, decreased vigilance, tachycardia, sickness and fatigue (Harrison, 2001; Janssens and Hens, 1997). The concentration of the carboxyhaemoglobin (CO-Hb-complex) in the blood is a good exposure indicator. No health effects are observed at CO-Hb concentration levels below 2 per cent. Cardiac effects of carbon monoxide include changes of the electro-physiological properties of the heart at quite low levels of carboxyhaemoglobin (5.5 per cent) and reduction of the threshold at which cardiac arrhythmia or arrest may occur. The ambient CO emissions in Flanders decrease due to the use of the catalytic converter. Strengthened CO-standards for motor vehicles will bring CO-emissions down, also in the future (VMM, 2002a).

$NO_x$, $SO_2$ and particulate matter also cause cardiovascular disorders. They combine with haemoglobin and consequently cause changes in the blood clotting mechanisms (Dockerey and Pope, 1994; Pope et al., 1995; Michelozzi et al., 2000; VMM, 2000; Harrison, 2001).

Lead (Pb) disturbs the production of haemoglobin (Deelstra et al., 1996). This explains why almost all organs and organ systems can be considered potential targets and why a wide range of biological effects is described. These effects include anemia and cardiovascular (such as increased blood pressure), liver, kidney, endocrine and gastrointestinal effects (Koren, 1996; VMM, 2000; VMM, 2002b). An extensive environment and health research in Flanders showed that increased Pb-concentrations in blood are related to decreased kidney functioning in adolescents and a decreased bone metabolism in adult women (Vlietinck, 2000). The increased use of unleaded petrol resulted in a sharp decrease in the lead emissions (VMM, 2000). The limit value in the Flemish legislation (2 $\mu g/m^3$) was not exceeded in any of the measuring stations in Flanders in 2001 (VMM, 2002a).

Epidemiological studies show that increased blood pressure and cardiovascular effects are associated with long-term exposure to high noise levels (65-70 dB(A)). The association is stronger for ischaemic heart disease than for hypertension (Babisch et al., 1999; WHO, 1999; Koszarny, 2000). About 1,800,000 people are exposed to such high noise levels in Flanders (VMM, 2002a).

Benzene mainly affects bone marrow, causing different types of white blood cell cancer (see 3.3.).

### 3.5.    Sleep disturbances

Effects on sleep, caused by traffic noise, are measurable at background noise levels of about 30 dB(A). The primary effects include the following: difficulties in falling asleep, awakenings and alterations of sleep stages or depth, especially a reduction in the proportion of the REM-sleep. Secondary effects include reduced sleep quality, increased fatigue, depressed mood or well-being and decreased performance (WHO, 1999). At levels experienced by at least 20 per cent of the Europeans, both deep sleep – which aids physical recovery – and dream sleep – which is important for mental recovery – are reduced (T&E, 1997). The potential number of noise-induced awakenings from road and railway traffic in Flanders is 2,129,000 each night. Traffic on regional roads, including motorways, is responsible for 273,000 awakenings; traffic on local roads for 1,643,000 awakenings. Railway traffic causes 213,000 awakenings (Ministerie van de Vlaamse Gemeenschap, 2001).

### 3.6.   Mental disorders

Mental disorders and effects on the nervous system are reported after exposure to high noise levels, lead and volatile organic compounds (BTEX-components).

Concentration problems, fatigue, uncertainty, lack of self-confidence, irritation, misunderstandings, decreased working capacity, problems with human relations and a number of stress reactions, such as frequent nervousness, have all been identified with high noise levels (Hecht and Maschke, 1997; WHO, 1999; Koszarny, 2000). Noise levels between 55 and 65 dB(A) directly interfere with auditory communication. This may cause annoyance, displeasure and dissatisfaction (Babisch et al., 1999; Koszarny, 2000; WHO, 1999). About 2,800,000 people are exposed to these noise levels in Flanders (Ministerie van de Vlaamse Gemeenschap, 2001). In noisy areas, it has been observed that there is an increased use of prescribed drugs such as tranquillizers and sleeping pills, and an increased frequency of psychiatric symptoms and mental hospital admissions (WHO, 1999).

The adverse effects of lead especially influence the nervous system. Effects on the nervous system are reported at lower concentrations in children than in adults (WHO, 1987; VMM, 2000). Lead causes a continuum of effects on the nervous system of children, varying from cognitive deficits to mental retardation (lower IQ), with delayed nerve conduction and behavioural disturbances (WHO, 1987; VMM, 2000; VMM, 2002b). Adults show pathological effects on attention, memory and psychomotor functions with long term exposure to low lead concentrations in blood (VMM, 2000). An adverse effect on psychomotor functioning among male adolescents has been reported in an extensive environment & health research in Flanders (Vlietinck, 2000).

The so-called BTEX-components (benzene, toluene, ethyl benzene and xylenes) also negatively affect the central nervous system with, among others, depression of the central nervous system, decreased co-ordination, spastic movements, tremor and heartbeat problems after acute exposure (Low, 1988; VMM, 1996; Greenberg, 1997). At lower concentrations following symptoms can be noticed: dizziness, agitation, staggering walk, irritation of the eyes, nose and/or throat, vomiting, stomachache and nausea (Koren, 1996).

**Box 6**. Noise nuisance from Brussels-National airport.

---

*History*

Brussels-National airport, located in Zaventem at 20 km from Brussels centre, is the largest airport in Belgium. The evolution of inhabitants exposed to high noise levels caused by airplanes from Brussels-National shows the following trends: the number of inhabitants living within the 60 dB(A) outline (people who are exposed to noise levels > 60 dB(A)) around the airport showed a limited decrease between 1990 and 1995 from 51,216 to 40,119. This was mainly caused by the ban on the noisiest airplanes during the night. From 1995 to 1999 an increase from 40,119 to 46,843 was noticed almost reaching the levels of 1990. This was caused by a changed distribution of aircraft movements across the runways and by an increase of aircraft movements (VMM, 2001).

*Aim*

Flanders policy handles in this context the following medium-term objective (2005-2010): decrease the number of inhabitants in Flanders exposed to air traffic noise $L_{Adn}$ > 60 dB(A) to 45,000 around the Brussels–National airport (VMM, 2002a).

*Importance for environmental health*

Considering the number of people affected by noise levels above 60 dB(A) by air traffic, the health impact is certainly not negligible. Many of the health effects caused by noise may be noticed at these noise levels. The importance of the health problem can also be shown by recalling the WHO guideline: 'To protect the majority of the population from serious (moderate) nuisance during daytime the noise level should not exceed 55 dB(A) (50 dB(A)). At night 45 dB(A) should not be exceeded, with a maximum noise level of 30 dB(A) in the bedroom, to avoid sleep disturbance' (WHO, 1999)

*Results and conclusions*

In response to a number of complaints by citizens living in the neighbourhood of the airport the Belgian federal authorities decided on measures in the framework of a global plan to decrease the noise nuisance from Brussels-National. The federal decision implies:
- A gradual and programmed restriction of noise peaks at night.
- Introduction of seasonal night quota.
- The disappearance of military flights from 1 January 2004.
- Modification of aviation procedures in order to direct flights over the least densely populated areas.

These measures were meant to encourage the replacement of noisy aircraft, with the result that the proposed medium-term objective would be achieved by 2003 (VMM, 2000).

As shown in Table 39 the medium-term objective was already achieved in 2000. The decreasing trend continues in the year 2001 (VMM, 2002a). In 2001, 23,967 people were exposed to very high noise levels, not only causing difficulties in falling asleep, reduced sleep quality and awakenings, but also cardiovascular effects (increased blood pressure, cardiovascular diseases) (VMM, 2002a). The decrease is mainly caused by the alteration of the composition of the aviation fleet flying into Brussels, e.g. the replacement of Boeing by Airbus and by the fact that a number of large cargo airplanes stayed away (VMM, 2001). The shift to quieter airplanes continued in 2001 because of the decision to impose further restrictions and because the so-called chapter II-airplanes (the noisiest airplanes) were banned from European airports from 1 April 2002 on. Moreover, the number of day and night flights decreased in 2001 as a consequence of the events of 11 September 2001 and the general malaise in the air traffic sector (e.g. the bankruptcy of the national aircraft company SABENA) (VMM, 2002a).

**Table 39**. Number of inhabitants in Flanders, exposed to air traffic noise within the $L_{Adn}= 60$ dB(A) outline around Brussels-National. * The inhabitants of the Brussels Region aren't included anymore since 1996. (VMM, 2002a).

| Year | Number of inhabitants exposed to $L_{Adn} > 60$ dB(A)* |
|------|---------------------------------------------------------|
| 1990 | 51 216 |
| 1995 | 40 119 |
| 1996 | 40 376 |
| 1997 | 41 824 |
| 1998 | 46 364 |
| 1999 | 46 843 |
| 2000 | 28 586 |
| 2001 | 23 967 |

Complaints of the affected people, however, led to the decision for a more widespread distribution of the night flights above Brussels, which in turn will affect more people, living in the neighbourhood of the airport, in the future.

## 4.    Conclusions

The emission of greenhouse gases, mainly composed of $CO_2$, continues to increase with the car fleet and the number of personkilometres. The short-term policy objective (reaching the $CO_2$ level of 1990 by 2005) will be most difficult to reach.

Since the ozone precursors ($NO_2$ and VOC) slightly increased again in 2001, additional measures will be necessary to limit the ozone concentrations in the future. Not only the number of days with exposure to concentration levels exceeding the EU-threshold value for the protection of human health increase, but also the background concentrations of ozone. The current ozone concentrations are responsible for about 900 extra deaths, 800 hospital admissions, 1 million asthma attacks and 24 million symptom days a year.

The decrease in the VOC and PAH concentrations in ambient air came to an end in 2001 in Flanders. This is a negative evolution in view of the carcinogenicity of these compounds. The benzene emissions from traffic are estimated to be responsible for almost 36 extra leukaemia patients per year. The current BaP concentrations in the ambient air cause at least 7 extra cancer deaths a year.

The potential number of noise-induced awakenings from road and railway traffic in Flanders amounted to 2,129,000 a night in 1998. Air traffic adversely affects the sleep and health quality of 26,400 people in 2001. In 1998, 951,000 Flemish inhabitants or 16.1 per cent were potentially seriously affected by the high noise levels from road traffic. In addition, about 80,000 people or 1.2 per cent experienced serious nuisance from railway traffic (Ministerie van de Vlaamse Gemeenschap, 2001). These levels can result in mental health effects including annoyance, irritation, nervousness, concentration loss, etc. About 1,800,000 people are exposed to noise levels exceeding 65 dB, which may result in health effects such as increased blood pressure and cardiovascular diseases. These figures indicate that traffic noise is an important health problem. The percentage of people exposed to noise levels above 65 dB(A) is still increasing, up to 30 per cent in 2001. The number of people affected by air traffic noise (> 60 dB(A)) decreased since 2000, but remains quite high. The different scenarios, 'business as usual' scenario and 'sustainable development' scenario, developed to predict the future evolutions in noise nuisance both show a negative evolution with an increase in the number of people affected by noise (Ministerie van de Vlaamse Gemeenschap, 2001).

The $PM_{10}$ concentrations show again an increasing trend since 1999, especially in urban areas. This is also a negative evolution since $PM_{10}$ is considered to be one of the most harmful air pollutants. It is, together with $SO_2$, $O_3$ and $NO_2$, associated with increased hospitalisation incidence and mortality. Patients with heart and lung diseases are particularly vulnerable. At least 1095 extra deaths are attributed to the current $PM_{10}$ concentrations. Lifetime exposure to current $PM_{10}$ concentrations results in the loss of one healthy life year per person. Estimations of the respiratory and cardiovascular health impacts of road traffic emissions are summarised in Table 40.

**Table 40.** Respiratory and cardiovascular health impact of road traffic emissions for 1998 in Flanders, mainly due to exposure to $PM_{10}$, $NO_2$ and $SO_2$ (ozone is not included). The results are expressed as the number of mortality or disease cases per year. The impacts 'chronic mortality' and 'restricted activity' are respectively expressed in number of lost life years and number of days (Ministerie van de Vlaamse Gemeenschap, 2001).

|  | Passenger cars | Lorries |
|---|---|---|
| Chronic mortality | 6700 lost life years | 5200 lost life years |
| Heart failure | 180 cases | 80 cases |
| Bronchodilator use (adult asthmatics) | 124 000 cases | 96 000 cases |
| Coughing (adult asthmatics) | 127 000 cases | 99 000 cases |
| Lower respiratory complaints (adult asthmatics) | 46 000 cases | 36 000 cases |
| Bronchodilator use (asthmatic children) | 25 000 cases | 19 000 cases |
| Coughing (asthmatic children) | 43 000 cases | 33 000 cases |
| Lower respiratory complaints (asthmatic children) | 33 000 cases | 25 000 cases |
| Chronic coughing (children) | 16 000 cases | 12 000 cases |
| Restricted activity | 758 000 days | 588 000 days |
| Chronic bronchitis | 700 cases | 600 cases |

As long as the $PM_{10}$, $O_3$ and $NO_2$ concentrations do not decrease significantly respiratory problems will remain important especially among people suffering from asthma or chronic lung diseases.

As the health impact of particulate matter is high, the composition of the car fleet is very important. Diesel cars emit higher concentrations of particulate matter than petrol cars. The potential decrease of health impacts will be strongly reduced by the increasing share of diesel cars within the car fleet. If the number of diesel cars continues to increase the health impact will follow this trend. Especially in cities, where the high traffic density is responsible for high concentrations and where many people are exposed to these high levels, the problem is important. For Flanders the main problems are situated in the regions of Antwerp and Brussels.

## References

Babisch, W., Ising, H., Gallacher, J.E., Sweetnam, P.M., and Elwood, P.C. (1999) Traffic noise and cardiovascular risk: the Caerphilly and Speedwell studies, third phase – 10-year follow up, *Arch. Environ. Health* 54(3), 210-6.

BIVV (2000) Jaarverslag Verkeersveiligheid 2000.

Deelstra, H., Massart, L., Daenens, P., and Van Peteghem, C. (1996) Vreemde stoffen in onze voeding, Monografie no. 35 van Stichting Leefmilieu, Antwerpen.

Dockery, D.W., and Pope III C.A. (1994) Acute respiratory effects of particulate air pollution, *Annual review of public health* 15,107-132.

Dora, C. (1999) A different route to health: implications of transport policies, BMJ 318, 1686-1689.

Duchamps, W., Botteldooren, D., and De Poorter, J. (1997) Het inventariseren van de geluidsniveaus veroorzaakt door lokaal wegverkeer in Vlaanderen, in opdracht van de Vlaamse Milieumaatschappij (VMM).

European Environment Agency (EEA) (1996) Environment and Health 1 – Overview and main European issues: Environmental Monograph No. 2.

European Environment Agency (EEA) (1997) Air and Health: a pamphlet for local authorities.

European Environment Agency (EEA) (2000) Are we moving in the right direction? Indicators on transport and environment integration in the EU. TERM 2000. Environmental issues series No. 12.

Greenberg, M.M. (1997) The central nervous system and exposure to toluene: a risk characterization, *Environ. Res.* 72(1), 1-7.

Harrison, R.M. (Ed.) (2001) Pollution. Causes, effects and control, fourth edition, Royal Society of Chemistry, London.

Hayes, R.B., Thomas, T., Dilverman, D.T., Vineis, P., Blot, W.J., Mason, T.J., Pickle, L.W., Correa, P., Fontham, E.T., and Schoenberg, J.B. (1989) Lung cancer in motor exhaust-related occupations, *Am. J. Ind. Med.* 16(6), 685-695.

Hecht, K., and Maschke, C. (1997) Health effects of traffic noise. Continuous nocturnal stress (interview by Bettina Schellong-Lammel), *Fortschr. Med.* 115(22-23), 8-10.

Int Panis, L., and De Nocker, L. (2001) The external costs of road transport in Belgium, VITO, Mol.

Janssens, P., and Hens, L. (1997) Mens en Milieu. Onze gezondheid bedreigd? Monografieën Stichting Leefmilieu 36 (in Dutch).

Koren, H. (1996) Illustrated dictionary of environmental health & occupational safety, CRC Press, Inc., Bora Roc.

Koszarny, Z. (2000) The effect of intensive traffic noise on well-being and self-assessed health status of urban population, *Rocz. Panstw. Zakl. Hig.* 51(2), 191-201.

Künzli, N., Kaiser, R., Medina, S., Studnicka, M., Chanel, O., Filliger, P., Herry, M., Horak, Jr F., Puybonnieux-Texier, V., Quénel, P., Schneider, J., Seethaler, R., Vergnaud, J-C., and Sommer, H. (2000) Public-health impact of outdoor and traffic-related air pollution: a European assessment, *Lancet* 356, 795-801.

Low, L.K., Meeks, J.R., and Mackerer, C.R. (1988) Health effects of the alkylbenzenes – toluene, *Toxicol. Indust. Health* 4, 49-73.

McClellan, R.O. (1987) Health effects of exposure to diesel exhaust particles, *Ann. Rev. Pharmacol. Toxicol.* 27, 279-300.

Medina, S., Plasència, A., Artazcoz, L., Quénel, P., Katsouyanni, K., Mücke, H-G., De Saeger, E., Krzyzanowsky, M., Schwartz, J. and the contributing members of the APHEIS group (2002) APHEIS Health Impact Assessment of Air Pollution in 26 European Cities, Second year report, 2000-2001, Institut de Veille Sanitaire, Saint-Maurice, September 2002; 225 pages.

Michelozzi, P., Forastiere, F., Perucci, C.A., Fusco, D., Barca, A., and Spadea, T. (2000) Acute effects of air pollution in Rome, Ann Ist Super Santina, 2000.

Ministerie van de Vlaamse Gemeenschap. AMINABEL – Cel Lucht. (2001) Onderzoeksopdracht 'Milieu-impactbepaling van het ontwerp Mobiliteitsplan Vlaanderen d.m.v. strategische m.e.r.'. 2001. Deelrapport 2 : Strategische MER van het ontwerp Mobiliteitsplan Vlaanderen. Wetenschappelijk consortium van RUG, Vakgroepen Informatietechnologie en Geografie, UIA, Departementen Biologie en Politieke en Sociale Wetenschappen, UFSIA, Onderzoeksgroep STEM, VITO en VUB, Vakgroep Menselijke Ecologie.

Neumeier, G. (1993) Occupational exposure limits, Criteria document for benzene, Commission of the European Communities, Directorate-General Employment, Industrial Relations and Social Affairs, Office for official publications of the European Community.

OECD (1988) Transport and the environment, OECD Publication Service, Paris.

Pope, C.A., Bates, D.V., and Raizenne, M.E. (1995) Health effects of particulate air pollution: time for reassessment? *Environmental Health Perspectives* 103(5), 472-480.

Pope, C.A. 3rd, Bumett, R.T., Thun, M.J., Calle, E.E., Krewski, D., Ito, K., and Thurston, G.D. (2002) Lung cancer, cardiopulmonary mortality, and long-term exposure to fine particulate air pollution, *JAMA* 287(9), 1132-1141.

Pilotto, L.S., Douglas, R.M., and Samet, J.M. (1997) Nitrogen dioxide, gas heating and respiratory illness, *Medical Journal of Australia* 167, 295-6.

Scheepers, P.T.J., and Bos, R.P. (1992) Combustion of diesel fuel from a toxicological perspective – Toxicity, *Int. Arch. Occup. Environ. Health* 64, 165-177.

Schwela, D. (2000) Air pollution and health in urban areas, *Rev. Environ. Health* 15(1-2): 13-42.

T&E. European Federation for Transport and Environment (1997) Traffic and health, By Barry Lynham, December 1997, T&E 97/7.

Van Larebeke, N. (2002) Literatuurstudie benzeen : het risico op kanker, Steunpunt Milieu en Gezondheid, Vakgroep Radiotherapie, Kerngeneeskunde en Experimentele Cancerologie, Universiteit Gent.

Vlietinck, R., Schoeters, G., Van Loon, H., and Loots, I. (2000) Ontwikkeling van een concept voor de opvolging en risico-evaluatie van blootstelling aan leefmilieupolluenten en hun effecten op de volksgezondheid in Vlaanderen, Algemene Koepeltekst, Eindrapport van het onderzoek milieu en gezondheid.

VITO (1999) Assessment of urban population exposure to benzene in Antwerp, VITO, Mol.

VMM (1996) Lozingen in de lucht: structuur en resultaten van de emissie-inventaris Vlaamse Regio in 1994, VITO, Mol.

VMM (1998) MIRA-T 1998, Milieu- en Natuurrapport Vlaanderen: thema's, Garant, Leuven-Apeldoorn.

VMM (1999) MIRA-T 1999, Milieu- en Natuurrapport Vlaanderen: thema's, Garant, Leuven-Apeldoorn.

VMM (2000). MIRA-S 2000. Milieu- en Natuurrapport Vlaanderen: scenario's, Garant, Leuven-Apeldoorn.

VMM (2001). MIRA-T 2001. Milieu- en Natuurrapport Vlaanderen: thema's, Garant, Leuven-Apeldoorn.

VMM (2002a). MIRA-T 2002. Milieu- en Natuurrapport Vlaanderen: thema's, Garant, Antwerpen-Apeldoorn.

VMM (2002b) Luchtkwaliteit in het Vlaamse Gewest, 2001, Erembodegem.

WHO (1987) Air quality guidelines for Europe, WHO Regional office for Europe (WHO Regional Publications, European Series No. 23), Copenhagen.

WHO (1995) Concern for Europe's tomorrow, Health and the Environment in the WHO European Region, WHO, Europe, Stuttgart.

WHO (1999) Guidelines for Community Noise, Edited by B. Berglund, T. Lindvall and D.H. Schwela, WHO, Geneva.

WHO (2000a) Transport, Environment and Health, Edited by C. Dora and M. Phillips, WHO Regional Publications, European Series, No. 89, Copenhagen.

WHO (2000b) Guidelines for Air Quality, WHO, Geneva.

## Appendix

**Table 41**. Summary of the health effects, the concentration levels in Flanders and the guidelines for the most important traffic pollutants (VMM, 2000; EEA, 2000; Koren, 1996; WHO, 2000b, VMM, 2002a; VMM, 2002b).

| Pollutant | Health effects | Concentration level in Flanders | Health impact in Flanders | Guidelines |
|---|---|---|---|---|
| Benzene | - Increased cancer risk: leukemia etc.<br>- Neurological effects, damage to the central and peripheral nervous system<br>- Haematological effects : anemia, damage to bone marrow<br>- Suppression of the immune system<br>- Chromosome aberrations<br>- Unconsciousness | $1.0\ \mu g/m^3$ (average benzene concentration in 2001) | 36 extra leukaemia patients a year | *EU 2005* :<br>$5\ \mu g/m^3$ (year) (limit value)<br><br>*WHO*:<br>$4.4 - 7.5\ 10^{-6}\ \mu g/m^3$ (guideline value) |
| Benzo(a)pyrene (BaP) | Carcinogen | $0.23 - 0.89$ $ng/m^3$ (small volume measurements) and $0.38 - 0.54$ $ng/m^3$ (large volume measurements) (2001) | about 7 extra cancer deaths a year | *EU-WHO*:<br>$< 1\ ng/m^3$ (background value) |
| Carbon monoxide (CO) | - The complex carbon monoxide-haemoglobin interferes with the metabolism, resulting in deficiency of the oxygen transport<br>- Dizziness, headache, decreased vigilance, tachycardia, sickness, fatigue, abnormal visual perception, co-ordination problems, breathing problems, unconsciousness, death, neurobehavioural effects<br>- Influence on cerebral and heart function<br>- Foetuses : risk of lower birth weight, congenital defects or non-livability | $0.41 - 0.60$ $mg/m^3$ (8 hours) (2002) | See Table 40 | *WHO*:<br>$100\ mg/m^3$ (10 minutes)<br>$60\ mg/m^3$ (30 minutes)<br>$30\ mg/m^3$ (hour)<br>$10\ mg/m^3$ (8 hours)<br><br>*Flanders (VLAREM)*:<br>$30\ mg/m^3$ (98 percentile) (limit value)<br><br>*EU 2005*:<br>$10\ mg/m^3$ (8 hours) (limit value) |

Table 41, continued.

| Pollutant | Health effects | Concentration level in Flanders | Health impact in Flanders | Guidelines |
|---|---|---|---|---|
| Lead (Pb) | - Haematological effects: disturbing the production of haemoglobin, anemia<br>- decreased bone metabolism among women<br>- Neurotoxical effects : cognitive deficits, mental retardation (lower IQ), delayed nerve conduction, behavioural disturbances, adverse effect on psychomotor functioning<br>- Kidney damage; cardiovascular, liver, endocrine and gastrointestinal effects<br>- Other: decreased fertility, obstructed development of the embryo | The limit value for Flanders isn't exceeded in any of the measuring stations in 2001. The annual average lead concentration in the ambient air was 0.5 $\mu g/m^3$ in 1996. | | *WHO :*<br>0.5 $\mu g/m^3$ (year) (air)<br>10 $\mu g/l$ (water)<br><br>*EU + Flanders (VLAREM) :*<br>2 $\mu g/m^3$ (limit value)<br><br>*EU 2005 :*<br>0.5 $\mu g/m^3$ (year) (limit value) |
| Ozone (O₃) | - Negative influence on the lung functions (decreased lung capacity, inflammations, hypersensitivity of the bronchial tubes, coughing, shortness of breath etc.)<br>- Increased asthma symptoms<br>- Irritation of the mucous membranes and eyes, nose and throat<br>- Headache, dizziness, nausea, general malaise and chest discomfort<br>- Increased hospitalisation and mortality | The EU-threshold value for the protection of human health (110 $\mu g/m^3$) was exceeded 38 days in 2001. In 2001 there were 11 ozone days (days with ozone concentrations > 180 $\mu g/m^3$). The highest ozone concentration in 2001 was 250 $\mu g/m^3$. | - 900 extra deaths<br>- about 800 hospital admissions<br>- 1 million asthma attacks<br>- 24 million symptom days<br>(all figures per year) | *WHO:*<br>120 $\mu g/m^3$ (8 hour value)<br><br>*EU + Flanders (VLAREM) :*<br>110 $\mu g/m^3$ (8 hour value) (threshold value for the protection of human health)<br>180 $\mu g/m^3$ (information threshold)<br>360 $\mu g/m^3$ (alert threshold)<br><br>*EU 2010:*<br>120 $\mu g/m^3$ (8 hours; max. 25 days with higher levels)(target value) |

Table 41, continued.

| Pollutant | Health effects | Concentration level in Flanders | Health impact in Flanders | Guidelines |
|-----------|---------------|--------------------------------|---------------------------|------------|
| Nitrogen dioxide ($NO_2$) | - Negative influence on the lung functions (inflammation, coughing, chronic lung infection, decreased lung function, paralysing impact on the smooth muscles of the cilia in the nose and bronchial tubes, etc.) <br> - Increased asthma symptoms <br> - Irritation of the respiratory system <br> - Cardiovascular effects <br> - Increased mortality and hospitalisation | The highest annual average concentrations vary between 42 and 48 $\mu g/m^3$ in 2001. The highest value for 2001 was 98 $\mu g/m^3$. | See Table 40 | *WHO* : <br> 200 $\mu g/m^3$ (hour) <br> 40 $\mu g/m^3$ (year) <br><br> *EU + Flanders (VLAREM)*: <br> 200 $\mu g/m^3$ (98 percentile) (limit value) <br> 135 $\mu g/m^3$ (98 percentile) (guideline value) <br><br> *EU 2010* : <br> 200 $\mu g/m^3$ (hour; max. 18 days with higher levels) <br> 40 $\mu g/m^3$ (year) (limit value) |
| Toluene | - Effects on central nervous system <br> - Headache, fatigue <br> - Eye, nose and throat irritation <br> - Kidney damage | Concentrations of toluene measured by VITO in ambient air: 5.0 $\pm$ 4 $\mu g/m^3$. | | *WHO*: <br> 1 $mg/m^3$ (30 minutes) <br> 260 $\mu g/m^3$ (week) |

Table 41, continued.

| Pollutant | Health effects | Concentration level in Flanders | Health impact in Flanders | Guidelines |
|---|---|---|---|---|
| Sulphur dioxide ($SO_2$) | - Negative influence on the lung functions (irritation, constriction, increased mucous secretion, breathing problems, coughing, decreased lung function) <br> - Increased asthma symptoms <br> - Irritation of the eyes and irritation of the mucous membranes of the nose and bronchial tubes <br> - Respiratory disorders and lung diseases (e.g. bronchitis and lung emphysema) <br> - Cardiovascular effects <br> - Increased mortality and hospitalisation | The EU-limit value (80 µg/m³ (50 percentile)) was respected in all stations in Flanders during the period 2001-2002. The highest P50-value was 23 µg/m³. The EU-limit value (250 µg/m³ (98 percentile)) was respected in all stations in Flanders during 2001-2002. The highest P98-value was 95 µg/m³. The guideline value for the annual average $SO_2$-concentrations (40 µg/m³) wasn't exceeded neither. The highest annual average was 29 µg/m³. | See Table 40 | *WHO:* <br> 500 µg/m³ (10 minutes) <br> 125 µg/m³ (day) <br> 50 µg/m³ (year) <br><br> *EU + Flanders (VLAREM)* : <br> 100-150 µg/m³ (day), <br> 40-60 µg/m³ (year) (guideline values); <br> 250-350 µg/m³ (98 percentile)(year), <br> 80-120 µg/m³ (50 percentile) (depending upon the concentration of particulate matter) (limit values) <br><br> *EU 2005* : <br> 350 µg/m³ (hour; max. 24 days with higher levels) <br> 125 µg/m³ (day; max. 3 days with higher levels) (limit values) |

Table 41, continued.

| Pollutant | Health effects | Concentration level in Flanders | Health impact in Flanders | Guidelines |
|---|---|---|---|---|
| Particulate matter ($PM_{10}$) | - Negative influence on the lung functions (decreased lung function, increased sensitivity to respiratory infections, breathing problems, inflammation of the bronchial tubes etc.)<br>- Increased asthma symptoms<br>- Increased occurrence of chronic lung problems (coughing, bronchitis, infections)<br>- Disturbance of mucus drainage<br>- Changes relating to the blood clotting mechanisms<br>- Cardiovascular effects<br>- Increased mortality and hospitalisation | During the meteorological year 2001-2002 the annual average $PM_{10}$-concentrations fluctuated between 28 and 42 $\mu g/m^3$. | - at least 1095 extra deaths a year (different calculations indicate between 1051 and 2101 extra deaths)<br>- one healthy life year a person is lost at lifetime exposure<br><br>see also Table 40 | *EU + Flanders (VLAREM)* :<br>100-150 $\mu g/m^3$ (day),<br>40-60 $\mu g/m^3$ (year) (guideline values);<br>250 $\mu g/m^3$ (98 percentile) (year),<br>80 $\mu g/m^3$ (50 percentile) (year) (limit values)<br><br>*EU 2005* :<br>50 $\mu g/m^3$ (day; max. 35 days with higher levels)<br>40 $\mu g/m^3$ (year) (limit values)<br><br>*EU 2010* :<br>50 $\mu g/m^3$ (day; max. 7 days higher)<br>20 $\mu g/m^3$ (year) (limit values) |
| Noise | - Mental health effects (annoyance, displeasure, dissatisfaction, concentration loss, stress, fatigue, repeated nervousness and irritation)<br>- sleep disturbance<br>- Increased risk on hearing loss and/or damage<br>- Increased blood pressure, cardiovascular effects<br>- Low birth weight | In 2001 approximately 30% of the general population is exposed to noise levels above 65 dB(A) (road traffic). | - 2,129,000 potential awakenings a night (road + railway traffic)<br>- 951,000 people potentially seriously affected by road traffic and 80,000 people with serious nuisance from railway traffic (mental health effects such as annoyance, irritation, concentration loss etc.)<br>- 1,800,000 people are exposed to noise levels from road traffic exceeding 65 dB, which may result in health effects such as increased blood pressure and cardiovascular diseases<br>- air traffic badly affects the sleep and health quality of 26,400 people | Guideline values in residential areas in the ambient air :<br><br>*Flanders (VLAREM)*:<br>45 dB(A) (day), 40 dB(A) (evening), 35 dB(A) (night)<br><br>*WHO* :<br>50 dB(A) (moderate nuisance) (day), 55 dB(A) (serious nuisance) (day), 45 dB(A) (night) |

# COUNTRY REPORT AUSTRIA ON MOBILITY, ENVIRONMENTAL ISSUES, AND HEALTH

H. MOSHAMMER[1], H.P. HUTTER[1] AND L. SCHMIDT[2]
[1]Institute for Environmental Health,
Medical University of Vienna, AUSTRIA
Kinderspitalgasse 15, 1095 Vienna, Austria
[2]SOMO. Social Sciences Mobility Research and Consultancy
Sonnenweg 5, 1140 Vienna, AUSTRIA

## Summary

Austria is a small country in the centre of the European continent with several important routes crossing the alps from North to South and at least one historically important route from East to West along the Danube river. The Austrian economy for centuries made use of this situation with a long history in freight transport and in tourism also. So motorised mobility has been at the core of Austrian self esteem and prosperity.

On the other hand the small alpine valleys leave little space for dwelling, agriculture and industry. Many living areas are situated near major roads and exposure towards noise and air pollutants is considerably high. Also the tourism industry and agriculture have recognised the values of a "clean" environment. Therefore there is a strong and long-lasting debate between interest groups that call for more and better roads and those opposing them.

Many measures have been implemented on various levels ranging from mobility plans of single enterprises and local cycling networks to national regulations (Austria was among the first countries in Europe to introduce unleaded petrol and the catalytic converter) but a scientific evaluation of most of the measures at local or regional level is still lacking. In spite of the success in reducing the emissions per vehicle-kilometre in the last years the increase in the number especially of diesel-driven light duty vehicles reversed the previously downward trend in air pollution exposure in several densely populated areas. Therefore nitrogen oxides and fine particles are among the key environmental problems in Austria.

*P. Nicolopoulou-Stamati et al. (eds), Environmental Health Impacts of Transport and Mobility, 223-237.*
© 2005 *Springer. Printed in the Netherlands.*

## 1. Introduction

The attitudes of the Austrians towards motorised mobility are very controversial. Maybe this is due to the historical and geographical features of the country. Between the wars the construction of new roads was seen as a token of independence of the small country and therefore very much supported by the Austrian fascist government. Car use was then restricted to a small elite only and the roads designed in those times were not planned for broader public use but for the entertainment of this elite and for tourists because tourism already then was considered financially important. Thus mountain roads like the "Großglockner Hochalpenstraße" were built in those times.

New roads were in fashion at the time with highways being constructed in Britain and the United States, but in Italy and Germany as well. The German Nazi Regime propagated the construction of new roads as a solution to the unemployment of the recession years. But motorization especially of the agricultural sector also had the goal to free labour forces for the army. The idea of a "Volkswagen" (i.e. a car for every citizen / family) was part of the regime's propaganda but this goal was not achieved while in contrast the Ford company in the States at the same time was successful in this target.

After the 2$^{nd}$ World War the reconstruction of the infrastructure was a necessity and included new roads as well (some of which had already been planned during the German occupation). Nevertheless society succeeded in promoting the feeling that free mobility for everybody was part of the new democratic and western way of life. It was generally considered that new and better roads would provide increasing opportunities for industry, tourism, and also for individuals.

Surprisingly the first groups to oppose new roads and especially local bypasses (which were greatly welcomed by the inhabitants of the villages) were the local inn keepers and shop owners who feared to their customers (which with the building of large out of town shopping centres in the end proved true). Environmental concerns were only raised many years later (beginning in the late 70s of the 20$^{th}$ century) but gained momentum rather fast.

Environmentalists in the beginning were defenders of nature and were concerned about habitats and wildlife. But very soon the general idea was formed that a demolished nature is also a threat to human health. "Environment" which was considered the sum of everything surrounding us and having an impact on us replaced "nature" as the prime topic. Biodiversity and threatened species were only of interest because their loss could in the long run mean a threat to humanity as well. Studies on the health effects of air pollution helped to link the picture of dead trees ("Waldsterben") with respiratory health. This close linkage between environment and health made environmentalism a popular theme in Austria. But new road projects were just one topic among many fought against by this movement.

Nowadays there is a strong environmental lobby in Austria opposing new roads and especially highways, new tunnel projects or even the expansion of existing roads. On the other hand the interest groups of car drivers, the freight transport enterprises and construction companies are also very powerful in our country.

Many politicians still seem to believe in the fairy tale of the 50s and 60s (then also taught at school) that new roads provide new jobs and new opportunities, although it is evident that there cannot be a linear association between number of roads and welfare but that there is certainly an optimal number of roads for a certain region. Until now there is a lack of integrated evaluation concerning this optimum. Various local, regional, and national responsibilities in the transport sector add to the difficulties of a coherent policy integrating transport and spatial planning.

On top of this Austria is a mountainous country with dense population in regions where the landscape allows settlement. Narrow alpine valleys and basins tend to concentrate air pollutants and noise is reflected from the slopes of the mountains. Few alpine passes bear a great part of the freight and passenger transport between the North and the South of Europe. While Austrian transport companies benefit from Austrian's key geographical position many dwellers of the alpine valleys along major roads are fed up with the heavy high duty traffic.

This chapter sets out to illustrate the current situation in Austria and to document trends and status of environmental problems that are at least in part caused by the transport sector. In the final subchapter ("case studies") some projects aiming at a more sustainable transport system will be discussed.

## 2.    Environmental problems

### 2.1.    $CO_2$ emissions

In Kyoto, Austria bargained the very ambitious goal to reduce by 2010 the emissions of total greenhouse gases to 87 % of the 1990 level. In fact total greenhouse gases in 2001 reached nearly 110 % of the 1990 level, $CO_2$ alone even 115 %. There is a steady (although lately rather slow) trend towards reduction of emissions in Austrian industry and an oscillating curve for annual emissions from heating of homes (depending on the mean winter temperatures of each year with a tendency to warmer winters due to climate change and therefore over the years rather lower emissions). This unsteady decrease is also supported by some achievements in insulation of houses. But (besides a smaller increase in the household use of electricity) all achievements are overruled by an increase in transport volume (both personal and freight transport) with increases especially in road transport but also (on a much lower level, but – at least until 9-11-01 – with a much steeper increase rate) in air traffic. Thus between 1990 and 2001 transport-related $CO_2$ emissions increased from 12.7 to 18.9 metric megatons annually, which makes a 48 % increase (data according to Umweltbundesamt, 2001a). Umweltbundesamt (2004) has now published the 2002 data also with no fundamental change.

**Table 42**. Austrian $CO_2$ emissions in megatons per year (Schleicher *et al.*, 2000). 2001 data based on different calculations (Umweltbundesamt, 2001a)

|      | Total | Electricity | District Heating | Industry | Transport | Consumers | Refinery | Others |
|------|-------|-------------|------------------|----------|-----------|-----------|----------|--------|
| 1980 | 63.81 | 7.53  | 1.51 | 10.98 | 11.52 | 15.63 | 2.46 | 14.19 |
| 1981 | 59.84 | 7.19  | 1.38 | 10.14 | 11.23 | 13.61 | 2.24 | 14.05 |
| 1982 | 57.39 | 7.12  | 1.57 | 9.57  | 11.28 | 13.20 | 1.91 | 12.74 |
| 1983 | 58.51 | 7.31  | 1.41 | 8.08  | 11.52 | 13.51 | 1.69 | 14.99 |
| 1984 | 58.38 | 8.03  | 1.47 | 7.75  | 11.24 | 14.07 | 1.79 | 14.03 |
| 1985 | 59.23 | 7.89  | 1.59 | 8.07  | 11.39 | 14.79 | 1.86 | 13.62 |
| 1986 | 59.51 | 7.74  | 1.60 | 7.68  | 11.81 | 14.72 | 1.84 | 14.11 |
| 1987 | 59.81 | 7.62  | 2.37 | 7.65  | 11.90 | 15.68 | 1.85 | 12.37 |
| 1988 | 57.21 | 6.35  | 2.26 | 7.14  | 12.75 | 13.92 | 1.72 | 13.07 |
| 1989 | 57.97 | 7.14  | 2.21 | 6.66  | 13.22 | 13.63 | 1.79 | 13.32 |
| 1990 | 62.13 | 9.92  | 2.47 | 7.43  | 13.57 | 13.31 | 2.02 | 13.42 |
| 1991 | 66.02 | 10.42 | 2.99 | 6.82  | 15.06 | 15.80 | 2.14 | 12.80 |
| 1992 | 60.15 | 7.28  | 2.60 | 6.95  | 15.05 | 14.35 | 2.24 | 11.69 |
| 1993 | 59.90 | 6.58  | 2.62 | 6.85  | 15.10 | 14.74 | 2.19 | 11.82 |
| 1994 | 61.75 | 6.88  | 2.60 | 6.66  | 16.16 | 14.71 | 2.26 | 12.48 |
| 1995 | 63.69 | 8.08  | 2.94 | 7.51  | 15.43 | 14.84 | 2.20 | 12.69 |
| 1996 | 65.91 | 9.63  | 1.88 | 7.65  | 15.38 | 14.45 | 2.59 | 12.13 |
| 1997 | 66.79 | 9.84  | 2.13 | 8.27  | 15.79 | 14.97 | 2.49 | 12.30 |
| 1998 | 66.60 | 9.64  | 2.09 | 8.15  | 16.75 | 14.89 | 2.64 | 12.49 |
| 2001 | 68,00 |       |      |       | 18,90 |       |      |        |

## 2.2.    *$SO_2$ and $NO_x$ emissions*

### *$SO_2$ and $NO_x$ emissions*

In 1990 national total $SO_2$ emissions amounted to 79,000 metric tons; emissions have decreased steadily since then and by the year 2001 emissions were reduced by 53% mainly due to lower emissions from residential heating, combustion in industries and energy industries. As can be seen in Table 43, the main source for $SO_2$ emissions in Austria with a share of 87% in 1990 and 76% in 2001 result from fuel combustion activities. Within this source residential heating has the highest contribution to total $SO_2$ emissions (41% in 1990 and 32% in 2001).

**Table 43.** $SO_2$ emissions per category, their trend 1990-2001 and their share in total emissions (Wieser *et al.*, 2003)

| Category | $SO_2$ Emissions [Gg] | | Trend 1990-2001 | Share in National Total | |
|---|---|---|---|---|---|
| | 1990 | 2001 | | 1990 | 2001 |
| Energy | 68.12 | 27.97 | -59% | 86.6% | 76.3% |
| Fuel Combustion Activities | 66.12 | 27.81 | -58% | 84.0% | 75.8% |
| Fugitive Emissions from Fuels | 2.00 | 0.16 | -92% | 2.5% | 0.4% |
| Including transport sector | 4.37 | 3.15 | -28% | 7% | 11% |
| Industrial Processes | 10.49 | 8.64 | -18% | 13.3% | 23.6% |
| Agriculture | 0.00 | 0.00 | -2% | 0.0% | 0.0% |
| Waste | 0.06 | 0.05 | -16% | 0.1% | 0.1% |
| **National Total** | **78.68** | **36.67** | **-53%** | **100%** | **100%** |

The 2010 national emission ceiling for $SO_2$ emissions in Austria as set out in Annex II of the Multi-Effects Protocol is 39,000 metric tons, which corresponds to a reduction of 50% based on 1990 emissions. Total emissions of 36,700 metric tons in 2001 emissions are already below the ceiling. Transport's share on $SO_2$ emissions is low although increasing relatively.

**Table 44.** $NO_X$ emissions per category, their trend 1990-2001 and their share in total emissions (Wieser *et al.*, 2003)

| Category | NOx Emissions [Gg] | | Trend 1990-2001 | Share in National Total | |
|---|---|---|---|---|---|
| | 1990 | 2001 | | 1990 | 2001 |
| Energy | 181.24 | 179.70 | -1% | 90% | 89% |
| Fuel Combustion Activities | 181.24 | 179.70 | -1% | 90% | 89% |
| Including transport sector | 101.99 | 101.46 | -1% | 56% | 56% |
| Industrial Processes | 17.41 | 14.64 | -16% | 7% | 9% |
| Agriculture | 5.20 | 5.02 | -3% | 3% | 3% |
| Waste | 0.04 | 0.03 | -32% | 0% | 0% |
| **National Total** | **203.88** | **199.40** | **-2%** | **100%** | **100%** |

In 1990 national total $NO_X$ emissions amounted to 204,000 metric tons; emissions have fluctuated since then, and in 2001 were only 2% below the level of 1990. As can be seen in Table 44, the main source for $NO_X$ emissions in Austria with a share of 90% in 1990 and 89% in 2001 results from fuel combustion activities. Within this source road transport has the highest contribution to total NO$_X$ emissions. About 50% of national total emissions arise from this source with 97.14 Gg in 2001.

The 2010 national emission ceiling for NO$_X$ emissions in Austria as set out in Annex II of the Multi-Effects Protocol is 107,000 metric tons (in the European National Emissions Ceiling Directive it is 103,000 metric tons), which corresponds to a reduction of 48% based on 1990 emissions (49% for the NEC Directive). With 199,000 metric tons in 2001, which is a reduction of 2% compared to 1990 levels, emissions in Austria are at the moment well above this ceiling.

**Table 45.** NOx emissions of the transport sector 1990 and 2001, their trend 1990-2001 (Wieser *et al.,* 2003)

| Category | NOx Emissions [Gg] | | Trend |
|---|---|---|---|
| | **1990** | **2001** | **1990-2001** |
| **Transport (total)** | **101.99** | **101.46** | **-1%** |
| Civil Aviation | 0.13 | 0.32 | 152% |
| Road Transportation | 98.93 | 97.14 | -2% |
| Passenger cars | 69.84 | 42.11 | -40% |
| Light duty vehicles | 2.31 | 5.29 | 129% |
| Heavy duty vehicles | 26.55 | 49.17 | 85% |
| Mopeds & Motorcycles | 0.20 | 0.54 | 170% |
| Railways | 1.95 | 1.69 | -13% |
| Navigation | 0.52 | 0.57 | 10% |
| Other | 0.46 | 1.74 | 281% |

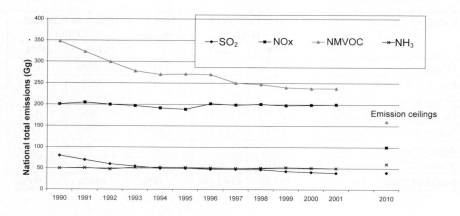

**Figure 42.** Trends in Austrian emissions of gaseous pollutants (Wieser *et al.,* 2003).

## 2.3    Particulate matter

Until recently the Austrian standard for air pollution monitoring on airborne dust only contained total suspended particles (TSP) which is a rather poor indicator for source-specific particulate pollutants. Beginning in the year 2000 dust of particle diameter less than 10 µm ($PM_{10}$) was introduced in Austrian ambient monitoring. $PM_{10}$ is also influenced by various sources, but it has been shown in several studies to be a good predictor of health outcome. From emission inventories and parallel $NO_2$ monitoring it was estimated that about 50% of the population's exposure in Austria towards $PM_{10}$ stems from motorised transport.

Current limit values ($PM_{10}$: no more than 35 daily mean values above 50 $\mu g/m^3$) were exceeded on many locations (both in urban areas and in narrow alpine basins). Highest daily mean values in 2002 were even above 200 $\mu g/m^3$! Even the limit value for the annual mean (40 $\mu g/m^3$) was exceeded at a few locations (Spangl and Nagl, 2003).

## 2.4.    Noise

Road traffic is the top reason for ambient noise exposure in Austria. From a very limited number of measurements and noise mapping in a few cities based on transport density data the number of Austrian households that were exposed (outside their window fronts) to environmental noise above certain levels was estimated for the year 1997 (Table 46, UBA, 2001).

Exposure to high noise levels has decreased substantially in some countries due to technological measures (e.g. reduction of emissions, change of road surfaces) and spatial measures such as noise barriers and spatial separation of transport and residential functions. Nevertheless, noise will remain a major problem due to the enormous growth in traffic (especially road and air) and the 24-hour economy. The OECD predicts an increase in motor vehicle kilometers of 40% in the next 20 years (OECD, 2001).

**Table 46.** Noise caused by road traffic, percentage of households above certain levels (Umweltbundesamt, 2001b)

| $L_{A,eq,day}$ or $L_{A,eq,night}$ + 10 dB | Communities with less than 20,000 inhabitants | Cities with more than 20,000 inhabitants | Vienna | Austrian average |
|---|---|---|---|---|
| $\geq 55$ | 61.6 | 51.1 | 67.0 | 60.9 |
| $\geq 60$ | 29.2 | 29.2 | 44.0 | 32.2 |
| $\geq 65$ | 5.0 | 18.5 | 17.1 | 9.8 |
| $\geq 70$ | 2.0 | 6.4 | 11.0 | 4.6 |
| $\geq 75$ | 0.0 | 2.0 | 3.4 | 1.0 |

## 3.    Health effects

### *3.1.    Impacts of air pollution*

Künzli *et al.* (2000) estimated the burden of disease due to air pollution in Austria. They only considered a few health endpoints. They used $PM_{10}$ exposure as the best indicator of total air pollution. Their estimates for Austria are summarised in Table 47.

Dose-response rates were derived from American cohort studies mainly. Exposure of Austrian population was calculated by means of the Austrian emission inventory and data were calibrated using the Austrian monitoring network for $NO_2$ (because no $PM_{10}$ monitoring network was in place in Austria then). The calculations were performed as part of an international study with teams from Austria, Switzerland, and France and the data were presented at the WHO ministerial conference in London 1999. Especially the figures on mortality (due to transport related air pollution) which are in the same range or even higher than the annual deaths caused by road accidents (approx. 1000 per year) gained much media coverage in Austria. In a later paper Künzli *et al.* (2001) argued that it was a pity that in 1999 there were no valid data on the actual loss of life years but that it could be estimated that each death due to air pollution would on average reduce life by about 6 months. But evidently the absolute numbers of attributable deaths made a great impact on the Austrian political debate. But neither $NO_2$ nor $PM_{10}$ have shown a clear downward trend in Austria since then.

**Table 47**. Annual health impact of air pollution in Austria (Künzli *et al.*, 2000)

| Health outcome | Estimated attributable number of cases or days (95% CI) | |
|---|---|---|
| | Total outdoor air pollution | Traffic related air pollution |
| Long-term mortality (adults) | 5600 (3400-7800) | 2400 (1500-3400) |
| Respiratory hospital admissions | 3400 (400-6500) | 1500 (160-2800) |
| Cardiovascular hospital admissions | 6700 (3500-10,000) | 2900 (1500-4300) |
| Chronic bronchitis incidence (adults) | 6200 (600-12,000) | 2700 (240-5300) |
| Bronchitis (children) | 48,000 (21,000-86,000) | 21,000 (9000-37,000) |
| Restricted activity days (adults) $(10^6)$ | 3.1 (2.6-3.6) | 1.3 (1.1-1.6) |
| Asthma attacks (children) | 35,000 (21,000-48,000) | 15,000 (9000-21,000) |
| Asthma attacks (adults) | 94,000 (46,000-143,000) | 40,000 (20,000-62,000) |

### 3.2.  Noise: annoyance and other health effects

Noise is generally distinguished from sound in that it is *undesirable* sound; i.e., sound which causes pain or discomfort, or interferes with human activities such as communication or sleep.

Census data (Statistics Austria) show that noise is the main reason of annoyance. The percentage of annoyed and highly annoyed persons declined over time (partly due to a change in the questionnaire after the 1994 census) but the absolute numbers remained fairly constant. The 1994 census data still give approx. 18% of annoyed or highly annoyed persons which is still rather a high number. Of these nearly 80% reported to be annoyed by transport (the remaining include industry noise, neighbours, building construction and others). The majority of the annoying transport noise stems from road traffic (Figure 43, Figure 44 and Figure 45).

Lately more and more studies are focusing not merely on simple dose-response relationships but also on the whole setting in which the noise is perceived (soundscape). A negative response to noise such as annoyance can be modified by both acoustical and non-acoustical parameters. People that feel unsafe due to road traffic are more annoyed than others. Improvements that include traffic safety measures can therefore be an effective way of reducing noise annoyance even though the actual noise level remains unchanged. Negative reactions to noise can influence effects of other exposures (e.g. air pollution).

The instance of transalpine traffic in the EU has shown that mountainous areas are especially sensitive to negative environmental effects (Lercher *et al.*,1995; Lercher and Kofler, 1996). Topography and meteorology unfavourably support the propagation of noise in such a way that the distant slopes of the valley often receive noise levels you see in the second or third lane along highways.

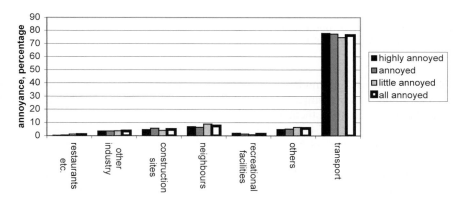

**Figure 43**. Noise sources. More than 70% of annoying noise is caused by transport (Statistik Austria, 2000).

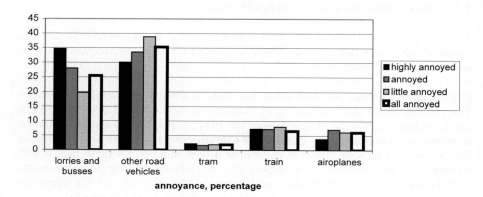

**Figure 44**. Noise from transport sources. Road transport is the leading cause of annoying transport noise (Statistik Austria, 2000).

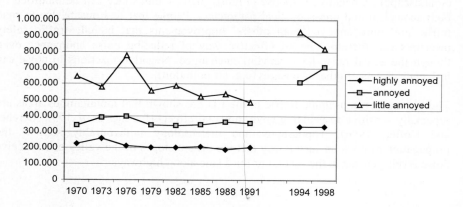

**Figure 45**. Number of households respectively persons annoyed stayed constant over the years (Statistik Austria, 2000).

Due to the low background levels in these distant areas the noise to background ratio increases and the noise impact on people is higher. For areas which resemble these characteristics that make them more vulnerable to the impact of environmental noise the term "sensitive area" has been used, which is often applied in a more restrictive sense only.

In a large Austrian field study child self-reported mental health was investigated in relation to road and railway noise. Self-reported mental health was only impaired in higher-exposed children with a history of low birth weight (Lercher *et al.*, 2002).

## 3.3.    Accidents

The risk of being involved in a road accident is strongly age and sex-dependent. Over the years there has been a decrease in the annual death rate (Figure 46). Now annually slightly less than 1000 people are killed on the roads (total population is about 8 millions) which is still far too much. This decrease happened not only because roads and vehicles are safer now especially for passengers, but also because of improved first aid and medical services of intensive care. Transport has not become better adapted to the needs of those most vulnerable (e.g. children) but children are kept away from the roads and are trained to adapt to transport needs. Nevertheless children as pedestrians suffer a high toll of injuries. But even more are injured as passengers in cars (Figure 47, Kuratorium für Verkehrssicherheit, 2004).

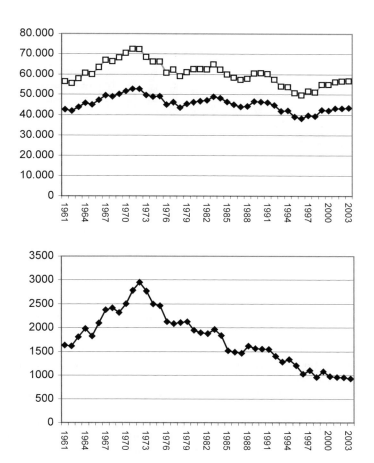

**Figure 46**. Accidents and casualties 1961–2003 (Kuratorium für Verkehrssicherheit, 2004).

In sum in 2003 a total of 4111 children (age 0 – 14) were injured. From these 1057 were pedestrians, 867 cyclists, 1812 passengers in cars and 375 participated otherwise in road transport.

The total number of small children killed on the roads per year is rather small. Half of them die as car passengers which is also due to the parents' belief that it is safer to deliver their children by car instead of allowing them to walk on their own. This leads to a sharp increase in fatalities and injuries when the youngsters are finally allowed to move on their own with motorised devices.

For all age groups injuries and deaths due to road accidents Austria ranks above the EU 15 average both when comparing on a per capita or a per vehicle basis.

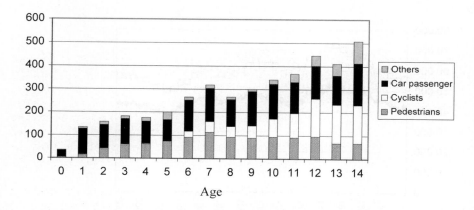

**Figure 47.** Children injured (fatalities and injuies) in 2003 (Kuratorium für Verkehrssicherheit, 2004).

### 3.4.   *Physical activity*

Children are less physically active than in former times. This is due to various reasons:

(1)  Increasing availability of home entertainment (TV, computer)
(2)  Perceived and real danger due to motorised traffic outdoors and lack of spaces for outdoor activities
(3)  No need for physical activity because of change in mobility patterns

There is compulsory physical education in most school types in Austria. But the amount of physical activity apart from these (usually 3 hours per week) classes is decreasing and physical activity is not an integrated part in everyday school life.

There are differences between age groups, gender, and regions. In a European-wide WHO survey (2000) exercise levels were still remarkably high in youngsters aged 11 and 13 but there was a drop in activity for 15 year old girls. In spite of doubts concerning the comparability of data from the different European countries this study showed that Austria is still in a rather good position concerning physical activities in children and adolescents. Tendencies for a change to less active life styles and a drop in activity when the youngsters grow older nevertheless call for attention.

On the other hand, overweight is quite common in Austria with a high prevalence also in children of lower social classes. There is a lack of data for the whole country. But studies on Vienna schoolchildren (10 to 14-year-olds) found differences according to gender (higher rates in boys) and to school type (higher rates up to 30 percent in secondary schools as compared to public schools with 14 percent; Widhalm, reported by Hecher 2004).

## 4.   Case studies

Many measures in Austria to reduce individual motorised traffic and to enhance more healthy modes of mobility form a patchwork all over the country. But too often still these patches do not fit with each other. Propagating public transport on the one side is at the same time accompanied by a reduction of public transport services due to financial constraints or financial interests. New roads and new railways are constructed in parallel with the digging of the tunnels for the railway stopped because of environmental impact considerations (danger for groundwater) but the tunnels for the motorways on the same route are still under construction.

Many fine solutions for transport problems have been implemented on the local and regional level. They include models of sustainable tourism ("holidays from the car") including a sophisticated infrastructure of cycling lanes, special service of inn keepers and hotels for cyclists (special prices, providing facilities for changing of clothes and showers, special menus, etc.), transportation of luggage by private and public transport enterprises, special services and ticket prices for tourists that leave their car at home, packages including rail ticket, hotel, and special entertainments like horse riding, horse-powered coaches or sleds, and much more.

Other measures were introduced by the management of enterprises that encouraged their workers to come to work by bike or with public transport facilities. Instead of investing in new parking space they paid for public transport tickets or provided facilities for the cyclists like showers and wardrobes. These activities were often accompanied by joint actions of the public transport authorities offering special bus lines that fit the working shift needs of the enterprise. Car sharing is also encouraged by other managements.

One famous project that tried to bring together many of the above-mentioned ideas in one special area is the Pilot study "Transport and Sensitive Areas – the Example of the Lake Neusiedl Region". This Lake Neusiedl Region in the very East of Austria for many years was placed at a "dead boarder" (to Hungary) hampering industrial development. This also enabled the survival of rare wildlife species and made the region a very valuable reservation not far from the urban centre of Vienna. Uncontrolled growth of tourism in some places and lack of jobs in many others were both recognised as serious problems of the region.

Therefore many programmes were jointly planned by Austrian and Hungarian authorities under the umbrella of this pilot study. Local authorities and other stakeholders were invited to participate and so several projects could be implemented including "ecotourism" (cycling, hiking, bird watching), improvements in cross boarder and local public transport, mobility services to encourage and assist commuters in car sharing activities, supports for local enterprises with their freight logistics, and many more.

One general problem with all the fine Austrian projects is a lack of evaluation of success. While many of the tourism projects are visibly successful in terms of tourists that come and ask for these facilities and services, it is not clear if more conventional recreational services would not have brought an equal economic benefit. And if "ecotourism" really has increased the number of tourists, there are no studies to measure the total environmental impact of that tourism. Even if the impact of the single tourist would be low it must be proven that the higher number of tourists does not outweigh these benefits. All the other projects, among those prominently the projects of the Lake Neusiedl study, have not implemented any benchmarking or monitoring of success. No quantitative goals have been published in the beginning, no baseline investigation on the satisfaction of the population of the region with the existing public transport facilities and other services, no ongoing integrative surveys have been carried out.

Therefore, as stated in the beginning, all these projects and measures remain a patchwork with no integrative national or even regional overall policy. Some turned out very fine, but none can be presented as a special case study which would serve as a model for broader implementation.

## References

Hecher, S. (2004) Juvenile Adipositas: Erstmals Prävalenzdaten, *Österreichische Ärztezeitung* 2004 4, 47-48

Künzli, N., Kaiser, R., Medina, S., Studnicka, M., Chanel, O., Filliger, P., Herry, M., Horak, F., Puybonnieux-Texier, V., Quenel, P., Schneider, J., Seethaler, R., Vergnaud, J.C., and Sommer, H. (2000) Public-health impact of outdoor and traffic-related air pollution: a European assessment, *Lancet* 356 (9232), 795-801.

Künzli, N., Kaiser, R., and Seethaler, R. (2001) Luftverschmutzung und Gesundheit: Quantitative Risikoabschätzung, *Umweltmed. Forsch. Prax.* 6, 202-212.

Kuratorium für Verkehrssicherheit (2004) Unfallstatistik 2003, Vienna, Kuratorium für Verkehrssicherheit, Institut für Verkehrstechnik und Unfallstatistik.

Lercher, P., Schmitzberger, R., and Kofler, W. (1995) Perceived traffic air pollution, associated behavior and health in an alpine area, *The Science of The Total Environment* 169, 71-74.

Lercher, P., and Kofler, W. (1996) Behavioral and health responses associated with road traffic noise along alpine through-traffic routes, *The Science of The Total Environment* 189/190, 85-89.

Lercher, P., Evans, G.W., Meis, M., and Kofler, W.W. (2002) Ambient neighbourhood noise and children's mental health, *Occup. Environ. Med.* 59, 380-6

OECD Organisation for Economic Co-operation and Development (2001) *Environmental Outlook 2020*, OECD, Paris.

Schleicher, S., Kratena, K., and Radunsky, K. (2000) Die österreichische $CO_2$-Bilanz 1998. Struktur und Dynamik der österreichischen $CO_2$-Emissionen, ACCC (Austrian Council on Climate Change, Österreichischer Klimabeirat), Graz.

Schmidt, L., and Schmidt, G.A. (2000) *Measures to improve transport safety as well as sustainability – two case studies*, Paper for the twelfth workshop of ICTCT in Corfu, October 05-07, 2000.

Spangl, W., and Nagl, Ch. (2003) Jahresbericht der Luftgütemessungen in Österreich 2002, UBA, Vienna (ISBN 3-85457-699-4)

Statistik Austria (2000) Umweltbedingungen, Umweltverhalten: Ergebnisse des Mikrozensus Dezember 1998, Vienna (ISBN: 3 7046 1531 5)

Umweltbundesamt (2001a) Sechster Umweltkontrollbericht des Bundesministers für Land- und Forstwirtschaft, Umwelt und Wasserwirtschaft an den Nationalrat Kapitel 3: Globaler Klimawandel, UBA, Vienna ISBN: 3-85457-593-9)
http://www.umweltbundesamt.at/umweltkontrolle/ukb2001/

Umweltbundesamt (2001b) Sechster Umweltkontrollbericht des Bundesministers für Land- und Forstwirtschaft, Umwelt und Wasserwirtschaft an den Nationalrat – Kapitel 16: Lärm, UBA, Vienna (ISBN: 3-85457-593-9).
http://www.umweltbundesamt.at/umweltkontrolle/ukb2001/

Umweltbundesamt (2004) Kyoto-Fortschrittsbericht Österreich 2004, Berichte BE-245, Vienna

WHO (2000) Health behaviour in school-aged children: a WHO cross-national study, International report, WHO Regional Office for Europe, Copenhagen.

Wieser, M., Poupa, S., Anderl, M., Wappel, D., Kurzweil, A., Halper, D., and Ritter, M. (2003) Austria's Informative Inventory Report 2003 - Submission under the UNECE Convention on Long-range Transboundary Air Pollution, UBA report BE-229, Vienna.

# HEALTH AND MOBILITY IN THE UK

J.A. NEWBY AND V.A. MOUNTFORD
*Developmental Toxico-Pathology Group*
*Department of Human Anatomy and Cell Biology*
*University of Liverpool, UK*

## Summary

This report addresses mobility and travel in the UK and the consequential cost to the environment, economy, quality of life and health. In 2001, the UK had a rapidly increasing population of 59,030,600 people who need effective, efficient modes of travel. We focus on road traffic accidents as well as the health-related and environmental impacts of road and air traffic, which are the modes of transport most associated with air and noise pollution. Engine noise and exhaust emissions are responsible for noise and air pollution, which cause environmental damage and health effects. Engine exhaust emissions contribute to global warming, poor air quality, ambient carcinogens, acidification and increased ground level ozone. Although total $CO_2$ emissions have reduced in the UK, road transport is the fastest growing source, almost doubling from 1970 to 2000. Nitrogen oxides, sulphur dioxide, nitrous oxides, volatile organic compounds, polycyclic aromatic hydrocarbons and particulate matter from engine exhaust emissions all contribute to poor air quality. This has serious implications for health, such as chronic and acute lung function conditions. Short-term exposure to air pollution may be bringing forward between 12,000 and 24,000 deaths and between 14,000 and 24,000 hospital admissions each year. Ozone, particulates and $SO_2$ are the main culprits. Emissions of diesel and petrol particulates, benzene, 1,3-butadiene, benzo[a]pyrene and benz[a]anthracine contribute to ambient carcinogens, add to lung cancer totals and may be implicated in childhood leukaemia. The UK seems, however, on course to meet most of its air quality objectives. We also address the noise-related health problems of living near to airports and main roads, including sleep disturbance, mental problems and effects on the cognitive performance of primary school children. In 2000, road traffic accidents killed around 3,500 people in the UK and 40,000 were seriously injured. The most vulnerable of UK citizens withstand the worst of transport-related health issues. More UK government-funded research is required to investigate the role of air pollution in the aetiology of cancer and respiratory disease. More noise reduction measures should be put in place near airports and traffic dense areas. More initiatives such as the Oxford Transport

*P. Nicolopoulou-Stamati et al. (eds), Environmental Health Impacts of Transport and Mobility, 239-276.*
© 2005 *Springer. Printed in the Netherlands.*

Strategy should be put into practice to promote alternatives to car use. Physical activity such as walking and cycling should be encouraged from an early age and the health benefits of such activity should be publicised.

## 1.  Introduction

### 1.1.  Overview

Mobility in human communities is essential. By mid 2001, the UK had a rapidly increasing population of 59,030,600 people (ONS, 2003a) who all need effective, efficient modes of travel. A disproportionately large population density occurs in the South East of England, causing higher stress on the transport system as well as on housing, as people commute long distances. Good road and rail networks are required for industry, commuting, efficient public transport and easy access to airports and seaports. However, effective mobility has a significant effect on the environment, human health and quality of life. This report addresses travel in the UK and the consequential cost to the environment and the health of the nation. It focuses mainly on the health-related and environmental impact of road and air traffic, which are transport modes associated with air and noise pollution, together with road accidents. This report does not address accidents during other modes of travel, or emissions from rail travel, because these are of relatively low concern, compared to other modes of travel.

The main health impacts of road and air traffic are through air and noise pollution and road traffic accidents. The air pollution comes from the emissions of engine exhausts, the major road traffic pollutants being:

- Carbon dioxide ($CO_2$), carbon monoxide (CO)
- Nitrous oxides ($NO_X$), e.g. NO and $NO_2$,
- Sulphur dioxide ($SO_2$)
- Particular Matter ($PM_{10}$)
- Volatile Organic Compounds (VOCs): hydrocarbons (alcanes, alcenes, aromatic monocyclic e.g. benzene and toluene), oxygenated compounds (aldehydes, acids, ketones, ethers)
- Polycyclic Aromatic Hydrocarbons (PAHs) e.g. benzo[a]pyrene, benzo[k]fluoranthene, benzo[b]fluoranthene, benzo[g,h,i]perylene, benz[a]anthracene
- Metals (lead)

(DEFRA, 2003a).

Aircraft exhausts also emit $CO_2$, VOCs and $NO_x$. Airports are among the greatest sources of local air pollution and noise pollution. (Holzman, 1997).

Air pollution is detrimental to human health and the environment in a number of ways. These are explored in more detail in other chapters of this book, but in summary:

- $CO_2$ is the main greenhouse gas;
- NOx and $SO_2$ are responsible for acidification;

- NOx also contributes to tropospheric ozone levels;
- Particulate matter is implicated in respiratory conditions such as asthma and
- VOCs and PAHs contribute to ambient carcinogens

(DEFRA, 2003a; IARC, 1989b).

An underestimated form of pollution, in terms of its health effects, is noise. Noise pollution from road and air traffic may be responsible for cardiovascular disease, sleep disturbance and mental health disorders such as depression (DEFRA, 2003a; DEFRA, 2003b; van Kempen, 2002; Holzman, 1997).

A direct health impact of road traffic is accidents. Every year in the UK, over 300,000 people are injured in road traffic accidents, 5% of them children (Donaldson 2002). In 2001, road traffic accidents accounted for over 40,000 pedestrians' deaths or injuries in the UK, a third of whom were children (DfT, 2003c).

### 1.2. Population statistics

There were about 59,030,600 people living in the UK in mid-2001. This was an increase since 1981 of 2.7 million people (4.7%). A year later, this had increased by another 0.3% to 59.2 million. The growth in the population of the UK is mainly due to net natural change (more births than deaths). (ONS 2003a)

In the UK, for the first time, people aged 60 and over form a larger part of the population than children form. There has also been a large increase in the number of people aged 85 and over, now over 1.1 million, or 1.9% of the population (ONS, 2003b). Unpolluted, safe, efficient modes of transport need to be a priority for elderly people, children and the sick, as these are the people who are likely to suffer the most from road and air traffic pollution.

### 1.3. Transport infrastructure and travel statistics

The UK has an expanding road network: motorways, principal and other road networks have expanded by more than 16% in the twenty-one years from 1980 to 2001, to over 380 thousand kilometres in total. The trunk road network, however, has decreased by 6% to 11.73 thousand kilometres in the same period (DfT, 2003a).

The number of vehicles using the road networks is increasing rapidly: Between 1991 and 2001, the number of cars rose 21% to 23,899,000 and the number of buses rose 23.6% to 89,000 (DfT, 2003b). UK total road traffic volume increased 18% from 1990 to 2002, reaching 486 billion vehicle kilometres in 2002. (UK Sust Dev, 2003).

In 2000, the modal split of UK road traffic (in billion vehicle kilometres) was:
- Car & taxi            378.7    (80.3%),
- Light vans            50.5     (10.7%),
- Goods vehicles        29.3     (6.2%),

- Buses & coaches   4.8      (1.02%),
- Motor cycles etc   4.4      (0.93%),
- Pedal cycles        4       (0.85%)

Adapted from (DEFRA 2003a).

Alternatives to motorised transport, particularly for local journeys, are cycling and walking. Cycle traffic on public roads fell dramatically from 23 billion passenger kilometres in 1952 to around 4 billion kilometres in the early 1970s. Despite rising 50% by the early 1980s, it returned to 4 billion kilometres in 1998. In 1996-1998, males aged 5 or over cycled an average of 112 km/yr on the public highway, while females cycled 23 km/yr. The peak age group was 11 to 17 year olds. Concern about road safety is a major reason for people not cycling (ONS 2002).

The number of households with regular use of a car has also increased slightly over the decade to 2001. By 2001, the percentage of households with regular use of one car was 46%, two cars 22% and three cars 5% (DfT, 2003a).

In 2002, the South East region of England had the largest share of English traffic on all roads, at about 18%, despite its relatively small area. The East of England and the North West had the next highest share with about 11% each. The North East carried the lowest amount of traffic on all roads with 4% (DfT, 2003d). The busiest motorway in 2002 was the M25 Western section with 147,000 vehicles per day. The M25 is the orbital motorway which encircles London. The average flow for the M25 as a whole was considerably higher than that of the next busiest motorways. These were the M1 (a major motorway serving the East of England connecting London to Leeds and continuing north to the A1 and the North East of England), M27 (a motorway which serves the south coast of England from the west of Southampton to Portsmouth), M60 (the Manchester orbital motorway) and southbound M6. The M6 motorway was the first motorway built in the UK and it is the longest motorway in England. It runs from a junction with the M1 in the South Midlands of England, past the major cities of Birmingham and Manchester, and finally to Carlisle, close to the border with West Scotland. (DfT, 2003c).

One of the main reasons for these traffic increases is the increasing mobility of the population. People travel further to work, shop and for leisure activities. Average annual distance travelled per person increased sharply in the late 1980s as shown in Table 48 (DfT 2003c).

**Table 48**. Changes in average distance travelled (Average miles/year). Data from (DfT 2003c).

| (miles/year) | 1985/86 | 1999/01 | % increase |
|---|---|---|---|
| Average distance travelled | 5,317 | 6,815 | +28% |
| Distance travelled by car | 3,796 | 5,354 | +41% |
| Distance walked | 244 | 189 | -23% |
| Distance travelled by bicycle or motorcycle | 95 | 68 | -28% |
| Distance by local bus | 297 | 245 | -18% |
| Distance by rail or tube train | 336 | 425 | +26% |

The increase in average distance travelled by car leapt by 41% and distance travelled by rail or underground train increased 26% while distance walked or by other road transport fell. (DfT 2003c)

There is considerable concern about the increasing trend towards car travel among schoolchildren, partly due to the increasing road congestion that this is causing, but also in terms of the loss of independence and the long-term health risks of taking less exercise. Various schemes are being devised to encourage fewer car journeys to and from schools, such as "Safe Routes to School", an initiative co-ordinated by Sustrans and adopted by many local authorities (Sustrans 2003). Safe Routes to Schools is a community approach to

- Encourage more people to walk and cycle to school safely
- Improve road safety and reduce child casualties
- Improve children's health and development
- Reduce traffic congestion and pollution (Sustrans 2003)

One example of "Best Practice" is the Oxford Transport Strategy (see box).

The growth in air transport in the last twenty years is far bigger than for other transport sectors. The number of passengers at UK airports (including those changing aircraft) increased more than threefold between 1980 and 2001, from 50 million to 162 million (DfT, 2003e).

The aim of this report is to educate and inform non-environmental health expert professionals about the heath and mobility issues in the United Kingdom and to provide an example of how these issues can be tackled to minimise the health impact.

## 2.   Environmental consequences of travel

### 2.1.   Road and air traffic emissions affecting air quality

As well as contributing to global greenhouse gases, road traffic emissions from engine exhausts (listed in the Introduction) have an adverse effect on air quality and have various health and environmental impacts.

In the UK, air quality is measured in two main ways: continuous automatic monitoring and monitoring using non-automatic equipment. Continuous automatic monitoring allows instantaneous measurements of air pollution concentrations at chosen sites. Measurements of air quality using non-automatic equipment gives concentration measurements over longer periods of time; for example, daily, weekly or monthly (DEFRA, 2003c).

There are three main networks funded by DEFRA to measure air quality: the Automatic Urban Network (AUN), the Automatic Rural Network (ARN) and the

Hydrocarbon Network. There were a total of 116 automatic monitoring sites around the UK in 2002; see Figure 48 (DEFRA, 2003c).

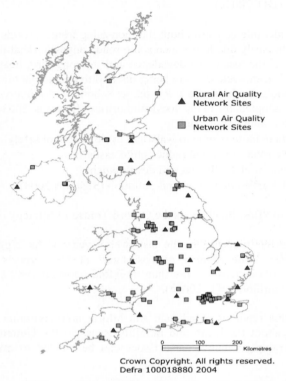

**Figure 48**. The automated air quality measuring sites in the UK (DEFRA, 2003c) (*Crown copyright material is reproduced with the permission of the Controller of HMSO and the Queen's Printer for Scotland.*)

The AUN urban network monitors $SO_2$, $NO_x$, CO, $O_3$ and particulate matter ($PM_{10}$). The urban sites are chosen as being representative of population exposure. In addition, so-called "hotspots" are monitored at urban kerbsides.

The ARN rural sites mainly monitor $O_3$ but also $SO_2$, $NO_x$. The Hydrocarbon Network monitors 25 volatile organic compounds, especially benzene, 1,3-butadiene and $O_3$ precursors at urban roadside, urban background and rural areas. DEFRA provide a public information service giving hourly updates of ambient $SO_2$, $NO_x$, $O_3$, CO and particulate matter concentrations via a free telephone line and a website (DEFRA, 2003c).

The health-based objectives for ambient air quality in the UK are set out in the Air Quality Strategy for England, Scotland, Wales and Northern Ireland (DEFRA, 2000). These objectives are set in order to protect human health from pollutants such as: $SO_2$, $PM_{10}$, $NO_2$, CO, Pb, benzene, 1,3-butadiene, $O_3$ and PAHs. There are also objectives for $SO_2$ and $NO_x$ to protect the environment from acidification.

## 2.2. $CO_2$ emissions (contribution to greenhouse effect)

Carbon dioxide ($CO_2$) is one of a "Basket" of six gases that comprise the so-called greenhouse gases. $CO_2$ is by far the most significant greenhouse gas, the other main offenders being methane ($CH_4$), nitrous oxide ($N_2O$), halocarbons (HCF, PFC) and sulphur hexafluoride ($SF_6$). Under the Kyoto protocol, the UK has agreed to reduce basket greenhouse gases by 12.5% below 1990 levels in the period 2000-2012. The UK already reduced emissions of basket greenhouse gases by 13% between 1990 and 2000, the challenge being to keep emissions from rising again as travel increases (DEFRA, 2003a).

In the pre-industrial atmosphere, the average concentration of $CO_2$ was thought to be about 280 ppm. During the present industrial time (1998), the average concentration of $CO_2$ is 365 ppm. The current rate of change of concentration of atmospheric $CO_2$ is 1.5ppm/year (DEFRA, 2002). $CO_2$ emissions in the UK from anthropogenic sources contribute about 2% to the global burden (DEFRA, 2003a). Global man-made emissions of $CO_2$ range between 6.2-6.9 billion tonnes of carbon per year (DEFRA, 2003a). However, The UK cut $CO_2$ emissions between 1970 and 2000 by 20% and has a goal of cutting $CO_2$ emissions by 20% below 1990 levels by 2010 (DEFRA, 2003a, DEFRA, 2002).

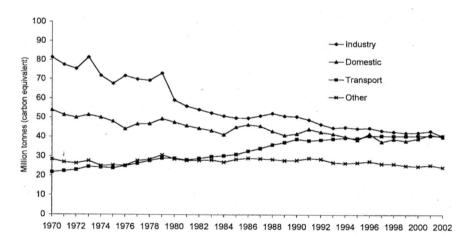

**Figure 49.** Carbon dioxide emissions by end user in the United Kingdom: 1970-2002 (DEFRA, 2003a) (*Crown copyright material is reproduced with the permission of the Controller of HMSO and the Queen's Printer for Scotland.*)

Road traffic is the fastest growing source of $CO_2$ emissions in the UK. Figure 49 shows that between 1970 and 2000, $CO_2$ emissions from road transport nearly doubled (from 20.7 million tonnes in 1970 to 38.6 million tonnes in 2000), becoming 24% of the total $CO_2$ emissions in 2000. Rail, shipping and civil aircraft each contribute 1% to the total UK $CO_2$ emissions (DEFRA, 2003a).

$CO_2$ emissions from cars have risen 81% since 1970 to 19,823 Ktonnes in 2001 (NAEI, 2002a).

Moreover, the cost of fuel does not seem to make a difference to forecast road traffic totals: Forecasts from the Department for Transport estimate that high cost fuel would do little to deter road traffic users (DfT, 2003d).

### 2.3.    Nitrogen oxides

Nitrogen oxides ($NO_x$) are acid gases that originate from fossil fuel combustion and road traffic emissions. As air pollutants, $NO_x$ are implicated in both chronic and acute lung function conditions, especially asthma. $NO_x$ also contribute to acid deposition and are ozone pre-cursors (DEFRA, 2003c).

In 2001, the total $NO_x$ emissions in the UK were 1,680,000 tonnes, of which 781,000 tonnes were attributable to road transport; 46% of the total. From 1970-1989, $NO_x$ emissions from road traffic rose from 764,000 to 1,290,000 tonnes, as a direct result of the increasing road traffic. However, despite road traffic continuing to rise after 1989, $NO_x$ emissions fell by 39% to the 2001 level, because of the introduction of catalytic converters. The emissions of $NO_x$ from civil aircraft have risen from 1,000 tonnes in 1970 to 4,000 tonnes in 2001. Road traffic diesel contributes more to total $NO_x$ than petrol. The UK has a target to reduce total $NO_x$ emissions to below 1,167,000 tonnes by 2010, under the EU National Emissions Ceiling Directive (DEFRA, 2003c).

To meet UK and EU health objectives by 2005, the concentration of $NO_x$ should not exceed an hourly mean of $200\mu g/m^3$ more than 18 times a year by 2005. This objective was met by 2002. In addition, the annual health objective of $40\mu g/m^3$ by 2005 has been achieved by around 70% of sites in 2002 (DEFRA, 2003c).

### 2.4.    Sulphur dioxide ($SO_2$)

Sulphur dioxide is also an acid gas and can affect health and vegetation in a similar way to $NO_x$. It particularly affects asthma and sufferers of chronic lung disease. In addition, $SO_2$ can affect the lining of the nose, throat and airways of the lung (DEFRA, 2003e).

The UK has a target to reduce $SO_2$ emissions below 585,000 tonnes by 2010, under the EU National Emissions Ceiling Directive (DEFRA, 2003e). In 2000, total $SO_2$ emissions were 1,125,000 tonnes, a reduction of 70% below 1990 levels. The $SO_2$ emissions attributed to road transport fell from 44,000 tonnes in 1970 to 3,000 tonnes in 2001 (DEFRA, 2003e).

Table 49 shows the target objectives for $SO_2$ and their achievement levels. All objectives were met in 2002 except at one site (DEFRA, 2003c).

**Table 49**. UK Health Objectives for $SO_2$ and achieved values (DEFRA, 2003c) (*Crown copyright material is reproduced with the permission of the Controller of HMSO and the Queen's Printer for Scotland.*)

|  | Target Objective ($\mu g/m^{3)}$ | Target Year | Permitted exceedences per year | Year (nearly) Achieved | No. of sites failing |
|---|---|---|---|---|---|
| 1 hour mean | 350 | ? | 24 | 2002 | - |
| 24 hour mean | 125 | 2004 | 3 | 2002 | 1 |
| 15 minute mean | 266 | 2005 | 35 | 2002 | 1 |

### 2.5.  Road transport - contribution to acidification

The main road transport emissions responsible for acidification are nitrous oxides, sulphur dioxide and ammonia. Acid deposition occurs by two processes: wet (polluted rainfall) and dry (when the chemicals fall directly back to the Earth due to gravity) (Air Quality Archive, 2003). The largest dry deposition of sulphur occurs in Northern areas of the UK, whereas nitrogenous dry deposition is highest in Southern England. Upland areas of UK suffer the highest levels of wet deposition of both sulphur and oxidised nitrogen due to high levels of rainfall in these areas (DEFRA, 2003c). Acid deposition is currently measured weekly at 32 sites across the UK. Five of these sites supply data to the European Monitoring and Evaluation Programme (EMEP) network (DEFRA, 2003c).

### 2.6.  Contribution to ground-level ozone

Both road and air traffic emissions contribute to ground level ozone pollution. The major contributing pollutants are nitric oxides and volatile organic compounds (VOCs) (DEFRA, 2003c). The majority of emissions of VOCs and $NO_x$ from air traffic occur during take off and landing cycles but idling aircraft at gates also contribute to air pollution. At airports, ground access traffic and aircraft pushing units contribute to approximately half of the total airport emissions. In addition, toxic glycols, used as a de-icer on aircraft, pollute local waterways (Holzman, 1997).

Ozone ($O_3$) can affect human health and is detrimental to vegetation and crops (DEFRA 2003c). Ground level ozone occurs naturally. However, levels increase when ozone is formed as a secondary pollutant as a result of chemical reactions between $NO_x$, oxygen and VOCs in the presence of sunlight (DEFRA, 2003c). Ground level ozone concentration has doubled from around $30\mu g/m^3$ to $60\mu g/m^3$ over the past century (DEFRA, 2003c). Ozone pollution from countries in mainland Europe can be blown across the UK by wind. Ozone pollution levels rise well over background levels in the summer months due to long periods of warm sunlight with low wind. Settled weather conditions can lead to ozone persisting for many days and being transported over long distances. Kent (2003) documented one such "classic ozone event" which occurred in the UK between May 27 and June 3, 2003.

The UK health objective for ground level ozone is that the daily maximum 8-hour running mean should not exceed a concentration of 100 $\mu$g/m$^3$ on more than 10 days a year at any site by 2005. This objective was met by nearly three quarters of sites in 2002.

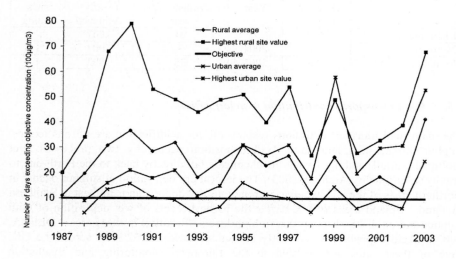

**Figure 50**. Ground level ozone exceedences: 1987-2003 United Kingdom. Adapted from Key Facts about: Air quality, Ground level ozone exceedences: 1987-2003 United Kingdom (DEFRA, 2003c) (*Crown copyright material is reproduced with the permission of the Controller of HMSO and the Queen's Printer for Scotland.*)

Ozone is usually much higher in rural areas than urban areas, as shown in Figure 50. This is because of higher urban levels of $NO_x$ from road traffic. The $NO_x$ reacts with the ozone to form $NO_2$ (DEFRA, 2003c). The fitting of catalytic converters after 1992 to all new petrol-engine vehicles has helped to reduce VOC emissions (DfT, 2003d).

Unfortunately, any local reduction in ozone emissions in the UK and Europe is likely to be outweighed by transboundary influx of globally produced ozone pollution, from developing countries. The Intergovernmental Panel on Climate Change (IPCC) predicts that a global-scale pool of ozone will grow and this, in turn, may increase surface $O_3$ concentrations in all the northern hemisphere continents through the cycle of constantly changing weather patterns. Local pollution episodes are built on top of these global baseline values. Thus, future global increases in tropospheric ozone may work against regional pollution control strategies designed to reduce exposure levels for both humans and vegetation (Derwent *et al.*, 2002).

## 2.7. Contribution to ambient carcinogens

Road and air traffic emissions to air include pollutants that are carcinogenic to humans such as particulate matter from diesel exhaust, Volatile Organic Compounds (VOCs), polycyclic aromatic hydrocarbons (PAHs) and heavy metals.

Volatile organic compounds are a diverse group of chemicals and are divided into methane and non-methane groups. Methane is primarily a greenhouse gas and will not be considered in this section. Most VOCs emitted from road traffic are not harmful to human health; however, benzene and 1,3-butadiene are toxic and carcinogenic. Total VOC emissions fell by 45% between 1990 and 2001 and road transport emissions fell by 66% (DEFRA, 2003c). Since 1970 Lead (Pb) emissions have fallen to virtually zero as a direct result of the introduction of unleaded petrol: leaded petrol was phased out from general sale at the end of 1999 (DEFRA 2003a, NAEI, 2004).

Benzene is one of the VOCs and is a group 1 carcinogen which causes various leukaemias (IARC, 1989a). Thirty-nine sites across the UK monitor benzene emissions in order to comply with EU directives on benzene. These sites use a combination of automatic and non-automatic monitoring methods. In 2001, road traffic was responsible for 6,140 tonnes of benzene emissions in the UK, 36% of the emission total (DEFRA, 2003c). Because benzene is a constituent of petrol, emissions arise from both evaporation and combustion of petrol (NAEI, 2002a). Benzene emissions have decreased since 1990 because of catalytic converters and a reduction in the benzene content in petrol. The UK objectives that the maximum running annual means for benzene should not exceed 16.25µg/m$^3$ by the end of 2003 were met at all monitoring sites. New health objectives for benzene levels have been agreed: in England and Wales the annual mean should not exceed 5µg/m$^3$ (DEFRA, 2003c).

Another potential contributor to ambient carcinogens is 1,3-butadiene, which is a group 2a suspected human carcinogen (IARC, 1999) emitted as a by-product of petrol combustion. Road traffic emissions form the bulk of total 1,3-butadiene emissions in the UK. At 3,480 tonnes in 2001, this was 78% of the UK total (NAEI, 2002a), which fell 64% between 1990 and 2001. Catalytic converters are helping to reduce the proportion of 1,3-butadiene emissions attributed to road transport.

Polycyclic aromatic hydrocarbons (PAHs) are emitted by motor vehicles due to incomplete combustion. They are highly toxic in high concentrations and some are carcinogenic. Benzo[a]pyrene (B[a]P) is recognised as one of the most carcinogenic of the group. It is a group 2a carcinogen along with Benz[a]anthracine (IARC, 1983 NAEI, 2002a). An addendum to the Air Quality Strategy in England, Scotland and Wales, issued in February 2003, has set out a new objective for PAH, for 0.25ng/m$^3$ as annual average, to be met by the end of 2010. B[a]P concentration is now used as the marker for the total mixture of PAH in the UK (DEFRA, 2003e). B[a]P and Benz[a]anthracine levels have reduced markedly in the UK. B[a]P emissions fell from 70 tonnes in 1990 to 10 tonnes in 2001 (DEFRA, 2003e). However, although

PAH emissions have decreased significantly since 1990, in 1990 the largest source of PAH was road transport combustion, which contributed 49% of the total emissions (NAEI, 2002a).

Road traffic engine exhausts contain thousands of gaseous and particulate substances. Particles emitted from petrol-burning engines are different from diesel engine exhaust particles, in terms of their size distribution and surface properties. Diesel engines produce two to forty times more particulate emissions than petrol engines, when both have catalytic converters and similar engine power output (IARC, 1989c). Aircraft particulate emissions contributed around 1% of the total emissions in 2001; this is a rise of 1% from 1990 levels (DfT, 2003e). The International Agency for Research on Cancer (IARC) studies found excesses of lung and skin tumours in animals exposed to diesel engine particulates. Diesel engine exhaust is classified as probably carcinogenic to humans (Group 2A) and petrol engine exhaust as possibly carcinogenic to humans (Group 2B) (IARC, 1989c).

There are areas of the UK where air quality is poorer than in others, as individual pollutant levels vary across the country. Air pollutant levels also vary between different places in the same area e.g. beside roads (DoH, 2002).

## 2.8.    Noise pollution

### 2.8.1.    Sources of noise pollution

Noise pollution from air and road traffic is underestimated. Noise pollution is implicated in a variety of physical disorders and mental illness. People living around airports and in close proximity to motorways and main roads are especially at risk from noise pollution.

Aviation is a major UK industry, carrying over 180 million passengers a year and over 2.1 million tonnes of freight (DfT, 2003e). If we consider Heathrow airport, Table 50 shows aircraft movements and noise exposure levels.

**Table 50**. Aircraft movements and noise exposure from Heathrow Airport 1990-2001 (DEFRA, 2003a). (*Crown copyright material is reproduced with the permission of the Controller of HMSO and the Queen's Printer for Scotland.*)

|                                           | 1990    | 2001    | % change |
|-------------------------------------------|---------|---------|----------|
| Number of aircraft movements [1]          | 362,000 | 463,000 | 28%      |
| Population exposed to $57_{leq}$ [2]      | 429,200 | 240,400 | -44%     |
| Population exposed to $63_{leq}$          | 109,600 | 54,900  | -50%     |
| Population exposed to $69_{leq}$          | 21,900  | 6,800   | -69%     |

[1] Aircraft movements = takeoffs and landings. [2] The equivalent continuous sound level (Leq) is an index of aircraft noise exposure. It is an averaged measure of the equivalent continuous sound level over a 16-hour day from 0700 to 2300 hours.

Table 50 shows that while aircraft movements rose 28% during the decade to 2001, the population exposed to high sound levels fell dramatically, due to quieter aircraft engines (DfT, 2003a). $57_{Leq}$ represents the onset of noise disturbance, $63_{Leq}$ moderate disturbances and $69_{Leq}$ high disturbances. The table shows that in 2001, there were still a quarter of a million people exposed to some noise disturbance from this airport alone. Thus aircraft noise still has major health implications and the complaints about noise nuisance are increasing (DEFRA, 2003a).

Figure 51 shows that complaints to Local Authorities about aircraft noise have increased more than fivefold since 1984/5. Road traffic noise complaints increased until 1995/6 then fell back to below 1984/5 levels. Caution must be taken in interpreting the air traffic statistics because many complaints about air traffic noise may not have been made to local authorities but to the Civil Aviation Authority, airports or the Department of Transport.

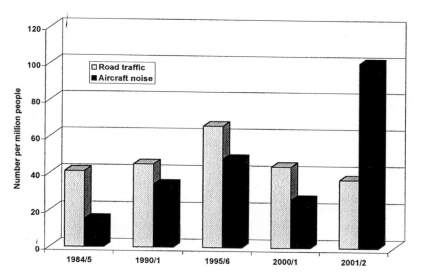

**Figure 51**. Noise complaints per million people for road and air traffic, 1984/5 – 2001/2 in the UK (DEFRA, 2003a) (*Crown copyright material is reproduced with the permission of the Controller of HMSO and the Queen's Printer for Scotland.*)

Air traffic restrictions between the hours of 23.00 and 07.00 can have a favourable effect on night-time noise levels. For example, very few people around Luton airport were exposed to disturbing levels of air traffic noise in 2001. (DEFRA, 2003a).

## 2.8.2. *Attitudes to noise pollution*

A survey carried out by the Building Research establishment in 2000, regarding the UK population's attitude to noise between 1991 and 1999 (National Survey of Attitudes to Environmental Noise) shows that the categories of noise people are most affected by are: road traffic, aircraft, neighbours and railways. The survey shows that between 1991 and 1999 there was a significant decrease in the proportion of respondents affected by road traffic noise, from 61% in 1991 to 56% in 1999, whereas there was no significant change in the percentage adversely affected by air traffic noise (DEFRA, 2003a).

A sub survey of The UK population shows that 22% of people that heard road traffic noise were significantly annoyed or disturbed by road traffic noise and 8% of people were extremely annoyed or disturbed by it. Aircraft noise is responsible for 7% of the UK population being significantly annoyed or disturbed and 2% of the UK population extremely annoyed or disturbed (DEFRA, 2003a).

Table 51 shows the effects of the different types of noise heard, on the UK population in 1999/2000. The table shows that road traffic noise interferes with sleep or rest in a third of respondents, with a similar proportion affected by aircraft noise. More than half of the respondents were irritated, fed up or angry about road traffic, and over 70% had similar emotions about aircraft noise, while fewer than 10% felt safe or relaxed about either.

**Table 51**. Attitudes to Noise. Percentage effects of different types of noise heard, UK 1999/2000 (DEFRA 2003a). (*Crown copyright material is reproduced with the permission of the Controller of HMSO and the Queen's Printer for Scotland.*)

|  |  |  | percentages |
|---|---|---|---|
|  | Noise source: | | |
| **Effect of noise:** | Road traffic | Aircraft | Neighbours/other people nearby |
| **The noise interferes with:** |  |  |  |
| Sleeping | 18 | 12 | 18 |
| Resting | 15 | 22 | 16 |
| Concentrating | 12 | 18 | 14 |
| Having windows/doors open | 19 | 17 | 13 |
| Reading or writing | 10 | 15 | 12 |
| Listening to TV, radio or music | 10 | 18 | 12 |
| Spending time in the garden | 9 | 21 | 9 |
| Having a conversation (including phone) | 7 | 18 | 8 |
| **Emotional reactions:** |  |  |  |
| Irritated | 30 | 45 | 25 |
| Fed up | 12 | 15 | 12 |
| Angry | 11 | 11 | 11 |
| Safe | 4 | 3 | 8 |
| Relaxed | 5 | 3 | 7 |
| Stressed | 6 | 10 | 7 |
| Upset | 3 | 7 | 5 |
| Worried | 7 | 11 | 3 |
| Frightened | 3 | 10 | 1 |

## 3.    Health effects of travel

As Figure 52 shows, respiratory diseases (including respiratory cancers) are the main cause of death in the UK (24% of the total), closely followed by coronary heart disease (21%) and non-respiratory cancers (19%) (BTS 2000). Pollution from transport can cause or exacerbate all of these types of disease.

In 1998, it was estimated that between 12,000 and 24,000 deaths and between 14,000 and 24,000 hospital admissions are brought forward by short-term exposure to air pollution each year, mainly in urban Great Britain (DoH, 1998).

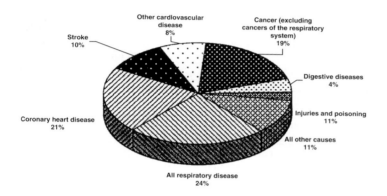

**Figure 52**. Overall breakdown of deaths by cause, 1999, UK. (BTS 2000)

### 3.1.    *Respiratory diseases, excluding cancers*

*3.1.1.    Overview*

The causal evidence linking transport emissions with respiratory health effects has been reviewed in other chapters of this book. However, it is difficult to assess, with any accuracy, the proportion of respiratory disease which may be attributed to such pollution.

Table 52 shows estimates made in 1998 by DoH, of health effects of short-term exposure to $PM_{10}$, $SO_2$ and ozone (DoH, 1998). The report also stated

> *".. in addition to the effects recorded here, it is likely that long-term exposure to air pollutants also damages health. At present there are insufficient UK data to allow acceptably accurate quantification of these effects..."*

**Table 52**. Numbers of deaths and hospital admissions for respiratory diseases affected per year by $PM_{10}$ and sulphur dioxide in urban areas of the UK, together with the numbers of deaths and hospital admissions for respiratory diseases affected per year by ozone in both urban and rural areas of the UK during summer only (DoH, 1998).

| Pollutant | Health outcome | UK Urban | |
|---|---|---|---|
| $PM_{10}$ | Deaths brought forward (all cause) | 8100 | |
| | Hospital admissions (respiratory) brought forward and additional | 10500 | |
| $SO_2$ | Deaths brought forward (all cause) | 3500 | |
| | Hospital admissions (respiratory) brought forward and additional | 3500 | |
| **Pollutant** | **Health outcome** | **UK threshold= 50 ppb** (urban and rural) | **UK threshold= 0 ppb** (urban and rural) |
| Ozone | Deaths brought forward (all causes) | 700 | 12500 |
| | Hospital admissions (respiratory) brought forward and additional | 500 | 9900 |

- Estimated total deaths occurring in urban areas of UK per year = c430,000
- Estimated total admissions to hospital for respiratory diseases occurring in urban areas of GB per year = c530,000

The British Thoracic Society published a report, (BTS, 2000), which gave a thorough and well-referenced review of the occurrence of respiratory disease in the UK, mainly using data from 1999. The foreword bluntly states:

> *"respiratory disease now kills more people [in the UK] than coronary heart disease - that is one in four people in the UK. ... it is the most common long-term illness among children, the most common illness responsible for an emergency admission to hospital, and costs the NHS more than any other disease area."*

The main points from the summary from this report follow (BTS, 2000). Respiratory cancer details have been omitted, as this is dealt with in section 3.4.

*3.1.2. Mortality, 1999.*

- In 1999, respiratory disease caused over 153,000 deaths: around 74,000 deaths in men and 79,000 in women.
- Diseases of the respiratory system are the cause of one in four deaths in the UK. More people die from respiratory disease in the UK than from coronary heart disease or cancer.

- Data from the World Health Organization shows that death rates from diseases of the respiratory system in the UK are well above the European average and around twice the European Union (EU) average.
- 44% of all deaths from respiratory disease are associated with social class inequalities.

(BTS, 2000).

### 3.1.3.   Disease levels, 1999

- The most commonly reported long-term illnesses in children are conditions of the respiratory system.
- Lung cancer is one of the most common cancers. See section 3.4 for details.

(BTS, 2000).

### 3.1.4.   Treatment, 1999

- Respiratory disease is the most common illness responsible for an emergency admission to hospital.
- Respiratory disease is the most common reason to visit the GP – almost a third of people will visit their GP at least once a year because of a respiratory condition.
- Cases of chronic obstructive lung disease and pneumonia (including acute lower respiratory infections) take up more than 2,800,000 hospital bed days a year.

(BTS, 2000).

### 3.1.5.   Costs

- Respiratory disease costs the UK National Health Service (NHS) more than any other disease area, amounting to £2,576 million (€3,830 million) in 2000 (BTS, 2000).
- The average cost for all ages for a respiratory hospital admission was £1,400 (€2,081) in 1999. Other NHS costs are likely but there is limited information on this (DoH, 1999a).

### 3.1.6.   Comparison to other countries, 1999

- The gap between the UK and other countries of the EU is particularly marked for women: females in Germany and France have death rates from respiratory disease around one-third the level of those found in the UK.
- WHO trend data show that between 1970 and 1997 death rates from respiratory disease fell by a third (32%) in the UK compared to bigger falls in Europe (46%) and the EU (44%).

(BTS 2000).

*3.1.7.    Prevalence of respiratory symptoms and asthma*

The Health Survey for England collected data on the prevalence of respiratory symptoms and doctor-diagnosed asthma in adults and children in England in 1995/6. At that time (BTS, 2000):

- Overall, around one third of adults (35% of men and 31% of women) and 28% of children aged 2-15 years (31% of boys and 26% of girls) had a history of wheezing.
- A fifth (21%) of adults and 18% of children had wheezed in the last twelve months.
- Just over one in ten adults (11% of men and 12% of women) and one fifth of children (23% of boys and 18% of girls) had doctor-diagnosed asthma.
- The prevalence of doctor-diagnosed asthma was higher in men than women and decreases with age in both sexes.
- In England, the prevalence of wheeze in adults increased from 16% in 1991 to 21% in 1995/96, an increase of around one third.

A number of studies in children suggest the prevalence of both wheeze and asthma have increased considerably during the past 25 years. Morbidity statistics from General Practice support this, showing that the proportion of children under five who are consulting their GP for asthma more than doubled during the 1980s. As asthma levels are generally increasing, especially in children, the current situation is likely to be rather worse.

### 3.2.    Accidents

*3.2.1.    UK overview for 1999*

The introduction to the White Paper "*Our Healthier Nation*" quotes the following statistics for 1999 (OHN, 1999):

- Accidents are responsible for 10,000 deaths a year across England.
- They are the greatest single life threat to children and young people.
- Accidents, particularly falls, are a major cause of death and disability in older people.
- Children from poorer backgrounds are more likely to die as the result of an accident than those from better off families.

*3.2.2.    Government policy regarding traffic accidents:*

In 1987, a UK target was set to reduce road casualties by one-third by 2000 compared with the 1981-85 baseline average. Pedestrians were set a higher target of reducing fatalities by 35% and serious injuries by 40%.

By 2000, road deaths had fallen by 39% and serious injuries by 45%. Towner reported in 2002 that overall child pedestrian deaths had fallen by 63% from the baseline average and overall child casualties by 30%. However, there had not been any such steep decline in the number of accidents, nor in the number of slight injuries. Over this period, traffic increased by 54% (DETR, 2000; Towner, 2002).

In *'Tomorrow's Roads – Safer for Everyone'*, the Government set road traffic casualty targets in 2000 (DETR 2000). The following targets to be attained by 2010 were set with respect to a baseline of 1994-1995:

- 40% reduction in the number of people killed or seriously injured (KSI) in road accidents.
- 50% reduction in the number of children (less than 16 years of age) killed or seriously injured.
- 10% reduction in slight casualty rate, expressed as the number of people slightly injured.

In October 2002, a cross-government taskforce report *'Preventing Accidental Injury - Priorities for Action'*, was launched (DoH, 2002). The Task Force adopted two population groups for priority attention. These are children and young adults, and older people.

The Task Force stated that the second highest burden of injury is from road accidents. They decided that the intervention areas which have the scope to make the biggest impact on road accidents in the short-term are:

- 20mph speed limits in areas of higher pedestrian activity
- Local child pedestrian training schemes and safe travel plans
- Systematic road safety intervention in inner city areas
- Advice and assessment programmes for elderly car drivers

### 3.2.3. Accidents arising from transport use

In 2000, around 3,500 people were killed every year on Britain's roads and 40,000 were seriously injured. In total, there were over 300,000 road casualties, in nearly 240,000 accidents, and about fifteen times that number of non-injury incidents (DETR 2000).

In terms of distance travelled, a pedestrian is roughly nineteen times more likely to be killed in a road accident than a car occupant is, and over 150 times more likely than a bus or coach passenger is (DETR 2000).

The following statistics, in Table 53, reveal the pattern of accidental death or injury by mode of transport:

- As Table 53 shows, more than half of all passengers/driver deaths per billion passenger kilometres occurred on two wheeled motor vehicles (mainly motor bikes), a quarter were pedestrians, and nearly a fifth were cyclists. Passenger death rates were low for travel by air, water (on merchant vessels), rail, bus or coach.
- Drivers & passengers killed or seriously injured (KSI) per billion passenger km, for the same decade, show a similar pattern.

- Considering deaths and all injuries, including lesser injuries; this shows a slightly different pattern, with pedal cyclists joining motor cycle drivers/passengers as the most accident-prone.

**Table 53**. Average annual passenger casualty rate per billion passenger kilometres, averaged for the decade 1991-2001. Data adapted from (DfT, 2003c).

| Per billion passenger kilometres | Deaths (% of total) | Killed or Seriously Injured (KSI) | Deaths & all injuries, incl. lesser injuries |
|---|---|---|---|
| Passengers | 211 (2.1% of total accidental deaths) | 3056 | 3056+ 11,200 |
| Of these: | (% of total passenger/driver deaths) | | |
| Motor bikes etc | 109 (51.3%) | 1490 (48.7%) | 5578 (39.0%) |
| Pedestrians | 57 (26.8%) | 633 (20.7%) | 2553 (17.9%) |
| Pedal cycles | 41 (19.4%) | 821 (26.9%) | 5465 (38.2%) |
| Car (include drivers) | 3.0 (1.4%) | 39 (1.3%) | 346 (2.4%), |
| Van | 1.2 (0.6%) | 16 (0.5%) | 121 (0.8%), |
| Bus/Coach | <1 | 14 (0.4%) | 197 (1.4%), |
| Rail[1] | <1 | ? | ? |
| Air | <1 | 0.1 (0%) | 0.07 (0%). |
| Water (merchant vessels) | <1 | 43 (1.4%) | ? |

[1] Reporting of rail injuries changed mid-decade, making comparisons difficult.

### 3.2.4.    Accidents to older people

Older people have a higher rate of accidental injury resulting in hospitalisation or death than any other age group. For both sexes, the age-specific death and admission rates for injury increase 'exponentially' with increasing age for women over 55 and men over 65. (DoH 2001).

Road traffic accidents cause 13% of all fatal injuries amongst older people. Transport accidents account for 3% of admissions to hospital for serious accidental injury for people aged 65 and over. Of pedestrians aged 60yrs or older, 5,832 were road accident casualties in 2000 (DoH, 2002).

### 3.2.5.    Accidents to children

Injury is the leading cause of child death in England and Wales. In the period 1998–2000 in England, 1003 children aged 0–14 years died as a result of accidental injury, of whom 287 were pedestrians, representing just over half of all road deaths in children.

In 2000, there were 320,283 road accident casualties in Great Britain, of whom 16,184 (5%) were child pedestrians (DoH, 2002).

There are steep social gradients (children in social class V are five times more likely to die from a pedestrian injury compared to children in social class I). There is also evidence of social gradients for non-fatal child pedestrian injuries. A range of evidence-based interventions including area-wide engineering measures, slowing vehicle speeds, education and enforcement aimed at vehicle drivers, education aimed at parents and children and child pedestrian skills training are available (Towner, 2002).

In the three-year period 1997-1999, the leading causes of unintentional injury deaths in childhood were motor vehicle traffic accidents (46%), with pedestrian deaths accounting for half of these (Towner, 2002).

**Table 54**. Accidental child road transport deaths aged 0-14, in 1999 in England and Wales. Source data adapted from Towner, 2002, who took it from ONS mortality statistics 1999.

|  | 1999 Boys (aged 0-14) | 1999 Girls (aged 0-14) | 1999 Total (Boys + Girls) |
|---|---|---|---|
| Total accidental deaths (all causes) | 220 | 127 | 347, 0 |
| Deaths from motor vehicle accidents | 95 | 58 | 153, 0 |
| Pedestrian deaths | 47 | 26 | 73, 0 |
| Cyclist deaths | 24 | 2 | 26, 0 |
| Passenger deaths (in motor vehicles) | 8 | 12 | 20, 0 |

Table 54 shows child transport death statistics for 1999. In 2000, there were 99 deaths of child pedestrians, with 2,921 seriously injured and 12,031 slightly injured (Towner, 2002).

There are relatively few cycle deaths in childhood but a large number of non-fatal injuries. Various measures are available to minimise these, including the promotion of bicycle helmets through educational and legislative approaches, engineering measures to separate cycles from other vehicles, reducing vehicle speeds and bicycle skills training.

The UK Accidental Injury Task Force noted that strategies to promote increased use of both walking and cycling were important but that these could increase children's exposure to injury risk. The injury priority areas needed to pay particular attention to increasing walking and cycling by providing safe environments in which these could be achieved (Towner, 2002).

### 3.3.    Health effects of noise pollution

#### 3.3.1.    Overview

Public perception of noise as a form of pollution from road traffic and air traffic is usually hidden under concerns about air and water pollution from industry, power production and transport. A survey carried out for DEFRA in 2001 titled "*Public*

*Attitudes to Quality of life and to the Environment"* recorded the public's attitude to 20 environmental issues. When asked about "pollution issues," respondents did not mention noise pollution as a major concern. When asked about "local issues," respondents (22%) did mention noise as a concern; however, noise pollution was the least important of the local issues (DEFRA, 2003c).

The World Health Organisation (WHO) has identified the following health risks of noise pollution:
- hearing impairment including tinnitus;
- pain and hearing fatigue (auditory effects);
- sleep disturbance: segregation of stress hormones;
- cardiovascular effects;
- reduced performance at work or school;
- interference with speech communication;
- interferences with social behaviour (aggressiveness, protest and helpfulness);
- annoyance (non-auditory effects).

(WHO, 2003).

The WHO programme on noise and health is studying the effects of noise in a dose-effect perspective, and is identifying the most vulnerable groups of people. Moreover, the programme is studying long-term night exposure to noise and its effect on health, especially long-term sleep disturbance and cardiovascular problems.

### 3.3.2.    Exposure-effect relationships

Different levels of noise pollution induce different health effects. For sleep disturbance, which is one of the most serious effects of environmental noise, exposure-response curves exist for awakenings in a laboratory setting and self-assessment questionnaires (WHO, 2003). Exposure-response curves can also be drawn for annoyance and loss of productivity in adults. Ambient noise affects memory recall and reading in children, although more evidence is needed to establish exposure-response relationships (WHO, 2003). Figure 53 shows the relationship between $L_{Amax}$ values (a measure of maximum sound levels) of Road traffic and aircraft noise and the probability of sleep disturbance, i.e. chance of awakening (Muzet, 2003).

Sleep disturbance can have a profound effect on an individual's health and quality of life. WHO guidelines say that for good sleep, sound levels of about 45 dB $L_{Amax}$ should not occur more than 10-15 times per night. Sleep disturbance can be measured objectively in a laboratory environment by a number of techniques. Subjective measuring such as interviewing can be used in such a way as to ascertain next day effects (Muzet, 2001). An individual waking in the early part of the night or just before the usual wakening time feels the most annoyance, suffers excessive daytime fatigue, is at a higher risk of accidents and has low work capacity (Muzet, 2001).

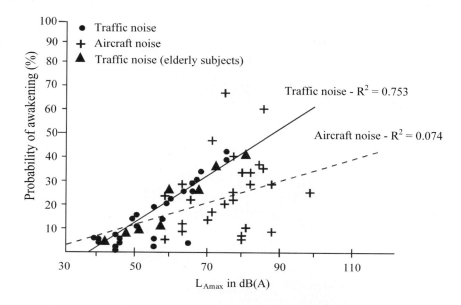

**Figure 53**. Relationship between $L_{Amax}$ values of road traffic and aircraft noises and the probability of awakening. Source: WHO taken from Hofman (1994 ) cited by Muzet (2003).

Road traffic noise is at a level which is detrimental to health, despite the number of complaints about road traffic noise having returned down to 1985 levels (see Figure 51). However, the effects of aircraft noise are better documented, perhaps indicating that this is perceived as being more of a problem.

Complaints about aircraft noise have risen over the last decade, see Figure 51. Thus, it is unsurprising that the highest numbers of complaints of sleep disturbance and annoyance come from people living around airports (WHO, 2003). Prescriptions of drugs are also higher around major airports (Muzet, 2001). Figure 54 is a photograph (taken during noise measurement) of a Boeing 747 Jumbo jet coming into land at Heathrow airport. The picture clearly shows the impact of having to experience aircraft landing nearby every two or three minutes throughout the day.

A cross-sectional study by Smith and colleagues looked at a possible association between aircraft noise and mental heath. Volunteers living in high and low aircraft noise areas around Manchester, East Midlands, Heathrow and Coventry answered questionnaires. The authors concluded that sleep disturbance due to aircraft noise was associated with mental health problems and other general health problems, but they warn that this may not be evidence of a causal relationship (Smith *et al.*, 2002).

**Figure 54**. Noise measurement of a Boeing 747-400 arriving at Heathrow. Taken from Aircraft Noise Monitoring Homepage, Civil Aviation Authority (CAA, 2003).

### 3.3.3.    Effects of noise on children

Children are one vulnerable group of people on whom noise has a profound effect. One study, on noise pollution and its effects on the cognitive performance of primary school children, found that average noise levels outside primary schools in all London boroughs exceeded the WHO guideline value for school playgrounds of 55dB. Road traffic noise was the predominant source of noise and aircraft noise was common. The average daily exposure of children in this study was 72db. Analysis of the Standard Assessment Tests (SATs) results indicated that chronic noise has a negative impact on children's academic performance. The children also reported that they were aware of, and annoyed by, environmental noise (Shield and Dockrell, 2002).

Another study, on West London schoolchildren (Stansfield *et al.,* 2003) confirmed most of the results from similar studies. Aircraft noise:

- is associated with impaired reading and annoyance;
- is weakly associated with hyperactivity and psychological morbidity;
- is associated with annoyance in parents and teachers.

However, Aircraft noise:

- was not associated with poorer performance of memory, sustained attention or overall reading score.
- was not associated with perceived stress, stressful life events, nor raised catecholamine or cortisol secretion.

(Stansfield *et al.,* 2003). These last two findings were not consistent with other studies on West London Schools.


## 3.4.    *The effect of motorised traffic on cancer incidence*

Cancer places an immense burden on the UK National Health Service. The cost of cancer care in the UK is £1,500,000,000 (€2,098,650,000) per annum (DoH, 1999). A government programme called the UK Cancer Plan sets out to achieve a reduction in the mortality rate of at least 20% by 2010. The latest figures show a 10.3% fall in death rates for cancer, so the plan is currently on track (DoH, 2003). However, cancer incidence rates are rising. At least one in three people in the UK will get cancer and at least one in four of the UK population will die from malignant disease (DoH, 2003; ONS, 2003c; Cancer Research UK, 2003). The UK government is spending an extra £570,000,000 (€797,487,000) on cancer services in 2003/4, which includes funding for cancer prevention measures mainly focused on smoking cessation, and diet (DoH, 2003). The problem of road traffic emissions, which are implicated in the aetiology of some cancers, is not addressed in the UK Cancer Plan.

Fuel and exhaust emissions from motorised forms of transport contain some cancer causing substances (carcinogens). The International Agency for research on Cancer (IARC) classifies all carcinogens into one of the groups shown in Table 55.

**Table 55**. Simplified IARC classification of carcinogens (IARC, 2000).

| Group 1 | Known human carcinogen |
|---------|------------------------|
| Group 2A | Probable human carcinogen |
| Group 2B | Possible human carcinogen |
| Group 3 | Not classifiable as to human carcinogenicity |
| Group 4 | Probably not carcinogenic to humans |

The carcinogens emitted by motorised transport fall into several of these categories, as shown in Table 56.

**Table 56**. Source, classification and target sites, of the main carcinogenic materials from motorised transport.

| Substance | Source of Carcinogen | Cancer Site | Carcinogen Group /Class | Source: |
|---|---|---|---|---|
| Benzene | Evaporation from and combustion of petrol (gasoline) | Leukaemia | Group 1 | (IARC, 1989a; NAEI, 2002a) |
| 1,3-butadiene | By-product of petrol combustion | Leukaemia, lymphomas | Group 2a | (IARC, 1999; NAEI, 2003c) |
| Benzo[a]pyrene (B[a]P) | Incomplete combustion of fuel | Lung, skin | Group 2a | (IARC, 1983; NAEI, 2003c) |
| Benz[a]anthracine | Incomplete combustion of fuel | Lung, skin | Group 2a | (IARC, 1983; NAEI, 2003c) |
| Diesel engine particulate matter | Incomplete combustion of fuel | Lung, bladder, leukaemia, Testes, skin | Group 2a | (IARC, 1989c) |
| Petrol engine particulate matter | Incomplete combustion of fuel | Lung, bladder | Group 2b | (IARC, 1989c) |
| Petrol (gasoline) | Occupational exposure | | Group 3 | (IARC, 1989b) |
| Diesel Fuel | Occupational exposure | Lung | Group 2b | (IARC, 1989b) |
| Marine diesel | Occupational exposure | Lung | Group 2b | (IARC, 1989b) |
| Jet fuel | Occupational exposure | | Group 3 | (IARC, 1989b) |

Exposure to carcinogens attributable to motorised transport can be via occupational exposure or from inhalation of polluted air. The latest world cancer report from the IARC estimates that between 1% and 4% of cancers can be accounted for by air pollution and up to 5% of lung cancers can be attributed to engine exhaust products. In recent years in developed countries, combustion engine exhaust emissions have contained fewer harmful substances because of the introduction of the catalytic converter. However, this is not the case in developing countries, where engine exhaust emissions will probably be a major public health concern in the future (Stewart and Kleihues, 2003).

Occupational exposure to carcinogens associated with mobility is mainly through contact with diesel and possibly petrol. People with occupations involving diesel production, storage, distribution and use, as well as occupations involving maintenance of diesel engines, are at risk of exposure through the skin and by inhalation. Occupational exposure to petrol occurs in a similar way (IARC, 1989b).

Petrol may contain 0-7%, typically 2-3%, benzene (IARC, 1989b), which causes leukaemia. However, exposure is not just limited to the work place. Benzene contributes to poor air quality and therefore people have little or no control over their exposure. Other constituents of vehicle exhaust emissions, such as 1,3-butadiene, Benzo[a]pyrene, Benz[a]anthracine, diesel engine particulate matter and petrol engine particulate matter also contribute to poor air quality and the public again have very little control over their exposure to these carcinogens.

Ultrafine particles from engine exhausts ($PM_{2.5}$) may be particularly damaging. Animal and cell culture studies show ultrafine particles cause oxidative stress and inflammation in lung tissue (Vrang *et al.*, 2002). In the UK, little monitoring of the mass or size of particulates per unit volume of air (gravimetric concentration) is undertaken (DoH, 2000b).

If we consider cancer risk in the UK from road traffic emissions, the DoH states:

> *"the available evidence strongly suggests that the risks associated with the levels found in the air in the UK are exceedingly small - that is, they are so small that they cannot be measured with any accuracy."* (DoH, 2002).

However, other studies have quantified cancer risk from road traffic exhaust emissions (Raaschou-Nielsen *et al.*, 2001; Nyberg *et al.*, 2000; Harrison *et al.*, 1999).

Figure 55 shows the ten most common forms of cancer in the UK (Cancer Research UK, 2003). The chart shows lung cancer as 14% of the total.

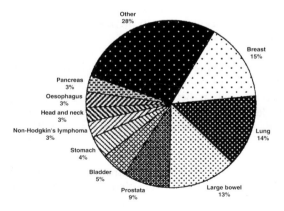

**Figure 55**. Pie chart showing the ten most common cancers diagnosed in the UK, excluding non-malignant melanoma (Cancer Research UK, 2003).

Figure 56 shows that in the UK, the trend for lung cancer cases for males is decreasing, whereas for females it has steadily increased between 1971 and 1999, though the increase has now slowed considerably (ONS, 2003d).

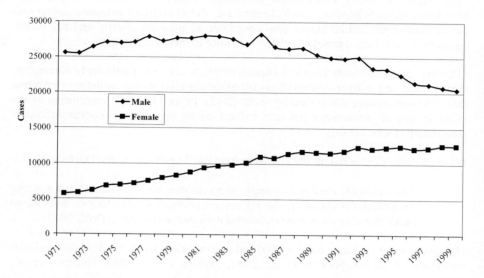

**Figure 56**. Lung cancer cases (all ages) in the UK, 1971-1999 (ONS, 2003d).

In 1999, there were around 23,500 cases of lung cancer in males, making it the second most common type. In females, there were around 14,700 cases, making it the third most common type. Survival rates for lung cancer are low. In both men and women, the one-year survival rate is 20% and the five-year survival rate is 5%. The only common cancer with lower survival rates in the UK is cancer of the pancreas (Cancer Research UK, 2003).

Do people, especially children, living in areas of high traffic density have an increased risk of cancer because of poor air quality? Not according to a study in California, that found no association between areas of high traffic density and childhood cancers (Reynolds *et al.*, 2002). However, studies in the UK, Denmark and Sweden, have found a small association between high traffic density and childhood cancer.

A study in Denmark, which looked at childhood exposure to high traffic density during childhood and pregnancy, found no association between high traffic density and childhood leukaemia, central nervous system tumours (CNS) and non-Hodgkin's lymphomas, even at the highest traffic density (exposure to >10,000 vehicles per day). $NO_2$ and benzene levels were calculated as markers of air pollution during the relevant period. The study found that there was an association between exposure *in utero* to increasing levels of $NO_2$ and benzene and Hodgkin's lymphoma. For a doubling of benzene and $NO_2$ concentration, there was an increased risk of 25% and 51% respectively for Hodgkin's lymphoma. This suggests a possible causal relationship (Raaschou-Nielsen *et al.,* 2001). The results for childhood leukaemia and CNS tumours in this study conflict with results from other studies.

A study in the West Midlands, UK, investigated whether there was an excess of childhood leukaemia in children living in close proximity to main roads and petrol stations. This study concluded that there is a small increase in risk for leukaemia but that further studies are required to establish the true risk. This study found no increased risk for solid tumours. The slightly increased risk of leukaemia was found in children living within 100m of a main road or a petrol station. It must be stressed that these results were not high enough to be statistically significant (Harrison, 1999). These results were consistent with an earlier case-referent study by Feychting *et al.,* carried out in Sweden (Feychting *et al.,* 1998).

If lung cancer is considered, the risk of developing lung cancer following exposure to city centre air pollution may be related to the distance from the source of the pollution. Biggeri and colleagues used a model based on distance from the source, which enabled estimation of risk gradients. They found that lung cancer was highly related to the city centre: relative risk at zero distance was 2.2 and there was a smooth decrease as distance from the source increased. The authors concluded that air pollution is a moderate risk factor of lung cancer (Biggeri *et al.,* 1996).

In a presentation reviewing the evidence for the links between transport pollution and cancer, Sasco (2003) concluded:

> *"there is an association between lung cancer in adults and some childhood tumours and air pollution, including but not limited to [that] coming from road traffic."*

Figure 57 reviews ten studies on the relative risk of lung cancer and exposure to diesel exhaust in railroad workers, all of which demonstrate a slightly increased occupational risk, although not all show a significant relative risk.

There are many confounding factors such as tobacco smoking, socioeconomic status, residential radon, and occupational exposures to be taken into account when considering the impact of road traffic emissions on cancer incidence. However, the available evidence seems to suggest that certain substances in exhaust emissions contribute to ambient carcinogens and to an increased cancer risk, especially to vulnerable groups of people such as children.

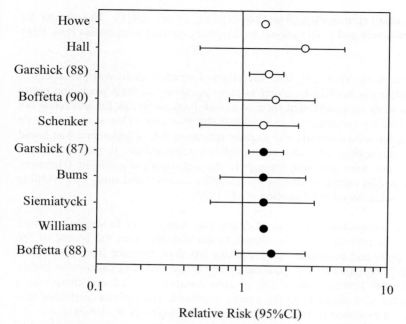

Relative Risk (95%CI)

● = RR adjusted for cigarette smoking; ○ = RR not adjusted for cigarette smoking. For the two studies by Howe and Williams, CIs were not reported and could not be calculated (from Cohen and Higgins 1995.)

**Figure 57.** Relative risk of lung cancer following exposure to diesel exhaust in railroad workers

## 4.    Conclusions

UK citizens pay high environmental, health and economic costs for, in the main, inefficient modes of travel. It is the most vulnerable UK citizens who bear the brunt of transport-related health problems, namely children and the elderly.

The UK addiction to car travel leads to a heavy price to pay in terms of pollution-related health costs as well as the contribution to global warming and other environmental degradation such as acid rain. Unfortunately, privatisation of rail and bus transport over the past decade is making it difficult to implement national or local strategies to ensure sufficient frequent reliable public transport – a prerequisite to persuading people out of their cars. Attempts are being made to redress years of underinvestment in the rail infrastructure.

In addition, the current government initially set out a transport plan which aimed to minimise new road building and reduce road travel, but lobbying caused a change in policy, reverting to attempting to solve traffic jams by building more roads, despite evidence that better roads generate more traffic, and hence more pollution.

More funding needs to be made available to local government, to safeguard and improve public transport where possible and to extend initiatives such as "Safe routes to school". This is one of various projects aimed at making cycling and walking more attractive and safe options. These should be encouraged and extended, together with more provision for safe use of buses and trains, especially for women at night. Physical activity such as walking and cycling should be encouraged from an early age and the health benefits of such activity should be publicised.

The UK seems to be meeting the majority of its own and European health objectives for $CO_2$ emissions and for other emissions that affect air quality. Further UK government-funded research is needed, to investigate the role of air pollution in the aetiology of cancer.

Noise is detrimental to health and noise reduction measures should be put in place in homes and schools near airports and traffic-dense areas.

**Box 7**. Example of best practice: Oxford Transport Strategy (OTS), Oxford, England.

---

**Aims**
To achieve significant environmental improvements, by improving air pollution and the ease of movement around the main streets in Oxford city centre, whilst catering for the increasing growth in demand for travel and allowing continued growth of the city centre economy.

**Methods**
Implement measures to reduce traffic in Oxford city centre, by promoting a shift from private car usage to more trips being made by bus, cycle and on foot.

**History of the project**

| | |
|---|---|
| 1992 - 1993 | Consultant's study, including extensive public consultation, studied the increasingly poor conditions in the city centre, particularly for pedestrians; the high level of traffic-related air pollution; the unacceptably high level of road accidents |
| 1993 - 1999 | Implementation of preliminary measures, including expansion of park and ride facilities; bus priority measures; capacity enhancements on the ring road; development of the cycle network. |
| 1999 | Implementation of key access restrictions to the city centre, including: selective increase in pedestrianisation; traffic restrictions to the main shopping streets; a bus priority route around the central area. |
| 1996 - 2001 | The EMITS study monitored traffic flows, economic vitality, air quality, public health, building stone erosion and public opinion, giving data from before and after OTS |

(OTS, 2000; Parkhurst, 2001).

**Health monitoring - Summary as at June 2003**

The health arm of the EMITS study aimed to assess the impact of OTS on the health of Oxford residents, by monitoring three health outcomes before and after the OTS implementation:

- Assessing changes in children's respiratory health,
- Assessing older adults' and parents' views on health issues before and after OTS,
- Assessing routine hospital admissions and mortality data for cardio-respiratory causes.

Health monitoring details were kindly provided by Dr Stephanie MacNeill, as final analysis is still ongoing prior to publication (MacNeill, 2003).

## 1. Children's health

Between 1998 and 2000, the respiratory health of 1389 children from seven Oxford primary schools was monitored though several multi-day visits per year:
- By peak expiratory flow (PEF);
- By questioning on self-reported symptoms of cough, wheeze, and runny nose;
- By parental questionnaires covering the child's medical history, exposures in the home (damp, cooking smoke and pets), and socio-economic and ethnic background.

This data is currently being analysed
- To assess differences in health outcomes before and after the implementation of OTS
- To determine whether local pollution levels have had an impact on children's respiratory health.

Results so far suggest that lung function improved slightly post-OTS, with a small but statistically significant increase in peak flows after controlling for potential confounding variables, and a statistically significant reduction in self-reported wheeze post-OTS.

## 2. Older adults and younger parents

The health and attitude to traffic related issues, of adults aged 65 years and older, was assessed before and after OTS through postal surveys, using a questionnaire enquiring on their respiratory health, and also through focus groups with a subset of the survey participants.

After the implementation of OTS, focus groups were also run with the parents of children participating in the study, to get the views of younger adults.

Discussion

Older adults: Those with pre-existing illnesses were more susceptible to traffic fumes. Air quality concerns concentrated on fumes (especially diesel) in the city centre, rather than noise or local air quality. More reported an improvement than deterioration in air quality post-OTS. However, air quality was regarded as less important to health than other traffic-related issues such as 'access', including issues of mobility, movement around the areas where people live, the accessibility of important amenities and safety. Concerns were raised about post-OTS ease of access, particularly to areas such as the bus station and railway station.

Bus services in the city were praised by the elderly; but criticised for their cost and extensiveness by the parents of young children.

Both older and younger age groups were concerned about the 'piecemeal nature' of the OTS. The younger group would have preferred a fully integrated transport system. Older people would find movement around Oxford much easier, more comfortable and less worrying if there were more and safer pedestrian crossings, without cyclists on the pavements, with easier access to the bus station and with more suitable seating.

### 3. Assessing cardio-respiratory data.

Routine hospital admissions and mortality data for cardio-respiratory causes between 1998 and 2000 were collected from the Oxfordshire Health Authority. These series were then matched with city-wide pollution levels and meteorological data. The data are currently being analysed to assess whether short-term fluctuations in air pollution have had a significant impact on daily health outcomes. (MacNeill, 2003).

### Results – traffic and pollution statistics.
Traffic statistics in the first 6 – 12 months after implementation showed:
- 20% reduction in traffic flows in central area, primarily by stopping through traffic
- 7% increase in park and ride bus usage
- 8 – 9% increase in bus passenger numbers, equating to approx. 2000 extra trips per day.
- 14% fewer cars parked in central area car parks
- No increase in traffic in other central area streets, i.e. no displacement of traffic.
- An increase in the number of people going into the centre area. Pedestrian surveys showed central area flows increased by 6000 people, including 8.5% more pedestrians on main shopping streets.
- Cycle flows were largely unchanged, although there has been a transfer to parallel (off main road) cycle routes. Nationally, Oxford still has one of the highest levels of cycling, with journey to work by cycle at 17%
- No significant change in accident levels in this short period.
- Broad improvement in air quality especially for carbon monoxide, with 75% improvement in one street. Particulate matter reduced by 20% on one shopping street which was closed to all traffic. Most sites showed reductions in nitrogen dioxide levels.
- The vacancy rate for retail units remained under 1%, while the sample of 9 retailers continuously monitored showed a decline in turnover, continuing a nationwide trend which had started in 1998, a year before the traffic changes.
- Modal split after implementation: Person trips: 39% by car; 44% by bus; 11% by cycle.
- Inner cordon vehicle composition: 44% Public service vehicle; 39% cars & taxis; 11% cycles; 4% light goods vehicles; 1% heavy goods vehicle; 1% motorcycle.

Monitoring was still ongoing, but further results were not available at the time of writing (OTS, 2000; Parkhurst, 2001).

### Results – health effects
- Primary schoolchildren's lung function generally improved slightly, post-OTS.
- The majority of surveyed older adults felt that air pollution had reduced but many felt that access to some areas was more of a concern.
- Analyses of cardio-respiratory data are ongoing.

(MacNeill, 2003).

### Conclusions
The desired aims of traffic reduction and improved air quality, whilst allowing increased travel in the city centre, appear to have been successful. Much of the assessment of health effects is ongoing, but those results so far available appear encouraging, though problems of access for the elderly may require further action.

# References

Biggeri, A., Barbone, F., Lagazio, C., Bovenzi, M., and Stanta, G. (1996) Air pollution and lung cancer in Trieste, Italy: spatial analysis of risk as a function of distance from sources, *Environ. Health. Perspect.* 104(7), 750-754.

BTS (2000) *The burden of lung disease,* British Thoracic Society Nov 2000, Downloaded September 2003 from http://www.brit-thoracic.org.uk/docs/BurdenofLungDisease.pdf

CAA (2003) *Aircraft noise monitoring: noise measurement of a Boeing 747-400 arriving at Heathrow,* Downloaded September 2003 from CAA website http://www.caa.co.uk/dap/environment/environment_noise/default.asp

Cancer Research UK (2003) *CancerStats incidence – UK,* Downloaded September 2003 from http://www.cancerresearchuk.org/aboutcancer/statistics/statsmisc/pdfs/cancerstats_incidence.pdf

Cohen, A.J, and Higgins, M.W.P. (1995) Health effects of diesel exhaust: Epidemiology. In: *Diesel Exhaust: A Critical Analysis of Emissions, Exposure, and Health Effects.* A Special Report of the Institute's Diesel Working Group. Health Effects Institute, Cambridge, MA; pp. 251-292.

DEFRA (2000) *The air quality strategy for England, Scotland, Wales and Northern Ireland,* DEFRA website, Originally DETR 2000, ISBN 0 10 145482-1, Downloaded July 2003 from http://www.defra.gov.uk/environment/airquality/strategy/

DEFRA (2002) *The environment in your pocket 2002,* DEFRA website, Downloaded June 2003 from: http://www.defra.gov.uk/environment/statistics/eiyp/pdf/eiyp2002.pdf

DEFRA (2003a) *e-Digest of environmental statistics,* DEFRA website, Index downloaded June 2003 from: http://www.defra.gov.uk/environment/statistics/index.htm

DEFRA (2003b) National emission ceilings directive: summary of responses to consultation, and implementing regulations. In *national strategy to combat acidification, eutrophication and ground-level ozone,* DEFRA website, Downloaded July 2003 from http://www.defra.gov.uk/environment/airquality/necd/

DEFRA (2003c) Index to: e-Digest of statistics about: air quality, in *e-Digest of environmental statistics,* DEFRA website, Downloaded June 2003 from: http://www.defra.gov.uk/environment/statistics/airqual/

DEFRA (2003d) *Air pollution-What it means for your health,* DEFRA website, Downloaded June 2003 from: http://www.defra.gov.uk/environment/airquality/airpoll/pdf/airpollution_leaflet.pdf

DEFRA (2003e) Index to: *Addendum to the air quality strategy,* Published February 2003 on DEFRA website, Downloaded July 2003 from http://www.defra.gov.uk/environment/airquality/strategy/addendum/index.htm

Derwent, R.G., Collins, W. J., Johnson, C.E., and Stevenson, D.S. (2002) Global ozone concentrations and regional air quality, *Env. Sci. Technol.* 36(19), 379A-382A.

DETR (2000) *Tomorrow's roads: safer for everyone. The government's road safety strategy and casualty reduction targets for 2010,* Department of Transport Environment and the Regions, London, Downloaded September 2003 from http://www.dft.gov.uk/stellent/groups/dft_rdsafety/documents/page/dft_rdsafety_504644.hcsp

DfT (2000) *Air traffic forecasts for the United Kingdom 2000,* DfT website, Downloaded July 2003 from: http://www.dft.gov.uk/stellent/groups/dft_aviation/documents/page/dft_aviation_503314.hcsp

DfT (2002) Public transport, Chapter 5 in *Transport statistics Great Britain: 2002 edition,* DfT website, Downloaded September, 2003 from: http://www.dft.gov.uk/stellent/groups/dft_transstats/documents/page/dft_transstats_023377.pdf

DfT (2003a) Index to: *Transport trends: 2002 edition,* DfT website, Downloaded June 2003 from: http://www.dft.gov.uk/stellent/groups/dft_control/documents/contentservertemplate /dft_index.hcst?n=7549&l=3

DfT (2003b) Transport trends 2002: section 1: tables, in *Transport trends,* DfT website, Downloaded June 2003 from: http://www.dft.gov.uk/stellent/groups/dft_control/ documents/contentservertemplate/dft_index.hcst?n=7556&l=4

DfT (2003c) Index to: *Transport statistics,* DfT website, Downloaded September, 2003 from: http://www.dft.gov.uk/stellent/groups/dft_transstats/documents/sectionhomepage/dft_tran sstats_page.hcsp

DfT (2003d) Index to: *Energy and the environment,* DfT website, Downloaded July 2003 from http://www.dft.gov.uk/stellent/groups/dft_control/documents/contentservertemplate/ dft_index.hcst?n=6930&l=2

DfT (2003e) Index to: *Aviation,* DfT website, Downloaded September 2003 from http://www.dft.gov.uk/stellent/groups/dft_aviation/documents/sectionhomepage/dft_aviati on_page.hcsp

DoH (1998) *COMEAP report: The quantification of the effects of air pollution on health in the United Kingdom,* Downloaded September 2003 from http://www.doh.gov.uk/comeap/ statementsreports/airpol7.htm

DoH (1999a) *Economic appraisal of the health effects of air pollution report,* Downloaded September 2003 from http://www.doh.gov.uk/airpollution/ecomain.htm

DoH (1999b) *Strategic priorities in cancer research and development,* Cancer Working Group 1999, Downloaded November 2003 from http://www.doh.gov.uk/research/ documents/rd3/cancer_final_report.pdf

DoH (2000a) *COMEAP advice: The health effects of air pollutants July 2000,* Downloaded September 2003 from http://www.doh.gov.uk/comeap/statementsreports %5Chealtheffects.htm

DoH (2000b) *Second report: sulphur dioxide, acid aerosols and particulates,* Advisory group on the medical aspects of air pollution episodes, Downloaded September 2003 from http://www.doh.gov.uk/airpollution/airpol3.htm

DoH (2001) *Priorities for prevention,* Accidental injury task force's working group on older people, December 2001, Downloaded September 2003 from http://www.doh.gov.uk/accidents/pdfs/olderpeople.pdf

DoH (2002) *Preventing accidental injury – priorities for action,* Report to the Chief medical officer from the accidental injury task force, ISBN 0-11-322477-X, Downloaded September 2003 from http://www.doh.gov.uk/accidents/pdfs/preventinginjury.pdf

DoH (2003) *The NHS cancer plan three year progress report maintaining the momentum,* Downloaded November 2003 from: http://www.doh.gov.uk/cancer/progressreport2003/ report.pdf

Donaldson, L. (2002) Letter introducing the report *preventing accidental injury - priorities for action,* The accidental injury task force, Downloaded September 2003 from http://www.doh.gov.uk/accidents/accinjuryreport.htm

Feychting, M., Svensson, D., and Ahlbom, A. (1998) Exposure to motor vehicle exhaust and childhood cancer, *Scand. J. Work. Environ. Health.* 24(1), 8-11.

Harrison, R., Leung, Pei-Ling., Somervaille, L., Smith, R., and Gilman, E. (1999) Analysis of incidence of childhood cancer in the West Midlands of the United Kingdom in relation to proximity to main roads and petrol stations, *Occup. Environ. Med.* 56, 774–780.

Hofman, W. F. (1994) *Sleep disturbance and sleep quality,* Doctoral Dissertation, University of Amsterdam, The Netherlands, cited by Muzet (2003).

Holzman, D., (1997) Plane pollution, *Environmental Health Perspectives* 105(12), 1300-1305.

IARC (1983) Benz[a]anthracene, in Polynuclear aromatic compounds, Part 1, chemical, environmental and experimental Data, *IARC Monographs on the evaluation of carcinogenic risks to humans* 32, Downloaded July 2003 from http://monographs.iarc.fr/ htdocs/monographs/vol32/benz%5Ba%5Danthracene.html

IARC (1989a) Benzene, *Overall evaluations of carcinogenicity: an updating of IARC monographs* Volumes 1 to 42 Supplement 7, Downloaded September 2003 from http://monographs.iarc.fr/htdocs/indexes/suppl7index.html

IARC (1989b) Occupational exposures in petroleum refining; crude oil and major petroleum fuels, *IARC monographs on the evaluation of carcinogenic risks to humans* 45, Downloaded September 2003 from http://monographs.iarc.fr/htdocs/indexes/ vol45index.html

IARC (1989c) Diesel and gasoline engine exhausts and some nitroarenes, *IARC monographs on the evaluation of carcinogenic risks to humans* 46, Downloaded July 2003 from http://monographs.iarc.fr/htdocs/indexes/vol46index.html

IARC (1999) 1,3-Butadiene, in Re-evaluation of some organic chemicals, hydrazine and hydrogen peroxide. *IARC monographs on the evaluation of carcinogenic risks to humans* 71, Downloaded July 2003 from http://www-cie.iarc.fr/htdocs/monographs/vol71/002-butadiene.html

IARC (2000) *Lists of IARC evaluations*, Downloaded September 2003 from http://monographs.iarc.fr/monoeval/grlist.html

Kent, A. (2003) *Air pollution forecasting: pollution episode report* (May/June 2003), from UK national air quality information archive website, Downloaded September 2003 from: http://www.airquality.co.uk/archive/reports/cat12/o3_episode_may2003.pdf

MacNeill, S. (2003) Personal communication with Dr Stephanie MacNeill, Imperial College, June 13th 2003.

Muzet, A. (2001) Aircraft noise and sleep, in the meeting report, technical meeting on aircraft noise and health, Bonn, Germany 29 – 30 October 2001, Downloaded September, 2003 from http://www.euro.who.int/Noise/Activities/20021203_1

Muzet, A. (2003) Noise exposure from various sources sleep disturbance, dose-effect relationships on adults at the WHO technical meeting on exposure-response relationships of noise on health, 19-21 September 2002 Bonn, Germany meeting report, Downloaded September, 2003 from http://www.euro.who.int/Document/NOH/ exposerespnoise.pdf

NAEI (2004) *Action on lead emissions*, National Atmospheric Emissions Inventory website, Downloaded December 2004 from: http://www.naei.org.uk/pollutantdetail.php

NAEI (2001) *Emissions data 1970-2001, all pollutants, all sources*, National Atmospheric Emissions Inventory website, Downloaded September 2003 from: http://www.naei.org.uk/emissions/emissions_2001/all_pollutants01_NFR_final_3.xls

NAEI (2002a) Index to: *Toxic air pollution*, National Atmospheric Emissions Inventory website, Downloaded September 2003 from http://www.naei.org.uk/issuedetail. php?issue_id=5

NAEI (2002b) *1999 Gothenberg protocol to abate acidification, eutrophication and ground-level ozone*, National Atmospheric Emissions Inventory website, Downloaded June 2003 from http://www.naei.org.uk/actiondetail.php?action_id=5

Nyberg, F., Gustavsson, P., Ja¨rup, L., Bellander, T., Berglind, N., Jakobsson, R., and Pershagen, G. (2000) Urban Air Pollution and Lung Cancer in Stockholm, *Epidemiology* 11, 487–495.

OHN (1999) White paper *'Saving lives: our healthier nation'*, The supporting website, including a summary of the report, is at http://www.ohn.gov.uk/ The full report is at http://www.archive.official-documents.co.uk/document/cm43/4386/4386.htm

ONS (2002) Distance travelled per person per year by bicycle: by age and gender, 1996-1998: *Social trends* 30, Downloaded September 26 2003 from http://www.statistics.gov.uk/STATBASE/xsdataset.asp?vlnk=1465

ONS (2003a) *Population estimates*, Office for national statistics website, Downloaded June 2003 from http://www.statistics.gov.uk/cci/nugget.asp?id=6

ONS (2003b) *Mid-2001 population estimates: estimated resident population by broad age-group and sex*, Office for national statistics website, Downloaded June 2003 from http://www.statistics.gov.uk/statbase/Product.asp?vlnk=601

ONS (2003c) *Cancer registrations in England, 2000*, Downloaded October 2003 from http://www.statistics.gov.uk/pdfdir/can0803.pdf

ONS (2003d) *Data for chart 7.13 death rates from selected cancers: by gender,* Downloaded October 2003 from http://www.statistics.gov.uk/StatBase/Expodata/Spreadsheets/D5226.xls

OTS (2000) *Oxford transport strategy, Assessment of impact, November 2000*, Downloaded July 2003 from   http://www.oxfordshire.gov.uk/ots_assessment-7.pdf

Parkhurst, G. (2001) *Monitoring of the Oxford transport strategy. Some questions for urban transport policy development,* ESRC TSU Publication 2001/1, January 2001.

Raaschou-Nielsen, O., Hertel, O., Thomsen, B. L., and Olsen, J. H. (2001) Air pollution from traffic at the residence of children with cancer, *Am. J. Epidemiol.* 153 (5), 433-443.

Reynolds, P., Behren, J., Gunier, R. B., Goldberg, D. E., Hertz, A., and Smith, D. (2002) Traffic patterns and childhood cancer incidence rates in California, United States, *Cancer Causes and Control* 13, 665–673.

Sasco, A.J., Olsson, A., and Chiron, C. (2003) *Traffic air pollution and cancer,* Presentation to the ASPIS conference on Mobility and Health, Kos, May 3-6 2003.

Sheild, B., and Dockrell, J. (2002) *The effects of noise on the attainments and cognitive performance of primary school children,* Department of Health, Downloaded September, 2003 from http://www.doh.gov.uk/noisepollution/cognitiveperf.htm.

Smith, A., Nutt, D., Wilson, S., Rich, N., Hayward, S., and Heatherley, S. (2002) *Noise and insomnia : a study of community noise exposure, sleep disturbance, noise sensitivity and subjective reports of health*, Department of Health, Downloaded September 2003 from http://www.doh.gov.uk/noisepollution/insomnia.pdf

Stansfeld, S., Haines, M., Brentnall, S., Head, J., and Roberts, R. (2003) *Final report, West London schools study: aircraft noise at school and children's cognitive performance and stress responses,* Department of Health, Downloaded September 2003 from http://www.doh.gov.uk/noisepollution/finrepwlondonschools.pdf

Stewart, B. W., and Kleihues, p. (eds) (2003) *World cancer report*, IARC Press, Lyon, p. 40.

Sustrans (2003) *Safe Routes to School*, Index to the website downloaded September 2003 from http://www.saferoutestoschools.org.uk/

Towner, E. (2002) *The prevention of childhood injury*, Background paper prepared for the UK Accidental Injury Task Force, September 2002, Downloaded July 2003 from http://www.doh.gov.uk/accidents/pdfs/childhoodinjury.pdf

UK NAQI Archive (2003) *Air pollution standards and banding*, UK National Air Quality Information Archive website, Downloaded July 2003 from http://www.airquality.co.uk/archive/standards.php#band

UK Sust Dev (2003) *Headline indicators of sustainable development, H11 road traffic*, Downloaded September 2003 from http://www.sustainable-development.gov.uk/indicators/headline/h11.htm.

UNECE (1999) Protocol to abate acidification, eutrophication and ground-level ozone, in *Convention on long-range transboundary air pollution,* United Nations Economic

Commission for Europe, Environment and Human Settlements Division, Downloaded July 2003 from: http://www.unece.org/env/lrtap/multi_h1.htm

van Kempen, E.E.M.M., Kruize, H., Boshuizen, H.C., Ameling, C.B., Staatsen, B.A.M., and de Hollander, A.E.M. (2002) The association between noise exposure and blood pressure and ischemic heart disease: a meta-analysis, *Environ. Health. Perspect.* 110(3), 307–317.

Vrang, M.L., Hertel, O., Palmgren, F., Wahlin, P., Raaschou-Nielsen, O., and Loft, S.H. (2002) Effects of traffic-generated ultrafine particles on health, *Ugeskr Laeger.* 164(34):3937-3941.

WHO (2003) *Noise and health,* World Health Organization Regional Office for Europe, Downloaded September 2003 from http://www.euro.who.int/eprise/main/WHO/Progs/ NOH/Home.

# EFFECTS OF MOBILITY ON HEALTH - AN OVERVIEW

P. NICOLOPOULOU-STAMATI[1], L. HENS[2], P. LAMMAR[2] AND
C.V. HOWARD[3]

[1]*National and Kapodistrian University of Athens
Medical School, Department of Pathology
75 Mikras Asias Street, 11527, Athens, GREECE*
[2]*Department of Human Ecology,
Faculty of Medicine and Pharmacy,
Vrije Universiteit Brussel, Laarbeeklaan 103,
B-1090 Brussels, BELGIUM*
[3]*University of Liverpool
Developmental Toxico-Pathology Research Group
Department of Human Anatomy & Cell Biology
L69 3GE Liverpool, UK*

## Summary

Contemporary mobility patterns necessitate a large use of resources and create considerable levels of pollution, both during the manufacture, utilisation and final disposal phases of motorised vehicle life cycles.

Widely studied air pollutants from exhausts, including particles, cause cardio-pulmonary problems and cancer. Noise is associated with annoyance, sleep disturbance and impairment of learning functions. Moreover, motorised vehicles contribute to a sedentary lifestyle. Sedentarism increases the risk of cardio-vascular problems, diabetes and obesity.

Most of the population is exposed to traffic-related noise and pollution, at levels that cause the above-mentioned effects. This is not only responsible for an important public health problem but also makes current mobility patterns the most important single source of environmental health problems in the European Union.

To ameliorate this situation, an integrated environmental health policy targeted at mobility is urgently required. This policy should include technical emission reduction measures in association with urban planning initiatives favouring the promotion of pedestrianisation and cycling. Reduction of car use and replacement with inexpensive or free public transport could be realised most effectively through pricing schemes and other fiscal measures. Awareness raising and education should

*P. Nicolopoulou-Stamati et al. (eds), Environmental Health Impacts of Transport and Mobility, 277-307.*

be targeted towards a more physically active lifestyle and the establishment of a societal acceptance of such an integrated policy.

In contrast to many environmental pollution policies, this approach to mobility-related health effects would bring some immediate benefits as well as longer term improvements in health and environment.


## 1.   Introduction

Modern society relies upon the mobility of its citizens for its normal functioning. However, it is clear from the information contained in this book that the ability to travel freely does not come about without consequences for health. These latter are closely related to the way the environment is used to allow current mobility patterns.

Global passenger car production not only reaches new record numbers year after year (Worldwatch Institute, 2001); during their production phase, cars, buses, planes and trains use high levels of resources, in terms of materials and energy. It has been estimated that approximately 10 per cent of the total energy consumption for a typical car will be used in its manufacture (ILEA, 1998). Moreover, airports, railway tracks and roads occupy increasing amounts of land and require large volumes of materials for their construction.

During the operation of these modes of transport, varying levels of energy use are required. The motorcar is an extreme case. A technological system that requires a ton of metal and plastic in order to move one person (weighing less than 100 kg) a couple of kilometers on a journey (to work or to take children to school) is grossly inefficient.

The waste production associated with the whole life-cycle of vehicle production and use is massive. The average car weighs about 1.14 tonnes. Most of it is steel, though increasing amounts of plastics are being used. However, during its production and utilisation, each car produces 25 tonnes of waste (UPI, 1999).

Mobility also results in health effects caused by the resource drivers required to produce and use energy and raw materials. Among these are oil extraction and its transport. Oil is, and will remain for the forseeable future, the primary energy resource for transport throughout the world. The world is now using over 60 million barrels of oil a day. Over the next two decades, most of the growth in the demand for oil will come from the transport sector, where high rates of growth are expected (IEA, 2001). A full materials analysis should also include the consequences of extracting and transporting crude oil. Accidents during the transportation of oil by sea occur frequently and have major ecological impacts. They threaten human health through the contamination of the food chain. On average, 25 per cent of the oil entering the sea through marine transportation activities arises from oil spills. The other 75 per cent is the result of discharges made by routine operations on ships (Göttinger, 2001). UPI (1999) calculated that the oil pollution of the oceans by accidental spillage plus routine washing out of tanks at sea amounts to 13 litres of crude oil for every car.

Land contamination, and the social problems related to it, are additional problems. These were a source of environmental and political crisis in the Ogoni lands of Southern Nigeria. In this case, the oil company was responsible for considerable land and water pollution. The Nigerian government executed the leaders of the environmental movement opposing such pollution.

All these activities lead to water, air and soil pollution; and have impacts on human health and wellbeing. These latter can be subdivided into four main groups:

1) Direct effects on health result from injuries related to road traffic accidents. Worldwide, in 1999, between 750,000 and 880,000 people died in road traffic accidents (TRL, 2000). More recent estimates give figures almost as high as 1.2 million people killed each year. The road traffic death toll represents only the 'tip of the iceberg' of the total waste of human and societal resources from road injuries. WHO estimates that, worldwide, between 20 million and 50 million people are injured or disabled each year in road traffic crashes (WHO, 2004). In the WHO European region approximately 120,000 die every year and more than 2,500,000 are injured every year as a result of road traffic collisions (Dora and Phillips, 2000).

2) Indirect effects result from exposure to air pollutants such as NOx, SOx, VOCs, ozone and (fine) particles. Road-traffic related air pollution has been calculated to be responsible for nearly 22,000 annual mortality cases (adults $\geq$ 30 years old) in Austria, France and Switzerland (Kunzli et al., 2000). In terms of acute effects, exposure to $SO_2$ and ozone causes harm to patients with respiratory disease, asthmatic subjects, and patients with chronic obstructive pulmonary disease (COPD) (Ayres, 1998). Epidemiological studies demonstrate a clear relationship between the levels of $PM_{10}$ and exacerbations of asthma and COPD, as well as deaths from cardiovascular causes, i.e. heart attacks and strokes. The basic mechanisms of these phenomena are only partially understood (Donaldson and MacNee, 1998). Effects associated with chronic long-term exposure to pollutants have also been demonstrated. These include cancer, prevalence of respiratory symptoms and decreased lung function, hospital admissions and premature mortality due to respiratory and cardiovascular diseases (Harrison, 2001; Nedellec et al., 2004). Traffic-related air pollution epidemiological data suggest an association between traffic exhaust and childhood cancer. Examples include an increased risk for leukemia and central nervous system tumours in children living near busy roads; increase in lymphomas associated with benzene and nitrogen dioxide exposure during pregnancy (Nedellec et al., 2004).

3) Noise produced by traffic is associated with morbidity (WHO, 1993). A rough estimate of the number of people in the pre-enlargement EU exposed to environmental noise ($L_{den}$) from road, rail and aircraft movements at a level > 55 dB(A) is 150 million or 40 per cent of the population. Adverse environmental

noise-induced health effects include annoyance, sleep disturbance, stress-related somatic effects such as increased blood pressure, effects on learning in children and, possibly, hearing damage.

4) Physical inactivity induced by the widespread availability of mechanised transport is a public health problem. It is known that sedentarism is associated with cardiovascular disease, diabetes, obesity, osteoporosis, gallstones, depression and different types of cancer. However, quantitative estimates of the extent of this problem are difficult to calculate. This is firstly because internationally agreed definitions of physical inactivity do not exist and secondly because epidemiological studies linking mobility-related sedentarism with health effects are scarce.

**Table 57**. Health damage caused by cars, Germany 1996, annual totals (UPI, 1999).

| Health effect | Number | Unit |
|---|---|---|
| Deaths from particulate pollution | 25,500 | deaths/year |
| Deaths from lung cancer | 8700 | deaths/year |
| Deaths from heart attacks | 2000 | deaths/year |
| Deaths from summer smog | 1900 | deaths/year |
| Deaths from road traffic accidents (RTAs) | 8758 | deaths/year |
| **TOTAL DEATHS** | 46,858 | deaths/year |
| Serious injuries (RTAs) | 116,456 | injured/year |
| Light injuries (RTAs) | 376,702 | injured/year |
| Chronic bronchitis (adults) | 218,000 | illnesses/year |
| Invalidity due to chronic bronchitis | 110 | invalids/year |
| Coughs | 92,400,000 | days/year |
| Bronchitis (children) | 313,000 | illnesses/year |
| Bronchitis | 1,440,000 | illnesses/year |
| Hospitalization (breathing problems) | 600 | hospitalizations/year |
| Hospitalization (breathing problems) | 9200 | days of care/year |
| Hospitalization (cardiovascular disease) | 600 | hospitalizations/year |
| Hospitalization (cardiovascular disease) | 8200 | days of care/year |
| Unavailable for work (not cancer) | 24,600,000 | days/year |
| Asthma attacks (days with attacks) | 14,000,000 | days/year |
| Asthma attacks (days with broncho-dilator) | 15,000,000 | days/year |

From these data it is obvious that mobility has health impacts. It is only recently that the full extent of transport's negative impact on health has become better substantiated. In an audit of the impact of cars on German society, UPI (1999) concluded that cars were responsible for 47,000 deaths each year and a range of the previously described health impacts. The quantitative data are shown in Table 57, which only summarises data related to cars and not to lorries or aircraft. This puts the European transport sector into a very serious public health context. Mobility should no longer be seen only in a context of transport, roads and highways, but increasingly as a public health problem. This chapter summarises the current state of knowledge of the health impacts. This is limited to consideration of the health effects of traffic-related air pollution, noise and physical inactivity. The findings are set in the context of the degree of scientific certainty of the effect being causal, associated with the findings. This allows us to relate the findings to mandatory policy initiatives to mitigate these effects. This chapter, although acknowledging the importance of traffic-induced injury, does not address this explicitly. It concentrates on impacts that are caused by environmental exposures.

## 2.    Health effects from traffic-related air pollution

### 2.1.    Non-particulate air pollutants

The health effects of air pollution have been studied intensely in recent years. Table 58 summarises the findings, which are discussed briefly below.

Exposure to air pollutants such as sulphur dioxide ($SO_2$), nitrogen oxides ($NO_x$) and ozone ($O_3$) have, just like airborne particulate matter, been associated with increases in mortality and hospital admissions due to respiratory and cardiovascular disease.

The most consistent results have been found for ozone. Ozone levels are associated with a significant increase in all causes, cardiovascular, and respiratory mortality. The effects are greater in the warm season in temperate climates and are independent of the effects of other pollutants (Touloumi et al., 1997; Hester and Harrison, 1998; Le Tertre et al., 2002). There is very good, consistent experimental evidence that ozone has an effect on health, with a consistent dose-response effect on a number of lung function parameters, at concentrations close to those seen in ambient air. A consistent and strong finding is a significant increase of daily admissions for respiratory diseases associated with elevated levels of ozone. This finding is stronger in the elderly, has a rather immediate effect, and is quite homogeneous over the different cities under study (Schwartz et al., 1996; Spix et al., 1998). The association between daily admissions for COPD and ozone is also generally quite high (Anderson et al., 1997). The studies are less consistent concerning the association between asthma admissions and ozone.

Sulphur dioxide is a potent bronchoconstrictor at high levels, and patients with asthma are much more sensitive to it than normal individuals. The association between daily hospital admissions for asthma and sulphur dioxide is highest among children (Sunyer et al., 1997). Because levels of sulphur dioxide and particulate matter co-vary closely, it has proved difficult to demonstrate effects of sulphur dioxide that are independent of the effects of particulate matter in epidemiological studies. It is likely that sulphur dioxide contributes to respiratory symptoms, reduced lung function, increases in hospital admissions and mortality, although the effects have been more consistently seen in Europe than in the USA (Harrison, 2001). In a large European study of short-term changes in mortality and hospital admissions (APHEA project) in relation to ambient air pollution, sulphur dioxide was significantly associated with both health outcomes. The effect was greater in Western European cities than those in Eastern and Central Europe (Katsouyanni et al., 1997; Spix et al., 1998).

**Table 58.** Studies linking non-particulate pollutants with health effects.

| Health effect | Type of estimate (95% CI) | SO₂ | Scientific certainty | NO₂ | Scientific certainty | Ozone (O₃) 1 hr max | Scientific certainty | Ozone (O₃) 8 hr max | Scientific certainty | Study location/description | Reference |
|---|---|---|---|---|---|---|---|---|---|---|---|
| Daily all cause mortality, per 50 µg/m³ increase pollutant | RR | 1.020 (1.015, 1.024) | M | | | | | | | Twelve European cities; time-series study | Katsouyanni et al. (1997) |
| Daily all cause mortality, per 10 µg/m³ increase pollutant | RR | | | 1.015 (1.008, 1.022) | M | | | | | Shanghai 2000-2001; time-series study | Kan & Chen (2003) |
| Daily all cause mortality, per 50 µg/m³ increase in O₃ | % increase | | | | | 2.9% (1.0, 4.9%) | H | | | Fifteen European cities; time-series study | Touloumi et al. (1997) |
| All non-accident mortalities, per 10th-90th percentile change in pollutant concentration (cool season) | RR | | | | | | | 1.04 (1.01, 1.06) | H | Hong Kong; time-series study | Wong et al. (2001) |
| Cardiovascular mortality, per 10th-90th percentile change in pollutant concentration (cool season) | RR | 1.07 (1.02, 1.11) | M | 1.10 (1.05, 1.16) | M | | | 1.05 (1.00, 1.11) | M-H | Hong Kong; time-series study | Wong et al. (2001) |

Table 58, continued.

| Health effect | Type of estimate (95% CI) | SO₂ | Scientific certainty | NO₂ | Scientific certainty | Ozone (O₃) 1 hr max | Scientific certainty | Ozone (O₃) 8 hr max | Scientific certainty | Study location/description | Reference |
|---|---|---|---|---|---|---|---|---|---|---|---|
| Respiratory mortality, per 10th-90th percentile change in pollutant concentration (cool season) | RR | 1.04 (1.00, 1.09) | M | 1.09 (1.02, 1.16) | M | | | 1.08 (1.02, 1.15) | H | Hong Kong; time-series study | Wong et al. (2001) |
| Daily respiratory admissions 15-64 Years, per 50 µg/m³ increase pollutant | RR | 1.009 (0.992, 1.025) | M | 1.010 (0.985, 1.036) | L | 1.019 (1.005, 1.033) | H | 1.031 (1.013, 1.049) | H | Five West European cities; time-series study | Spix et al. (1998) |
| Daily respiratory admissions 65+ Years, per 50 µg/m³ increase pollutant | RR | 1.020 (1.005, 1.046) | M | 1.019 (0.982, 1.060) | L | 1.031 (1.015, 1.047) | H | 1.038 (1.018, 1.058) | H | Five West European cities; time-series study | Spix et al. (1998) |
| Hospital admissions - COPD, per 50 µg/m³ increase pollutant | RR | 1.02 (0.98, 1.06) | M | 1.02 (1.00, 1.05) | L | | | 1.04 (1.02, 1.07) | M-H | Six European cities; time-series study | Anderson et al. (1997) |
| Hospital admissions - asthma < 15 Years, per 50 µg/m³ increase pollutant | RR | 1.075 (1.026, 1.126) | H | 1.026 (1.006, 1.049) | M | 1.006 (0.976, 1.037) | L-M | 0.989 (0.941, 1.038) | L-M | Four European cities (Barcelona, Helsinki, Paris and London); time-series study | Sunyer et al. (1997) |
| Hospital admissions - asthma > 15-64 Years, per 50 µg/m³ increase pollutant | RR | 0.997 (0.961, 1.034) | L-M | 1.029 (1.003, 1.055) | M | 1.015 (0.955, 1.078) | L-M | 1.035 (0.937, 1.144) | L-M | Four European cities (Barcelona, Helsinki, Paris and London); time-series study | Sunyer et al. (1997) |

Table 58, continued.

| Health effect | Type of estimate (95% CI) | SO₂ | Scientific certainty | NO₂ | Scientific certainty | Ozone (O₃) 1 hr max | Scientific certainty | Ozone (O₃) 8 hr max | Scientific certainty | Study location/description | Reference |
|---|---|---|---|---|---|---|---|---|---|---|---|
| Cough without infection / dry cough at night, in 1st year of life | OR | | | 1.40 (1.12, 1.75) / 1.36 (1.07, 1.74) | M | | | | | German part of TRAPCA on childhood asthma; cohort study | Gehring *et al.* (2002) |
| Reduced lung function (Forced Expiratory Volume in 1 sec, FEV₁) in children aged 10-18 | $ | | | -101.4 (-164.5, -38.4) | M | -44.5 (-138.9, 50.0) | L | | | Los Angeles, cohort of children, studied 1993-2001; cohort study | Gauderman *et al.* (2004) |
| Lower respiratory symptoms (assoc w 5-day mean) | RR | 2.25 (1.42, 3.55) | M | 1.79 (1.39, 2.30) | M | | | | | Children in the Netherlands; panel study | Boezen *et al.*, (1999) |
| Elevated total IgE / SPT reactivity to any allergen / SPT reactivity to indoor allergens | OR | | | 3.12 (1.81, 5.38) / 1.70 (1.03, 2.81) / 1.94 (1.13, 3.33) | L-M | | | | | Dutch Schoolchildren; community-based study | Janssen *et al.* (2003) |

(L: low, M: moderate, H: high); RR = relative risk; OR = odds ratio; CI = confidence interval; SPT = Skin Prick Test, IgE = Immunoglobulin E,
$ = difference in average growth in lung function over the eight-year study period from the least to the most polluted community.

The evidence from challenge and from epidemiological studies generally does not show significant health effects for nitrogen dioxide, an oxidant pollutant. It is likely, however, that longer-term exposures may have an impact in terms of chronic effects (Hester and Harrison, 1998). The evidence is less consistent than for sulphur dioxide, ozone and particulate matter. However, European studies in the framework of the APHEA project indicate a significant association between nitrogen dioxide and hospital emergency visits and hospital admissions for asthma (Dab *et al.*, 1996; Sunyer *et al.*, 1997; Tenias *et al.*, 1998), as well as significant associations with mortality (Touloumi *et al.*, 1997; Le Tertre *et al.*, 2002).

Another gaseous non-particulate pollutant is carbon monoxide, which binds very avidly to haemoglobin, thereby reducing the oxygen-carrying capacity of the blood. There has been a paucity of recent research into the potential health effects of low-level carbon monoxide exposure. Some epidemiological evidence suggests that in certain susceptible individuals with heart disease, ambient levels can be associated with health effects (Hester and Harrison, 1998; Harrison, 2001).

Other non-particulate air pollutants originating from traffic, such as VOCs and PAH, are known carcinogens. The best studied are benzene and benzo(a)pyrene, which are respectively mainly linked with causing leukaemia and lung cancer (Hester and Harrison, 1998; Harrison, 2001).

## 2.2    *Particulate air pollution*

There is strong and widely accepted scientific evidence that exposure to particulate aerosols causes increased mortality rates leading to premature deaths, as shown in Table 59. This is largely attributable to increases in cardiovascular and respiratory illnesses (Laden *et al.,* 2001). In addition, increased lung cancer deaths are associated with $PM_{2.5}$ and sulphates (Laden *et al.*, 2001; Pope *et al.*, 2002).

Over the past two decades, an increasing number of epidemiological studies showed a significant association between the mass concentration of particulate matter in the ambient air and adverse respiratory and cardiovascular health effects. The main effects include hospital admissions, emergency room and clinician visits for respiratory and cardiac diseases, increased use of medication and time lost from work and school.

Cardiovascular diseases have been shown to increase with the amount of $PM_{10}$ attributed to highway vehicles (Lippmann *et al.*, 2003). $PM_{2.5}$ exposure is associated with an increase in exhaled nitric oxide (NO), a marker of pulmonary inflammation. This suggests that $PM_{2.5}$ at ambient concentrations can act as an inflammatory agent (Koenig *et al.*, 2002). Moreover, $PM_{10}$ exposure is associated with the occurrence of asthma symptoms. Recently, diabetics have been identified as an important susceptible population. PM-associated hospital admissions for heart disease in diabetics is double that in the general population (Zanobetti and Schwartz, 2002). Individuals with other pre-existing diseases are also at higher risk. Respiratory diseases, for example, modify the risk of cardiovascular admissions. Heart failure modifies the risk of hospital admissions for chronic obstructive pulmonary disease (Zanobetti *et al.*, 2000).

**Table 59.** Some studies linking particulate aerosols with health effects. A non-exhaustive list.

| Health effect, and pollution increment (interpercentile range) | Type of estimate (95% CI) | UFPs | Scientific certainty | PM$_{2.5}$ | Scientific certainty | PM$_{2.5-10}$ | Scientific certainty | PM$_{10}$ | Scientific certainty | Other | Scientific certainty | Study location/description | Reference |
|---|---|---|---|---|---|---|---|---|---|---|---|---|---|
| Daily all cause mortality (5-95), per 10 μg/m³ increase pollutant | RR | | | 1.058 (1.042, 1.074) | H | 1.02 (0.992, 1.049) | H | | | | | Six eastern US cities; time-series study | Schwarz et al. (1996) |
| Daily mortality, per 100μg/m³ increase in black smoke | RR | | | | | | | | | 1.19 (1.02, 1.38) | H | Amsterdam; time-series study | Verhoeff et al. (1996) |
| Respiratory mortality, in cool season, per 10th-90th percentile change in pollutant concentration | RR | | | | | | | 1.06 (1.00, 1.13) | H | | | Hong Kong; time-series study | Wong et al. (2001) |
| Respiratory mortality, assoc with PM13, per 50 μg/m³ increase | RR | | | | | | | | | 1.04 (1.00, 1.09) | H | Lyons, France, 1985-1990; time-series study | Zmirou et al. (1996) |
| Lung cancer mortality per 10 μg/m³ increase PM$_{2.5}$ | RR | | | 1.14 (1.04, 1.23) | H | | | | | | | US-wide Cancer prevention II Study; cohort study | Pope et al. (2002) |
| Cardiopulmonary mortality assoc with living near a major road | RR | | | | | | | | | 1.95 (1.09, 3.52) | H | Netherlands; cohort study | Hoek et al. (2002) |
| Daily respiratory admissions (25-75) | RR | | | 1.037 (1.015, 1.059) | M | 1.023 (1.010, 1.036) | M | | | | | Toronto 1992-1994 (Summers); time-series study | Burnett et al. (1997) |
| Daily cardiac admissions (25-75) | RR | | | 1.031 (0.997, 1.066) | M | 1.036 (1.015, 1.057) | M | | | | | Toronto 1992-1994 (Summers); time-series study | Burnett et al. (1997) |

Table 59, continued.

| Health effect, and pollution increment (interpercentile range) | Type of estimate (95% CI) | UFPs | Scientific certainty | PM$_{2.5}$ | Scientific certainty | PM$_{2.5-10}$ | Scientific certainty | PM$_{10}$ | Scientific certainty | Other | Scientific certainty | Study location/description | Reference |
|---|---|---|---|---|---|---|---|---|---|---|---|---|---|
| Chronic bronchitis, assoc. with 10 μg/m³ TSP | OR | | | | | | | | | 1.07 (1.02, 1.12) | M-H | Fifty-three urban areas of USA; prevalence study (cross-sectional field health survey) | Schwartz (1993) |
| Lower respiratory symptoms (25-75) | OR | | | 1.33 (1.11, 1.58) | H | 1.14 (0.98, 1.66) | M | | | | | Harvard Six Cities; cohort study | Schwartz & Neas (2000) |
| Attack of shortness of breath & wheezing, per increase of one interquartile range | OR £ | 1.26 (1.08, 1.48) | M | | | | | | | | | Asthmatics in Erfurt, Winter 1996/7; panel study | Von Klot et al. (2002) |
| Waking with breathing problems, per increase of one interquartile range | OR £ | 1.26 (1.13, 1.41) | M | | | | | | | | | Asthmatics in Erfurt, Winter 1996/7; panel study | Von Klot et al. (2002) |
| Shortness of breath, per increase of one interquartile range | OR £ | 1.24 (1.11, 1.40) | M | | | | | | | | | Asthmatics in Erfurt, Winter 1996/7; panel study | Von Klot et al. (2002) |
| Cough (lag 2) (25-75) | OR | | | 1.13 (1.01, 1.26) | M | 1.15 (1.04, 1.27) | M | | | | | Kuopio; panel study | Tiittanen et al. (1999) |
| Cough, per increase of one interquartile range | OR £ | 1.20 (1.06, 1.35) | M | | | | | | | | | Asthmatics in Erfurt, Winter 1996/7; panel study | Von Klot et al. (2002) |
| Cough without infection / dry cough at night, assoc. with PM$_{2.5}$ & PM$_{2.5}$ absorbance in 1st yr of life | OR | | | 1.34 (1.11, 1.61) / 1.31 (1.07, 1.60) | M | | | | | 1.32 (1.10, 1.59) / 1.27 (1.04, 1.55) | M | German part of TRAPCA on childhood asthma; cohort study | Gehring et al. (2002) |

Table 59, continued.

| Health effect, and pollution increment (interpercentile range) | Type of estimate (95% CI) | UFPs | Scientific certainty | $PM_{2.5}$ | Scientific certainty | $PM_{2.5-10}$ | Scientific certainty | $PM_{10}$ | Scientific certainty | Other | Scientific certainty | Study location/description | Reference |
|---|---|---|---|---|---|---|---|---|---|---|---|---|---|
| Chronic phlegm, assoc. with outdoor air pollution level | OR | | | | | | | | | 4.2 (1.1, 16.9) | M | Krakow schoolchildren; prevalence study (cross-sectional field health survey) | Jedrychowski & Flak (1998) |
| Reduced lung function ($FEV_1$) in children aged 10-18 | $ | | | -79.7 (-153.0, -6.4) | H | | | -82.1 (-76.9, 12.8) | M | -87.9 (-146.4, -29.4 elemental carbon) | H | Los Angeles, cohort of children, studied 1993-2001; cohort study | Gauderman et al. (2004) |
| Hay fever ever / SPT reactivity to outdoor allergens | OR | | | 2.28 (1.13, 4.57) / 1.90 (1.06, 3.40) | M | | | | | | | Dutch Schoolchildren; community-based study | Janssen et al. (2003) |
| Elevated total IgE associated with soot | OR | | | | | | | | | 2.67 (1.16, 6.12) | M | Dutch Schoolchildren; community-based study | Janssen et al. (2003) |
| Very low birthweight baby, assoc with TSP & $SO_2$ | OR | | | | | | | | | 2.88 (1.16, 7.13) | L | Georgia district 9; population-based case-control study | Rogers et al. (2000) |

UFPs = ultrafine particles; IgE = Immunoglobulin E, $FEV_1$ = Forced Expiratory Volume in one second, RR = relative risk; OR = odds ratio; CI = confidence interval; £ = assoc with 14-day mean; $ = difference in average growth in lung function over the eight-year study period from the least to the most polluted community.

## 3.   Noise

### 3.1.   Overview

Table 60 summarises the state of knowledge from field studies of research into health effects of environmental noise. There is sufficient scientific evidence that noise exposure can induce hearing damage (although unlikely to occur at typical levels of community noise exposure), annoyance and sleep disturbance (especially in adults). It also affects children's learning (cognition) and motivation. The evidence of cardiovascular risk, as a health effect of noise, is more limited. However, hypertension and ischaemic heart disease seem to be more and more associated with noise exposure. With respect to the causal role of environmental noise in mental health impacts, no definite conclusion can be drawn. A significant association was observed between self-reported noise exposure and depression as well as cognitive failures. Nevertheless, a number of studies applying psychometric questionnaires to assess psychological morbidity yielded inconsistent results. There is no convincing evidence of a direct effect of noise exposure on congenital abnormalities, birth weight or disorders related to the immune system. A few studies show an increased risk of low birth-weight in children of aircraft-noise exposed mothers, but the influence of important confounders such as socioeconomic status and smoking was not taken into account (Staatsen et al., 2004).

**Table 60**. Long-term effects of exposure to environmental noise, observed thresholds, relative risk (RR) and degree of scientific certainty (DOC) of the effect being causal (L: low, M: moderate, H: high; *: significant).

| Health effect | Observation threshold | | $RR_{5\ dB(A)}$ | DOC | References |
|---|---|---|---|---|---|
| | Metric | Value in dB(A) | | | |
| Hypertension | $L_{den}$ | 70 | | M-H | HCN (1994); Stansfeld et al. (2002) |
| | $L_{Aeq,\ 6\text{-}22\ hr}$ (road traffic) | < 55-80 | 0.95 (0.84, 1.08) | | Van Kempen et al. (2002) |
| | $L_{Aeq,\ 7\text{-}19\ hr}$ (air traffic) | 55-72 | 1.26 (1.14, 1.39)* | | Van Kempen et al. (2002) |
| Ischaemic heart disease | $L_{den}$ | 70 | 1.1-1.5 | M-H | HCN (1994); Babisch (2000) |
| | $L_{Aeq,\ 6\text{-}22\ hr}$ (road traffic) | 51-70 | 1.09 (1.05, 1.13)* | | Van Kempen et al. (2002) |
| Myocardial infarction | $L_{Aeq,\ 6\text{-}22\ hr}$ (road traffic) | 51-80 | 1.03 (0.99, 1.09) | M | Van Kempen et al. (2002) |
| | $L_{Aeq}$ | > 65 vs ≤ 60 (men) | OR: 1.18 (0.93, 1.49) | M | Babisch (2004) |
| | $L_{Aeq}$ | > 65 vs ≤ 60 (men living ≥10 years at present address) | OR: 1.33 (1.00, 1.76)*- 1.45 (1.03, 2.05)* | M | Babisch (2004) |

Table 60, continued.

| Health effect | Observation threshold | | $RR_{5\,dB(A)}$ | DOC | References |
|---|---|---|---|---|---|
| | Metric | Value in dB(A) | | | |
| Angina pectoris | $L_{Aeq,\ 6\text{-}22\ hr}$ *(road traffic)* | 51-70 | 0.99 (0.84, 1.16) | M | Van Kempen *et al.* (2002) |
| | $L_{Aeq,\ 7\text{-}19\ hr}$ *(air traffic)* | 55-72 | 1.03 (0.90, 1.18) | M | Van Kempen *et al.* (2002) |
| Cardiovascular risk | $L_{day}$ | 65 | | M | Ising and Kruppa (2004) |
| Annoyance | $L_{den}$ | for %HA[#] 42 | | H | Miedema and Oudshoorn (2001); Ouis (2002); Stansfeld and Matheson (2003) |
| Hearing impairment | $L_{night}$ | 50 | | H | Hoeger *et al.* (2002) |
| | $L_{Aeq,\ 24h}$ | 70 | | H | Passchier-Vermeer and Passchier (2000) |
| Performance of children | $L_{Aeq,\ school}$ | 70 | | H | Passchier-Vermeer and Passchier (2000) |
| Sleep disturbance - sleep stages | SEL | 35 | | H | Passchier-Vermeer and Passchier (2000) |
| - sleep pattern | $L_{night}$ | < 60 | | H | Passchier-Vermeer and Passchier (2000) |
| Behavioural awakening | SEL | 55 | | H | Passchier-Vermeer and Passchier (2000) |
| Self-reported sleep disturbance | $L_{night}$ | for %HSD§ < 45 | | H | Miedema and Vos (2003) |
| Increased motility, Motoric unrest | $L_{max\_i\ (night)}$ | 32 | | H | Passchier-Vermeer (2003) |
| Hormone levels (cortisol) | $L_{max\_i\ (night)}$ | 33-52 | | L | Ising and Ising (2002) Babisch *et al.* (2001) Maschke and Hecht (2004) |
| Increased allergies | $L_{night}$ | | | L | Ising *et al.* (2003) |
| Decreased birthweight | $L_{dn}$ | 70-75 | OR: 1.324 (1.183,1.482) | L | Matsui *et al.* (2003) Kawada (2004) |
| Mood next day | $L_{Aeq,\ night}$ | >60 | | H | HCN (1994); Passchier-Vermeer and Passchier (2000) |
| Psychological symptoms | $L_{dn}$ | 70 (mental instability, depression) | | L | Hiramatsu *et al.* (2000) Stansfeld and Matheson (2003) |

# HA – Highly Annoyed, § HSD – Highly Sleep Disturbed, $L_{den}$ - equivalent sound level over 24-hour period (day, evening, night), $L_{dn}$ - equivalent sound level over 24-hour period (day, night), $L_{night}$ - equivalent sound level over night time, $L_{day}$ - equivalent sound level over daytime, $L_{Aeq,T}$ - equivalent sound level over period of time T, SEL – sound exposure level, $L_{max\_i}$ - maximal indoor sound level during an aircraft noise event, OR – odds ratio.

## 3.2.   *Annoyance*

Annoyance, as a direct outcome of environmental noise exposure, has the lowest threshold and is the most thoroughly documented of all outcomes. Annoyance is defined as any feeling of resentment, displeasure, discomfort and irritation, occurring when noise intrudes into someone's thoughts and moods when contemplating or interferes with an activity.

Quantitative exposure-response relationships are well established (Miedema and Oudshoorn, 2001). Aircraft noise has a higher impact on annoyance than road traffic. Railway traffic has the least effect on annoyance. The effect of road traffic noise does not differ between day and night. In contrast, annoyance reactions are higher at night for railway and air traffic noise. For aircraft noise, nighttime annoyance rises faster than annoyance during the day, especially when above an equivalent sound level of 50 dB(A).

Severe annoyance is estimated to be responsible for 400 to 2700 Disability Adjusted Life Years (DALYs) per million inhabitants (Staatsen *et al.,* 2004).

## 3.3.   *Sleep disturbance*

The auditory system continuously analyses auditory stimuli, which are in turn interpreted by a variety of brain structures. Noise is an aspecific stressor that arouses the autonomic nervous and endocrine systems, in particular the pituitary and adrenal glands. Therefore, it is not surprising that exposure to traffic noise causes sleep disturbances and changes in hormonal balance.

Sleep disturbance may manifest itself in various ways. Noise may affect sleeping behaviour (e.g. increasing the time one is awake during the night), sleep pattern, physiological responses or it may cause chronic changes. The following effects can be noted:
- Primary effects like difficulties in falling asleep, awakenings, sleep stage changes and instantaneous arousal effects during sleep (temporary increase in blood pressure, heart rate, vasoconstriction, release of stress hormones in the blood, increased motility);
- Secondary or 'after effects' measured the next day: decrease of perceived sleep quality, increased fatigue and decrease in mood and performance;
- Long-term effects on well being: increased medication use or chronic annoyance (Staatsen *et al.,* 2004).

Sleep disturbance can be described with physiological, hormonal and motility measures and on the basis of self reported observations and evaluations. Passchier-Vermeer and Passchier reviewed the literature on sleep disturbance as measured by motility, motoric unrest, self-assessed awakenings and self-reported sleep disturbances. They concluded that self-reported sleep disturbance is a dose-related effect that strongly depends upon the type of noise and frequency of awakening.

Aircraft noise in the nighttime was found to be the most damaging to health (Passchier-Vermeer and Passchier, 2000; Passchier-Vermeer *et al.*, 2003).

Behavioural awakening is usually assessed by pressing a button on a counting device on each occasion of awakening during sleeping hours. A review of the literature reveals that aircraft noise again emerges as having the highest impact.

Motility is an indication of the number of awakenings during the night. Motility studies have demonstrated that the human body reacts to nighttime noise at very low intensities, even when the noise exposure cannot be recalled. Aircraft noise induces increases in motility start from maximum equivalent noise levels of 32 dB(A) (Passchier-Vermeer *et al.*, 2003).

Although children appear to be less disturbed during their sleep than adults (with respect to awakenings and sleep quality), there is evidence of 'hidden effects' occurring during sleep that, in the long term, might add to the risk of cardiovascular disease (Staatsen *et al.*, 2004).

Noise increases the changes between sleep stages and the number of awakenings during the night, starting from SEL levels of about 35 and 60 dB(A), respectively. Reported sleep quality is likely to be affected at nighttime noise levels above 40 dB(A). In most studies, an effect of nighttime noise on performance and mood the next day is only seen at levels above 60 dB(A). Nighttime noise exposure may increase heart rate during the night; habituation to this effect does not seem to occur. The observation threshold is a SEL value of 40 dB(A). There are indications that noise-induced sleep stage changes are associated with elevated (stress) hormone levels (Staatsen *et al.*, 2004).

Severe sleep disturbance is estimated to be responsible for 150 to 1300 DALYs per million inhabitants (Staatsen *et al.*, 2004).

### 3.4.  *Cardiovascular diseases*

Noise as a stressor is also known to affect biological risk factors such as blood pressure and cardiac output which in turn might contribute to an increase in cardiovascular diseases. The epidemiological evidence is, however, still limited. In laboratory conditions, exposure to traffic noise is known to cause changes in heart rate and EEG patterns. Studies on environmental noise exposure suggest an association with hypertension (Stansfeld and Matheson, 2003). In a recent meta-analysis, a significant association was observed between hypertension and air traffic noise exposure, but not with road traffic noise exposure (Van Kempen *et al.*, 2002).

In cross sectional studies, road traffic noise exposure has been related to an increased risk of myocardial infarction and total ischaemic heart disease. Prospective studies regarding ischaemic heart disease suggest an increase in risk of outdoor noise levels above 65 to 70 dB(A) during the daytime (Babisch, 2000; Babisch, 2004). Air traffic noise exposure is associated with the consultation frequency of medical doctors, the use of cardiovascular medicines and sleep medication and angina pectoris (van Kempen *et al.*, 2002). A consistent trend towards an increased cardiovascular risk is observed at equivalent exposure levels of 65 dB(A) (Ising and Kruppa, 2004).

Of nine studies investigating the effects of air traffic, road traffic and rail traffic noise on blood pressure in children aged 3-16 years, six studies found blood pressure elevations associated with noise exposure. In three studies a statistically significant noise-related increase in blood pressure has been observed (Staatsen *et al.*, 2004).

Effects of road, rail and air traffic noise exposure on stress responses, e.g. cortisol, adrenaline and noradrenaline, have been studied. The results of these studies were inconclusive: if any associations were observed, the effects were only small (Staatsen *et al.*, 2004). Stress hormone levels (epinephrine) were higher in children exposed to aircraft noise at the old Munich airport (Hygge *et al.*, 1996). After the airport was moved, the levels of epinephrine rose among children living under the flight paths of the new airport. Children exposed to road and rail noise levels of more than 60 dB(A) had raised urinary cortisol levels but no difference in urinary (nor)adrenaline, compared to lower-exposed children ($L_{dn}$ 50 dB(A)) (Evans *et al.*, 2001). Ising observed raised cortisol levels in children exposed to road traffic noise (indoor levels Lmax 33-52 dB(A)) during the first, but not the second, half of the night (Ising and Ising, 2002).

Mortality due to hypertension attributed to noise is estimated to be responsible for a maximum of 700 DALYs per million inhabitants (Staatsen *et al.*, 2004).

### 3.5.   *Psychological impacts*

Apart from annoyance, noise exposure is associated with impaired learning ability (cognition) and success at school (motivation)(Staatsen *et al.*, 2004).

The following results have been found in children exposed to high levels of noise (aircraft, train and road), as compared to children in quieter schools:
-   deficits in sustained attention and visual attention;
-   difficulties in concentration;
-   poorer auditory discrimination and speech perception;
-   memory impairment for tasks that require high processing demands;
-   poorer reading ability and school performance on national standardised tests.

For reading ability, consistent results are observed, indicating a negative association between chronic (long-term) noise exposure and reading acquisition. Studies looking at the association between noise exposure and attention deficits vary in results (Staatsen *et al.*, 2004).

The multi-centre RANCH study found that aircraft noise exposure was associated with impaired reading comprehension. A 5 dB(A) increase in noise was associated with a one to two month impairment in reading age. The findings with regard to road traffic noise are inconsistent. While some studies indicate that the effects on reading may be reversible, if the noise ceases, the long-term developmental consequences, of exposure that persists throughout the child's education, remain yet to be determined. However, intervention measures (outdoor and indoor), such as reducing the noise levels in classrooms, have been shown to result in improving long-term memory and reading ability of children (Clark, 2004; van Kamp, 2004).

Children highly exposed to chronic environmental noise seem to be less motivated when placed in situations where task performance is dependent on persistence. Associations were found between noise exposure and reduced persistence on challenging puzzles. Teachers in high noise areas report greater difficulties in motivating children, as compared to their colleagues in low noise areas (Stansfeld *et al.*, 2000).

In a London study of psychological distress in children exposed to chronic aircraft noise, no differences were found attributable to noise exposure, adjusting for socioeconomic status (Haines *et al.*, 2001a). However, in a further larger study, higher psychological distress scores were found in children exposed to chronic noise (Haines *et al.*, 2001b). In adults, Japanese results suggest that high levels of military aircraft noise may have effects on mental health; in a cross-sectional study of 5,963 inhabitants around two air bases in Okinawa, those exposed to noise levels of Ldn 70 or above had higher rates of 'mental instability' and depression (Hiramatsu *et al.*, 2000). Those who were more annoyed showed higher risk of mental or somatic symptoms.

## 4. Sedentarism

Sedentarism or physical inactivity is a major public health problem. It is, however, difficult to link this directly to everyday mobility behaviour, because reported physical activity is only rarely evaluated in health effects studies by asking for everyday mobility behaviour. Except for walking only stair-climbing have been asked in some studies. Usually more intensive leisure time sports are required or occupational physical activities. From these findings and from the studies that investigate the consequences of cardio-respiratory fitness the health effects shown in Table 61 are supported. But because different studies use different measures of physical activity it is very difficult to estimate the possible quantitative effect of walking and cycling from these other activity measures. For some end points the effect might be rather indirect only or could even be confounded by an inverse causality: better health leads to a higher activity level. Usually no data on thresholds are available in this field. Generally any physical activity adds on already existing activity levels (which presumably for most outcomes are above the theoretical threshold level anyway). For cardio-respiratory fitness and some other outcomes the dose-response-curve is steeper at low activity levels. Although the quantity of the effect is not certain, there is good external evidence for most of the effects in Table 61 concerning causality. That's why all heatlth effects get a high scientific certainty, except colon cancer.

Cardio-respiratory fitness is more strongy correlated to health benefits than reported physical activity. Physical activity in the experimental design clearly increases cardio-respiratory fitness. This has also been shown for non-motorised mobility patterns (Hendriksen et al., 2000): cycling to work (mean single trip distance = 8.5 km) during six months more than three times a week led to a significant increase of the maximal performance on the cycle ergometer by 13 per cent (W per kg body weight) both in male and female cyclists while the control group remained constant.

Literature shows that regular physical activity increases life expectancy, stress tolerance and independence in old age. It decreases the risk of developing cardiovascular disease, Type II diabetes, obesity, colon and breast cancer, osteoporosis, gallstones and depression (see Table 61). Other effects like control of body weight, reductions in symptoms of anxiety and stress, and risk reduction for prostate cancer are mentioned as well. It has been shown that physical activity equivalent to 30 minutes of walking on all or most days of the week provides protective and preventive benefits for this range of health conditions (ISDE, 2002; Mason, 2000; Bassuk and Manson, 2003).

Reduced physical activity has been shown to impact mood and psychomotor development of children (Huttenmoser and Degan-Zimmermann, 1995). Sedentary life is also a risk factor for neuro-degenerative diseases (Trejo et al., 2002).

In two prospective cohort studies, obesity significantly increased the risk of pancreatic cancer. Physical activity appears to decrease the risk of pancreatic cancer, especially among those who are overweight (Michaud et al., 2001).

Being overweight is an important element in the top ten of health risks worldwide. In industrialised countries, it ranks within the top five risk factors. Obesity begins early in life; obesity in childhood is a strong predictor of becoming overweight in adulthood. Longitudinal studies from the USA, the Netherlands and Finland reported that the decline in physical activity was most marked between the ages of 12/13 and 15/16. It can clearly be concluded that overall physical activity peaks at the age of 12/13 years, after which time a marked decline can be observed.

Obesity has serious long-term consequences. The incidence of Type II diabetes in US children has increased in parallel with the rise in incidence of obesity. Furthermore hypertension, hypercholesterolaemia, heart disease, asthma, sleep apnoea, depression, low self-esteem and orthopaedic disorders have been linked to obesity (Schmidt, 2003; Galvez et al., 2003).

Although physical activity seems to promote self-esteem, no firm conclusions can be drawn on mental health, because factors which are perhaps more crucial than the physical activity itself are rarely accounted for (Martin et al., 2004).

**Table 61**. Health effects of sedentarism, relative risk (RR) and degree of scientific meetraint of the effect being causal (M: moderate, H: high).

| Health Effect | Intensity | RR for the disease | RR for mortality | Scientific Certainty | Reference |
|---|---|---|---|---|---|
| Life expectancy | 5 vs 15 km/week | | 1.49 | H | Paffenbarger *et al.* (1986, 1994) |
| Cardiovascular disease | | 1.84 | 1.43 | H | Martin *et al.* (2001) |
| Cardiovascular disease: death | 1000 vs > 2000 kcal/week | | 1.84 | | Paffenbarger *et al.* (1984) |
| Cardiovascular disease: hospitalisation | Walking <1hr vs >4hrs/week | 1.45 | | | LaCroix *et al.* (1996) |
| Type II diabetes | | 1.88 | 3.00 | H | Martin *et al.* (2001) |
| Obesity | | | | H | Arluk *et al.* (2003) |
| Colon cancer | | 1.90 | 1.68 | M | Martin *et al.* (2001) |
| Breast cancer | | 1.39 | 1.00 | H | Martin *et al.* (2001) |
| Osteoporosis | | 2.00 | | H | Martin *et al.* (2001) |
| Symptomatic gall stone disease | Men watching television >40 hrs vs <6 hrs/week | 3.32 (older men); 1.58 (younger men) | | H | Leitzmann *et al.* (1998) |
| | Decreased physical activity | 1.72 (men <65y); 1.33 (men ≥65y) | | | Leitzmann *et al.* (1998) |
| | Women spending 41-60 hrs/week sitting vs <6 hrs sitting | 1.42 | | | Leitzmann *et al.* (1999) |
| | Women spending >60 hrs/week sitting | 2.32 | | | Leitzmann *et al.* (1999) |
| Depression | | 3.15 | | H | Martin *et al.* (2001) |
| | <1000 vs 1000-2499 kcal/week | 1.20 | | | Paffenbarger *et al.* (1994) |
| | <1000 vs >2500 kcal/week | 1.39 | | | |
| Back pain | | 1.36 | | H | Martin *et al.* (2001) |
| Hypertension | | 1.47 | 1.00 | H | Martin *et al.* (2001) |

Many obese children suffer from a metabolic syndrome. This is characterised by the presence of excessive abdominal fat, high blood pressure, increased triglyceride levels in blood, low high density lipoprotein (HDL) cholesterol levels and high blood sugar. Childhood obesity exacerbates the risk of potentially fatal health problems later in life. For example, autopsies on obese children have shown higher levels of atherosclerotic plaque, which is a risk factor for stroke and myocardial infarction. They also suffer from non-alcoholic fatty liver disease, which promotes hepatitis, fibrosis and ultimately sclerosis in adults (Schmidt, 2003).

The global estimates of WHO indicate that physical inactivity causes about 10-16 per cent of cases each of breast cancer, colon and rectal cancers and diabetes mellitus, and about 22 per cent of ischaemic heart disease, resulting in 1.9 million deaths and 19 million DALYs (WHO, 2002).

A significant relationship exists between child obesity and maternal body mass index (BMI) (Arluk *et al.*, 2003).

Levels of activity are correlated with a number of socio-cultural parameters. The contribution of school-based physical education seems to be particularly important in groups with low activity levels (Martin *et al.*, 2004).

Obesity and its related health effects are strongly influenced by diet and lifestyle. For instance, contemporary major modes of entertainment, such as television, cinema and computer-based ones, are primarily sedentary in nature. It is unclear, however, what proportion of sedentarism is caused by mechanised transport. On the other hand, there is convincing evidence that walking and cycling decrease the risk of diseases associated with sedentarism.

## 5.   Discussion of scientific findings

Environmental health effects resulting from societal mobility patterns are multiple and diverse. The preceding tables show that the spectrum of conditions involved includes physiological, endocrinological, neurological and immunological mechanisms.

Nevertheless, common patterns of effects emerge. Noise, particles, non-particle air pollution and even sedentarism all appear to target the cardio-pulmonary system. Noise and obesity to some extent target the same neuro-endocrine homeostatic mechanisms. Particles and other air pollutants and sedentarism have all been shown to be positively correlated with increased cancer incidence.

The degree of overlap of any common mechanisms that may be triggered by different environmental impacts, however, remains to be fully elucidated. There are clearly, however, common underlying systems in the body which indicate the biological plausibility for considering that additive or synergistic effects can occur, e.g. aromatase/oestrogen/obesity.

Adverse health effects resulting from environmental pollution and sedentarism related to traffic movements occur in the general population. It has been estimated that 32 per cent of Europeans are exposed to road traffic noise levels above 55 dB(A), 19 per cent to levels between 55 and 65 dB(A), 11 per cent between 65 and 75 dB(A) and 2 per cent over 75 dB(A) (EEA, 2000).

It is estimated that in the UK alone, 8,100 deaths are caused annually by $PM_{10}$ exposure. The number of respiratory hospital admissions brought forward exceeds 10,000 (Department of Health, 1998). It is estimated that 50 per cent of the population's exposure arises from motorised transport (Newbery, 1998).

The WHO (2002) global estimate for physical inactivity among adults is 17 per cent. In a questionnaire by the EU (Eurobarometer), 47 per cent of citizens reported to have done no physical activity of moderate intensity in the last seven days (EC, 2003).

These selected data demonstrate that there is a major public health problem resulting from the widespread availability and use of motorised transport. No other single disease entity causing these levels of mortality and morbidity would be tolerated by society and there would be demands for research and action to reduce this burden.

Next to these direct health effects, the indirect impacts on health of global warming need to be considered. The main traffic-related greenhouse gases are $CO_2$ and $N_2O$. Road traffic is the fastest growing source of $CO_2$ emissions in most countries in the EU. In the UK, for instance, $CO_2$ emissions from road transport nearly doubled in the thirty years between 1970 (20.7 million tons) and 2000 (38.6 million tons). In this way, road traffic generated 24 per cent of the total $CO_2$ emissions in 2000. Rail, shipping and civil aircraft each contributed 1 per cent to the total UK $CO_2$ emissions for that period (DEFRA, 2003).

The current rises in temperature, induced by climate change, will lead to more heat waves, storms, droughts, floods and fires. These will cause additional deaths. In the summer of 1995, a heat wave struck the Eastern and Midwestern United States, leaving more than 500 deaths in Chicago alone (Hardy, 2003). In 2003 hundreds of extra deaths were recorded in France as a result of high temperatures (Dhainaut *et al.*, 2004). Malaria and other vector-borne diseases, such as dengue fever, schistosomiasis, and Leishmaniasis are expected to extend their ranges northwards into Europe, as a result of the changing climate (WHO Europe, 2003). For malaria alone, 2020 million people worldwide are at risk in response to climate change (Kovats *et al.*, 2000).

## 6.    Policy implications

Scientific data provide sufficient grounds for policy action on the following basis: Citizens and, in particular children, have a right to clean air, a quiet environment, a safely habitable urban environment and a healthy lifestyle. These rights should take priority over the right to personalised motor transport.

In different EU countries, a pollutant reduction policy is being implemented. Reductions in $SO_2$, lead and benzene already date from the 1980s. Plans to reduce tropospheric ozone, $CO_2$ and particulate aerosols are much more recent.

In view of the interrelated character of the health effects and the underlying mechanisms of traffic-related pollution and lifestyle, it should be advocated that an integrated approach mobility and health be established. Such a policy should include the following elements:

i) **A reduction of emissions** in particular of ozone precursors, particulate aerosols and noise. This reduction should be reached by better technology of cars, trucks and planes, but also by reducing the amounts of kilometres we drive.

ii) **Education, information and awareness-raising** to promote walking and cycling at both the individual and societal levels. At the level of the individual and in households, important decisions can be made, as half of all journeys made in cars in Europe are of less than 3km, an ideal distance for walking or cycling. In addition, 40 to 50 per cent of car use is for leisure activities (ISDE, 2002). Information related to the negative impacts of physical inactivity should also create a societal basis for this policy. The lessons learnt from the anti-smoking campaigns should be heeded. It took a remarkably short time for society to view smoking as an anti-social habit. Society needs to develop similar feelings towards short distance journeys by car. The professions allied to medicine should advocate the advantages of a physically active lifestyle. This is urgent because, currently, nearly two thirds of the adult population in the Western world are sedentary or not physically active enough to gain health benefits (ISDE, 2002).

iii) **Urban planning and design** currently tends to promote the use of motor vehicles more than walking and cycling. Too many places have unsafe sidewalks or lack cycle paths. Concerns for physical safety lead parents to keep their children indoors or to transport them to school by car. The planning and organisation of public areas should, as a priority, incorporate infrastructure for walking and cycling in safety.

iv) **Road pricing schemes** to discourage the use of motor vehicles in highly populated areas. A recent example of this has been introduced successfully in central London. An additional important option is to allocate the revenue raised to fund public transport by direct hypothecation. This can also be combined with tax incentive schemes or the provision of low cost or free public transport to encourage people to travel to work by bus, tram, train, cycling or walking. In Flanders (northern Belgium) public bus transport is free for pensioners and students (in their respective university towns). To alleviate the traffic pressure on Brussels, the capital, civil servants are provided with free public transport. The general introduction of these price measures has resulted in a huge increase in the number of bus passengers in Flanders since 2000. In 2003, the bus company 'De Lijn' transported 362.2 million passengers, an increase of 13.8 per cent compared to 2002. The number of passenger-kilometers increased parallel with the number of passengers, amounting to 176.8 million passenger-kilometers in 2003. Use of the train is also becoming more popular with - for the sixth consecutive year - an increase in the number of passengers and person-kilometers (VRIND, 2003). Although this will have a positive environmental effect, the precise impact is not yet known because it remains to be calculated how many cars are left at home as a result of the increased use of public transport (MIRA, 2003).

v) The levels of scientific certainty in some areas concerning mobility and health act as a bar to progressive policies. While there is sufficient knowledge for action on the above mentioned points, research should be instituted to further reduce those areas of scientific uncertainty that remain. These are, among others:

a) Biomonitoring and analysis of non-primary effects of mobility-related pollutants, including sedentarism.
b) Mechanisms underlying these effects, with special reference to unifying models that lead to an understanding of the interrelations between the effects caused by the different pollutants.
c) Data leading to quantification of the impacts on the population.
d) Health economic analyses of the impacts of mobility-related pollutants on health to determine the true cost of the impacts to society.
e) Validation methods for the effectiveness and efficiency of policy measures related to health and mobility.

This policy will be characterised by both long- and short-term benefits. Lowering of the cancer incidence after reducing exposure of the population to traffic-related carcinogens will take decades to be realised. However, reduction of exposure to noise will have an immediate effect. Increase in personal physical activity will also bring health benefits in a short time scale. For example, the adoption of as little as 30 minutes of brisk walking each day can reduce the risk of coronary artery disease by 50 per cent, the equivalent of giving up smoking. This makes such a policy attractive because decision makers are likely to be able to realise some direct benefits within their term of office.

## 7.    Conclusions

The health outcomes of mobility-related pollution are, in part, pollutant-specific, for example sleep disturbance by noise or intellectual impairment through exposure to lead. For the other part, there is a remarkable similarity in health outcomes from apparently diverse exposures. It appears that the cardio-pulmonary system is the prime target from a number of different mobility-related influences. Exposure to $NO_x$, particles and the effects of sedentarism are examples. In addition, the endocrine system seems to be targeted by a number of mobility-related pollutants. Noise and some VOCs have been recognised as capable of changing hormonal balances. Sedentarism increases the risk of osteoporosis, a condition under hormonal control. Among mobility-related pollutants are numerous carcinogens, such as benzene, benzopyrene and particles.

This is a major public health problem. In the EU the majority of the population, 150 million people, are exposed on a daily basis to a complex mixture of traffic-related pollutants and noise. In addition, two thirds of the population are sedentary. The largest single cause of all these problems is the use of motorised vehicles.

This should make mobility, environment and health the core elements of an integrated policy, which should be targeted to: reduction of the need for individual car use by society, promotion of walking, cycling and public mass transport and reduction of traffic emissions.

While many types of environmental pollution are long-term problems, for example POP and nuclear waste, the problems associated with traffic and mobility will respond very quickly to regulatory measures to reduce impacts. Traffic reduction brings an immediate associated reduction in noise, particle and gas emissions.

## References

Anderson, H.R., Spix, C., Medina, S., Schouten, J.P., Castellsague, J., Rossi, G., Zmirou, D., Touloumi, G., Wojtyniak, B., Ponka, A., Bacharova, L., Schwartz, J., and Katsouyanni, K. (1997) Air pollution and daily admissions for chronic obstructive pulmonary disease in 6 European cities: results from the APHEA project, *Eur. Respir. J.* 10(5), 1064-1071.

Arluk, S.L., Branch, J.D., Swain, D.P., and Dowling, E.A. (2003) Childhood obesity's relationship to time spent in sedentary behaviour, *Mil. Med.* 168(7), 583-586.

Ayres, J.G. (1998) Health effects of gaseous air pollutants, in R.E. Hester and R.M. Harrison (Eds.), *Air Pollution and Health*, The Royal Society of Chemistry, Letchword, U.K., 1-20.

Babisch, W. (2000) Traffic noise and cardiovascular disease: epidemiological review and synthesis, *Noise Health* 2(8), 9-32.

Babisch, W. (2003) Stress hormones in the research on cardiovascular effects of noise, *Noise Health* 5(18), 1-11.

Babisch, W. (2004) Chronischer Lärm als Risicofaktor für den Myokardinfarkt Ergebnisse der "NaRoMi"-studie, Umweltbundesamt, *WaBoLu-Hefte*, Nr. 02/2004.

Bassuk, S.S., and Manson, J.E. (2003) Physical activity and cardiovascular disease prevention in women: how much is good enough? *Exerc. Sport. Sci. Rev.* 31(4), 176-181.

Boezen, H.M., van der Zee, S.C., Postma, D.S., Vonk, J.M., Gerritsen, J., Hoek, G., Brunekreef, B., Rijcken, B., and Schouten, J.P. (1999) Effects of ambient air pollution on upper and lower respiratory symptoms and peak expiratory flow in children, *Lancet* 353(9156), 874-878.

Burnett, R.T., Cakmak, S., Brook, J.R., and Krewski, D. (1997) The role of particulate size and chemistry in the association between summertime ambient air pollution and hospitalization for cardiorespiratory diseases, *Env. Health. Perspect.* 105(6), 614-620.

Clark, C. (2004) Aircraft and road traffic noise and children's cognition and health: exposure-effect relationships from the RANCH project, 3rd International Conference on Children's

Health and the Environment, organised by ISDE, PINCHE and INCHES, 31 March to 2 April 2004, London, UK.

Dab, W., Medina, S., Quenel, P., Le Moullec, Y., Le Tertre, A., Thelot, B., Monteil, C., Lameloise, P., Pirard, P., Momas, I., Ferry, R., and Festy, B. (1996) Short term respiratory health effects of ambient air pollution: results of the APHEA project in Paris, *J. Epidemiol. Community Health* 50(suppl 1), s42-s46.

DEFRA (2003) *e-Digest of environmental statistics,* DEFRA website, Index downloaded June 2003 from: http://www.defra.gov.uk/environment/statistics/index.htm

Department of Health (1998) *Quantification of the effects of air pollution on health in the United Kingdom,* Committee on the Medical Effects of Air Pollution (COMEAP), HMSO: London.

Dhainaut, J.F., Claessens, Y.E., Ginsburg, C., and Riou, B. (2004) Unprecedented heat-related deaths during the 2003 heat wave in Paris: consequences on emergency departments, *Crit. Care* 8(1), 1-2.

Donaldson, K., and MacNee, W. (1998) The mechanism of lung injury caused by $PM_{10}$, in R.E. Hester and R.M. Harrison (Eds.), *Air pollution and health*, The Royal Society of Chemistry, Letchword, U.K., 21-32.

European Commission (EC) (2003) Eurobarometer, Physical activity, Special Eurobarometer 183-6 / Wave 58.2 – European Opinion Research Group EEIG. http://europa.eu.int/comm/public_opinion/archives/ebs/ebs_183_6_en.pdf

EEA (2000) *Are we moving in the right direction? Indicators on transport and environment integration in the EU, TERM 2000*, Environmental Issue Report No 12, EEA, Copenhagen, Denmark.

Evans, G.W., Lercher, P., Meis, M., Ising, H., and Kofler, W.W. (2001) Community noise exposure and stress in children, *J. Acoust. Soc. Am.* 109(3), 1023-1027.

Galvez, M.P., Frieden, T.R., and Landrigan, P.J. (2003) Obesity in the 21st century, *Env. Health Perspect.* 111(13), A684-685.

Gauderman, W.J., Avol, E., Gilliland, F., Vora, H., Thomas, D., Berhane, K., McConnell, R., Kuenzli, N., Lurmann, F., Rappaport, E., Margolis, H., Bates, D., and Peters, J. (2004) The effect of air pollution on lung development from 10 to 18 years of age, *N. Engl. J. Med.* 2004 351(11), 1057-1067.

Gehring, U., Cyrys, J., Sedlmeir, G., Brunekreef, B., Bellander, T., Fischer, P., Bauer, C.P., Reinhardt, D., Wichmann, H.E., and Heinrich, J. (2002) Traffic-related air pollution and respiratory health during the first 2 yrs of life, *Eur. Respir. J.* 19, 690-698.

Göttinger, H.W. (2001) Monitoring pollution accidents: the case of oil spills: Preface and introduction, *Int. J. Environment Pollution* 15(3), 243-248.

Haines, M.M., Stansfeld, S.A., Job, R.F., Berglund, B., and Head, J. (2001a) Chronic aircraft noise exposure, stress responses, mental health and cognitive performance in school children, *Psychol. Med.* 31(2), 265-277.

Haines, M.M., Stansfeld, S.A., Brentnall, S., Head, J., Berry, B., Jiggins, M., and Hygge, S. (2001b) The West London Schools Study: the effects of chronic aircraft noise exposure on child health, *Psychol. Med.* 31(8), 1385-1396.

Hardy, J.T. (2003) *Climate change. Causes, effects and solutions*, John Wiley and Sons, Ltd., Hoboken, NY.

Harrison, R.M. (ed.) (2001) *Pollution. Causes, Effects and Control,* Fourth Edition, The Royal Society of Chemistry, London, UK.

HCN - Health Council of the Netherlands (1994) *Noise and Health,* Report of a committee of the Health Council of the Netherlands, Report No. 1994/15E, The Hague, 15 September, 1994.

Hendriksen, I.J.M., Zuiderveld, B., Kemper, H.C.G., and Bezemer, P.D. (2000) Effect of commuter cycling on physical performance of male and female employees, *Medicine and Science in Sports and Exercise* 32, 504-510.

Hester, R.E., and Harrison, R.M. (eds.) (1998) Air pollution and health, in *Issues in Environmental Science and Technology,* The Royal Society of Chemistry, London, UK.

Hiramatsu, K., Minoura, K., Matsui, T., Miyakita, T., Osada, Y., and Yamamoto, T. (2000) An analysis of the general health questionnaire survey around airports in terms of annoyance reaction, *Proc. Internoise 2000*, 2089-2093, 2000.08 (Nice).

Hoeger, R., Schreckenberg, D., Felscher-Suhr, U., and Griefahn, B. (2002) Night-time noise annoyance: state of the art, *Noise Health* 4(15), 19-25.

Hoek, G., Brunekreef, B., Goldbohm, S., Fischer, P., and van den Brandt, P.A. (2002) Association between mortality and indicators of traffic-related air pollution in the Netherlands: a cohort study, *Lancet* 360, 1203-1209.

Huttenmoser, M. and Degan-Zimmermann, D. (1995) *Lebensraume fur Kinder*, Zurich: Swiss Science Foundation.

Hygge, S., Evans, G., and Bullinger, M. (1996) The Munich airport noise study: cognitive effects on children from before to after the change-over of airports, in F.A. Hill and R. Lawrence (eds.), Proceedings from Inter-Noise; 1996 jul 31-aug 2; Liverpool, UK; 1996, 2189-2194.

IEA - International Energy Agency (2001) *World Energy Outlook: 2001*, Paris, http://www.iea.org

ILEA - Institute for Life Cycle Environmental Assessment (1998) *Automobiles: Manufacture vs. Use*, Downloaded July 2004 from http://www.ilea.org/lcas/macleanlave1998.html.

ISDE - International Society of Doctors for the Environment (2002) *Transport - Environment – Health*, ISDE, 2002.

Ising, H., and Ising, M. (2002) Chronic cortisol increases in the first half of the night caused by road traffic noise, *Noise Health* 4(16), 13-21.

Ising, H., Lange-Asschenfeldt, H., Lieber, G.F., Weinhold, H., and Eilts, M. (2003) Respiratory and dermatological diseases in children with long-term exposure to road traffic emissions, *Noise Health* 5(19), 41-50.

Ising, H., and Kruppa, B. (2004) Health effects caused by noise: evidence in the literature from the past 25 years, *Noise Health* 6(22), 5-13.

Janssen, N.A.H., Brunekreef, B., van Vliet, P., Aarts, F., Meliefste, K., Harssema, H., and Fischer, P. (2003) The relationship between air pollution from heavy traffic and allergic sensitization, bronchial hyperresponsiveness, and respiratory symptoms in dutch schoolchildren, *Environ. Health Perspect.* 111, 1512–1518.

Jedrychowski, W., and Flak, E. (1998) Effects of air quality on chronic respiratory symptoms adjusted for allergy among preadolescent children, *Eur. Respir. J.* 11, 1312-1318.

Kan, H., and Chen, B. (2003) Air pollution and daily mortality in Shanghai: a time-series study, *Arch. Environ. Health* 58(6), 360-367.

Katsouyanni, K., Touloumi, G., Spix, C., Schwartz, J., Balducci, F., Medina, S., Rossi, G., Wojtyniak, B., Sunyer, J., Bacharova, L., Schouten, J.P., Ponka, A., and Anderson, H.R. (1997) Short term effects of ambient sulphur dioxide and particulate matter on mortality in 12 European cities: results from time series data from the APHEA project, *BMJ* 314(7095), 1658-1663.

Kawada, T. (2004) The effect of noise on the health of children, *J. Nippon Med. Sch.* 71(1), 5-10.

Koenig, J.Q., Jansen, K., Mar, T.F., Kaufman, J., Sullivan, J., and Liu, L.-J.S. (2002) Measurement of offline nitric oxide in an air pollution health effect study, *Am. J. Respir. Crit. Care Med.* 165, A306.

Kovats, R.S., Menne, B., McMichael, A.J., Corvalan, C., and Bertollini, R. (2000) *Climate change and human health: impact and adaptation*, WHO/SDE/OEH/00.4, Geneva, Switzerland.

Kunzli, N., Kaiser, R., Medina, S., Studnicka, M., Chanel, O., Filliger, P., Herry, M., Horak, F.Jr., Puybonnieux-Texier, V., Quenel, P., Schneider, J., Seethaler, R., Vergnaud, J.C., and Sommer, H. (2000) Public-health impact of outdoor and traffic-related air pollution: a European assessment, *Lancet* 356(9232), 795-801.

LaCroix, A.Z., Leveille, S.G., Hecht, J.A., Grothaus, L.C., and Wagner, E.H. (1996) Does walking decrease the risk of cardiovascular disease hospitalizations and death in older adults? *Journal of the American Geriatrics Society* 44, 113-120.

Laden, F., Schwartz, J., Speizer, F.E., and Dockery, D.W. (2001) Air pollution and mortality: a continued follow-up in the Harvard Six Cities study, *Epidemiology* 12, S81.

Leitzmann, M.F., Giovannucci, E.L., Rimm, E.B., Stampfer, M.J., Spiegelman, D., Wing, A.L., and Willett, W.C. (1998) The relation of physical activity to risk for symptomatic gallstone disease in men, *Ann. Intern. Med.* 128(6), 417-425.

Leitzmann, M.F., Rimm, E.B., Willett, W.C., Spiegelman, D., Grodstein, F., Stampfer, M.J., Colditz, G.A., and Giovannucci, E. (1999) Recreational physical activity and the risk of cholecystectomy in women, *N. Engl. J. Med.* 341(11), 777-784.

Le Tertre, A., Quenel, P., Eilstein, D., Medina, S., Prouvost, H., Pascal, L., Boumghar, A., Saviuc, P., Zeghnoun, A., Filleul, L., Declercq, C., Cassadou, S., and Le Goaster, C. (2002) Short-term effects of air pollution on mortality in nine French cities: a quantitative summary, *Arch. Environ. Health* 57(4), 311-319.

Lippmann, M., Frampton, M., Schwartz, J., Dockery, D., Schlesinger, R., Koutrakis, P., Froines, J., Nel, A., Finkelstein, J., Godleski, J., Kaufman, J., Koenig, J., Larson, T., Luchtel, D., Liu, L.J., Oberdorster, G., Peters, A., Sarnat, J., Sioutas, C., Suh, H., Sullivan, J., Utell, M., Wichmann, E., and Zelikoff, J. (2003) The U.S. Environmental Protection Agency Particulate Matter Health Effects Research Centers Program: a midcourse report of status, progress, and plans, *Environ. Health Perspect.* 111(8), 1074-1092.

Martin, B.W., Beeler, I., Szucs, T., Smala, A.M., Brügger, O., Casparis, C., Allenbach, R., Raeber, P.A., and Marti, B. (2001) Economic benefits of the health-enhancing effects of physical activity: first estimates for Switzerland. Scientific position statement of the Swiss Federal Office of Sports, Swiss Federal Office of Public Health, Swiss Council for Accident Prevention, Swiss National Accident Insurance Organization (SUVA), Department of Medical Economics of the Institute of Social and Preventive Medicine and the University Hospital of Zurich and the Network HEPA Switzerland, *Schweiz. Z. Sportmed. Sporttraumatol.* 2001; 49(3), 131-133.

Martin, B., Martin-Diener, E., Balandraux-Olivet, M., Mäder, U., and Ulrich, U. (2004) Transport-related health effects with a particular focus on children. Topic Report Physical Activity. Contribution to the UNECE-WHO Transport, Health and Environment Pan-European Programme – THE PEP. Transnational project and workshop series of Austria, France, Malta, the Netherlands, Sweden and Switzerland.

Maschke, C., and Hecht, K. (2004) Stress hormones and sleep disturbances – electrophysiological and hormonal aspects, *Noise Health* 6(22), 49-54.

Mason, C. (2000) Transport and health: en route to a healthier Australia? *Med. J. Aust.* 172(5), 230-232.

Matsui, T., Matsuno, T., Ashimine, K., Hiramatsu, K., Osada, Y., and Yamamoto, T. (2003) The Okinawa study: Effect of chronic aircraft noise exposure on birth weight, prematurity and intrauterine growth retardation, in *Proceedings of 'The 8th International Congress on Noise as a Public Health Problem'*, 29 June- 3 July 2003, Rotterdam, The Netherlands.

Michaud, D.S., Giovanucci, E., Willett, W.C., Colditz, G.A., Stampfer, M.J., and Fuchs, C.S. (2001) Physical activity, obesity, height, and the risk of pancreatic cancer, *JAMA* 286(8), 921-929.

Miedema, H.M., and Oudshoorn, C.G. (2001) Annoyance from transportation noise: relationships with exposure metrics DNL and DENL and their confidence intervals, *Environ. Health Perspect.* 109(4), 409-416.

Miedema, H.M., and Vos, H. (2003) Noise sensitivity and reactions to noise and other environmental conditions, *J. Acoust. Soc. Am.* 113(3), 1492-1504.

MIRA (2003) Milieu- en natuurrapport Vlaanderen, Achtergronddocument 2003, 1.5 Verkeer & vervoer, Ina De Vlieger, Erwin Cornelis, Luc Int Panis, Caroline De Geest en Els Van Walsum, Vlaamse Milieumaatschappij, http://www.milieurapport.be

Nedellec, V., Mosqueron, L., Desqueyroux, H., Boudet, C., Le Moullec, Y., Annessi-Maesano, I., and Medina, S. (2004) Transport-related health effects with a particular focus on children. Topic Report Air Pollution. Contribution to the UNECE-WHO Transport, Health and Environment Pan-European Programme – THE PEP. Transnational project and workshop series of Austria, France, Malta, the Netherlands, Sweden and Switzerland.

Newbery, D. (1998) *Fair payment from road users: a review of the evidence on social and environmental costs,* The Automobile Association: Basingstoke, Hants.

Ouis, D. (2002) Annoyance caused by exposure to road traffic noise: an update, *Noise Health* 4(15), 69-79.

Paffenbarger, R.S.Jr., Hyde, R.T., Wing, A.L., and Steinmetz, C.H. (1984) A natural history of athleticism and cardiovascular health, *JAMA* 252, 491-495.

Paffenbarger, R.S.Jr., Hyde, R.T., Wing, A.L., and Hsieh, C-C. (1986) Physical activity, all-cause mortality, and longevity of college alumni, *N. Engl. J. Med.* 314, 605-613.

Paffenbarger, R.S.Jr., Lee, I-M., and Leung, R. (1994) Physical activity and personal characteristics associated with depression and suicide in American college men, *Acta. Psychiatr. Scand. Suppl.* 377, 16-22.

Passchier-Vermeer, W., and Passchier, W.F. (2000) Noise exposure and public health, *Environ. Health Perspect.* 108 (suppl 1), 123-131.

Passchier-Vermeer, W., Miedema, H.M.E., and Vos, H. (2003) Aircraft noise and sleep: study in The Netherlands, in Proceedings of 'The 8th International Congress on Noise as a Public Health Problem', 29 June- 3 July 2003, Rotterdam, The Netherlands.

Peden, M., Scurfield, R., Sleet, D., Mohan, D., Hyder, A.A., Jarawan, E., and Mathers, C. (Eds.) *World report on road traffic injury prevention,* World Health Organization, Geneva, 2004.

Pope, C.A. 3rd, Burnett, R.T., Thun, M.J., Calle, E.E., Krewski, D., Ito, K., and Thurston, G.D. (2002) Lung cancer, cardiopulmonary mortality, and long-term exposure to fine particulate air pollution, *JAMA* 287(9), 1132-1141.

Rogers, J.F., Thompson, S.J., Addy, C.L., McKeown, R.E., Cowen, D.J., and Decoufle, P. (2000) Association of very low birth weight with exposures to environmental sulfur dioxide and total suspended particulates, *Am. J. Epidemiol.* 151, 602-613.

Schmidt, C.W. (2003) Obesity: a weighty issue for children, *Env. Health Perspect.* 111(13), A700-707.

Schwartz, J. (1993) Particulate air pollution and chronic respiratory disease, *Environ. Res.* 62, 7-13.

Schwartz, J., and Neas, L.M. (2000) Fine particles are more strongly associated than coarse particles with acute respiratory health effects in schoolchildren, *Epidemiology* 11(1), 6-10.

Schwartz, J., Spix, C., Touloumi, G., Bacharova, L., Barumamdzadeh, T., Le Tertre, A., Piekarski, T., de Leon, A.P., Ponka, A., Rossi, G., Saez, M., and Schouten, J.P. (1996) Methodological issues in studies of air pollution and daily counts of deaths or hospital admissions, *J. Epidemiol. Community Health* 50(suppl 1), S3-S11.

Spix, C., Anderson, H.R., Schwartz, J., Vigotti, M.A., Le Tertre, A., Vonk, J.M., Touloumi, G., Balducci, F., Piekarski, T., Bacharova, L., Tobias, A., Ponka, A., and Katsouyanni, K.

(1998) Short-term effects of air pollution on hospital admissions of respiratory diseases in Europe: a quantitative summary of APHEA study results, *Arch. Environ. Health* 53(1), 54-64.

Staatsen, B.A.M., Nijland, H.A., van Kempen, E.M.M., de Hollander, A.E.M., Franssen, A.E.M., and van Kamp, I. (2004) Transport-related health effects with a particular focus on children. Topic Report Noise. Contribution to the UNECE-WHO Transport, Health and Environment Pan-European Programme – THE PEP. Transnational project and workshop series of Austria, France, Malta, the Netherlands, Sweden and Switzerland.

Stansfeld, S., Haines, M., and Brown, B. (2000) Noise and health in the urban environment, *Rev. Environ. Health* 15(1-2), 43-82.

Stansfeld, S.A., and Matheson, M.P. (2003) Noise pollution: non-auditory effects on health, *Br. Med. Bull.* 68, 243-257.

Sunyer, J., Spix, C., Quenel, P., de Leon, A.P., Ponka, A., Barumandzadeh, T., Touloumi, G., Bacharova, L., Wojtyniak, B., Vonk, J., Bisanti, L., Schwartz, J., and Katsouyanni, K. (1997) Urban air pollution and emergency admissions for asthma in four European cities: the APHEA project, *Thorax* 52(9), 760-765.

Tenias, J.M., Ballester, F., and Rivera, M.L. (1998) Association between hospital emergency visits for asthma and air pollution in Valencia, Spain, *Occup. Environ. Med.* 55(8), 541-547.

Tiittanen, P., Timonen, K.L., Ruuskanen, J., Mirme, A., and Pekkanen, J. (1999) Fine particulate air pollution, resuspended road dust and respiratory health among symptomatic children, *Eur. Respir. J.* 13(2), 266-273.

Touloumi, G., Katsouyanni, K., Zmirou, D., Schwartz, J., Spix, C., de Leon, A.P., Tobias, A., Quennel, P., Rabczenko, D., Bacharova, L., Bisanti, L., Vonk, J.M., and Ponka, A. (1997) Short-term effects of ambient oxidant exposure on mortality: a combined analysis within the APHEA project. Air Pollution and Health: a European Approach, *Am. J. Epidemiol.* 146(2), 177-185.

Trejo, J.L., Carro, E., Nunez, A., and Torres-Aleman, I. (2002) Sedentary life impairs self-reparative processes in the brain: the role of serum insulin-like growth factor-I, *Rev. Neurosci.* 13(4), 365-374.

TRL-Transport Research Laboratory (2000) *Estimating global road fatalities*, Crowthorne, UK. http://www.trl.co.uk

UPI-Umwelt und Prognose Institut (1999) Öko-Bilanz von Fahrzeugen, *UPI Berichte* Nr. 25, 6. Auflage, May, Heidelberg, Germany.

van Kamp, I. (2004) Effects of aircraft noise and road traffic noise on annoyance, sleep quality and perceived health in children; The RANCH study. 3rd International Conference on Children's Health and the Environment, organised by ISDE, PINCHE and INCHES, 31 March to 2 April 2004, London, UK.

van Kempen, E.E., Kruize, H., Boshuizen, H.C., Ameling, C.B., Staatsen, B.A., and de Hollander, A.E. (2002) The association between noise exposure and blood pressure and ischaemic heart disease: a meta-analysis, *Environ. Health Perspect.* 110(3), 307-317.

Verhoeff, A.P., Hoek, G., Schwartz, J., and van Wijnen, J.H. (1996) Air pollution and daily mortality in Amsterdam, *Epidemiology* 7, 225-30.

von Klot, S., Wölke, G., Tuch, T., Heinrich, J., Dockery, D.W., Schwartz, J., Kreyling, W.G., Wichmann, H.E., and Peters, A. (2002) Increased asthma medication use in association with ambient fine and ultrafine particles, *Eur. Respir. J.* 20, 691-702.

VRIND (2003) *Vlaamse Regionale Indicatoren*, Administratie Planning en Statistiek, Vlaamse Overheid, Brussels.

WHO-World Health Organisation (1993) *Community noise, environmental health criteria document*, WHO European Office, Copenhagen, Denmark.

WHO-World Health Organisation (2002) *The World Health Report 2002 – Reducing risks, promoting healthy life*, WHO, Geneva, Switzerland.

WHO-World Health Organisation Europe (2003) *Methods of assessing human health vulnerability and public health adaptation to climate change*, Health and Global Environmental Change, Series No. 1.

WHO-World Health Organisation (2004) World report on road traffic injury prevention, M. Peden, R. Scurfield, D. Sleet, D. Mohan, A.A. Hyder, E. Jarawan, and C. Mathers (eds.), Geneva.

Wong, C., Ma, S., Hedley, A.J., and Lam, T. (2001) Effect of air pollution on daily mortality in Hong Kong, *Environ. Health Perspect.* 109, 335-340.

Worldwatch Institute (2001) *Vital Signs 2001*, Earthscan, London.

Zanobetti, A., Schwartz, J., and Gold, D. (2000) Are there sensitive subgroups for the effects of airborne particles? *Environ. Health Perspect.* 108(9), 841-845.

Zanobetti, A., and Schwartz, J. (2002) Cardiovascular damage by airborne particles: are diabetics more susceptible? *Epidemiology* 13(5), 588-592.

Zmirou, D., Barumandzadeh, T., Balducci, F., Ritter, P., Laham, G., and Ghilardi, J.P. (1996) Short term effects of air pollution on mortality in the city of Lyon, France, 1985-1990, *J. Epidemiol. Community Health* 50 Suppl 1, S30-35.

# LIST OF ABBREVIATIONS

| | |
|---|---|
| A | Annoyed |
| ACCC | Austrian Council on Climate Change |
| ACSM | American College of Sports Medicine |
| AIDS | Acquired Immunodeficiency Syndrome |
| APHEA | Air Pollution Health Effects: a European Approach |
| APHEIS | Air Pollution and Health: a European Information System |
| ARN | Automatic Rural Network |
| Asl. | Above sea level |
| AUN | Automatic Urban Network |
| BALF | Bronchoalveolar avage fluid |
| BaP | Benzo(a)pyrene |
| BAT | Best Available Technology |
| BSRB | Bulgarian Statistical Reference Book |
| BTEX | Benzene, Toluene, Ethyl benzene and Xylenes |
| CAA | Civil Aviation Authority |
| CALIPSO | Atmospheric Lidar |
| CDC | Centers for Disease Control |
| CEHAP | Child Environmental Health Action Plans |
| $CH_4$ | Methane |
| CLRTAP | Long-Range Transboundary Air Pollution |
| CNES | Centre National d'Etudes Spatiales |
| CO | Carbon monoxide |
| $CO_2$ | Carbon dioxide |

| | |
|---|---|
| COHb | Carboxyhemoglobin |
| CO-Hb | Carboxyhaemoglobin |
| COPD | Chronic Obstructive Pulmonary Disease |
| $CO_x$ | Carbon oxides |
| CTM | Chemical Transport Model |
| DALYs | Disability-Adjusted Life Years |
| dB(A) | Measure for the noise level (decibel) |
| DEFRA | Department for Environment, Food and Rural Affairs |
| DEP | Diesel Exhaust Particles |
| DEPSC-SM | Department of Environment, Public Services and Cleanness at Sofia Municipality |
| DETR | Department of Environment, Transport and Regions |
| DfT | Department for Transport |
| DLCO | Carbon monoxide diffusion capacity coefficient |
| DOAS | Differential Optical Absorption Spectroscopy |
| DOC | Degree of scientific certainty |
| DoH | Department of Health |
| DPM | Diesel Particulate Matter |
| DPTA | Diethylenetriamine pentaacetic acid |
| EAD | Particle Aerodynamic Diameter |
| EARLINET | European Aerosol Research Lidar Network to Establish an Aerosol Climatology |
| EC | European Commission |

| | | | |
|---|---|---|---|
| EEA | European Environment Agency | JPD | Johannesburg Political Declaration |
| EMECAS | Spanish Multicenter Study on Health Effects of Air Pollution | JPoI | Johannesburg Plan of Implementation |
| EPA | Environmental Protection Agency | KfV | Kuratorium für Verkehrssicherheit (Austrian Road Safety Board) |
| Eq | Equivalent | KSI | Killed or Seriously Injured |
| ET | Extrathoracic | LA | Little Annoyed |
| ETP | Employer Transport Plan | $L_{adn}$ | A-weighed day/night noise level |
| EU | European Union | | |
| FEV 1 | Forced Expiratory Volume in the first second | LIDAR | Light Detection And Ranging |
| FVC | Forced Vital Capacity | LITE | Lidar In-space Technology Experiment |
| GDP | Gross Domestic Product | LRD | Lower Respiratory Disease |
| GLAS | Geosciences Laser Altimeter System | LRT | Long-Range Transport |
| GP | General Practitioner | M | Motorway |
| GPS | Global Positioning System | MAC | Maximum Allowable Concentrations |
| HA | Highly Annoyed | | |
| Hb | Haemoglobin | MI | Myocardial Infarction |
| HCN | Health Council of the Netherlands | MIA | Ministry of Internal Affairs |
| | | MINA-plan | Flemish Environment and Nature plan |
| HCs | Hydrocarbons | | |
| HEPA | Health-Enhancing Physical Activity | MIRA | Report on the environment and nature in Flanders |
| HIA | Health Impact Assessment | MIRA-S | Report on the environment and nature in Flanders: scenarios |
| HIV | Human Immunity Virus | | |
| HSD | Highly Sleep Disturbed | | |
| IARC | International Association for Research on Cancer | MIRA-T | Report on the environment and nature in Flanders: themes |
| IEA | International Energy Agency | MoEW | Ministry of Environmental and Waters |
| IEH | Institute for Environment and Health | $N_2O$ | Nitrous oxide |
| IHD | Ischaemic Heart Disease | NAAQS | National Ambient Air Quality Standards |
| ILEA | Institute for Life Cycle Environmental Assessment | NAEI | National Atmospheric Emissions |
| IQ | Intelligence Quotient | NAIE | National Air Quality Information Archive |
| ISDE | International Society of Doctors for the Environment | NASA | National Agency for Space Administration |
| ISO | International Organization for Standardization | NEHAP | National Environmental Health Action Plan |

| | | | |
|---|---|---|---|
| NF-kappa B | Necrosis Factor kappa B | RTA | Road Traffic Accident |
| NGO | Non-Governmental Organisation | SABENA | Societé Autonyme Belge d'Exploitation de la Navigation Aerienne |
| $NH_3$ | Ammonia | | |
| NHS | National Health Service | SD | Sustainable Development |
| NIS | Newly Independent States | SEL | Sound Exposure Level |
| NLCS | Netherlands Cohort Study on Diet and Cancer | $SO_2$ | Sulphur dioxide |
| | | TB | Tracheobronchial |
| NMHC | Nonmethane hydrocarbons | THE PEP | Transport, Health and Environment Pan European Programme |
| NMMAPS | National Mortality, Morbidity and Air Pollution Studies | | |
| | | TNF-$\infty$ | Tumor necrosis factor alpha |
| NMVOC | Non Methane Volatile Organic Compound | TRL | Transport Research Laboratory |
| NO | Nitrogen monoxide | TSH | Thyrotropin |
| $NO_2$ | Nitrogen dioxide | TSP | Total Suspended Particles |
| $NO_X$ | Nitrogen oxides | UBA | Umweltbundesamt (Austrian Federal Environmental Protection Agency) |
| NSI | National Statistical Institute | | |
| $O_2$ | Oxygen | | |
| $O_3$ | Ozone | | |
| OECD | Organisation of Economic Co-operation and Development | ufCB | ultrafine carbon black |
| | | UK | United Kingdom |
| | | UNECE | United Nations Economic Commission for Europe |
| ONS | Office for National Statistics | | |
| | | UPI | Umwelt und Prognose Institut |
| OTS | Oxford transport Strategy | | |
| P50 | 50 percentile | US | United States |
| P98 | 98 percentile | UT | Universal Time |
| PAH | Polyaromathic Hydrocarbons | VITO | Flemish Institute for Technological Research |
| PAR | Population Attributable Risk | VLAREM | Legislative document containing the environmental quality objectives for Flanders |
| PBL | Planetary Boundary Layer | | |
| PCBs | Polychlorinated Bipheniles | | |
| PEF | Peak Expiratory Flow | VOC | Volatile Organic Compounds |
| PM | Particulate Matter | | |
| PMN | Polymorphonuclear | WEHAB | Water, Energy, Health, Agriculture and Biodiversity |
| $PM_X$ | Particulate Matter (x=particle diameter) | | |
| | | WHO | World Health Organisation |
| ppm | Parts per million | WSSD | World Summit on Sustainable Development |
| REM | Rapid Eye Movement | | |
| RR | Relative Risks | WTC | World Trade Center |

# LIST OF UNITS

## Prefixes to Units

| | | | | | | |
|---|---|---|---|---|---|---|
| da | deca | $(10^1)$ | d | deci | $(10^{-1})$ |
| h | hecto | $(10^2)$ | c | centi | $(10^{-2})$ |
| k | kilo | $(10^3)$ | m | milli | $(10^{-3})$ |
| M | Mega | $(10^6)$ | $\mu$ | micro | $(10^{-6})$ |
| G | Giga | $(10^9)$ | n | nano | $(10^{-9})$ |
| T | Tera | $(10^{12})$ | p | pico | $(10^{-12})$ |
| P | Peta | $(10^{15})$ | f | femto | $(10^{-15})$ |

## Units

| | | | | |
|---|---|---|---|---|
| °C | degree Celcius or centigrade | pa | per annum |
| d | day | pH | acidity |
| Drachmas | Greek currency unit | ppb | parts per billion |
| Euro | European currency unit | ppm | parts per million |
| g | gram | s | second |
| h | hour | t | ton |
| $kg_{bw}$ | kilogram body weight | te | ton emission gas |
| kgpa | kilogram per annum | tpa | ton per annum |
| l | litre | US$ | US Dollar |
| m | metre | y | year |
| $Nm^3$ | Normalised cubic metre | | |

# INDEX